Library of
Davidson College

George P. Richardson

FEEDBACK THOUGHT in Social Science and Systems Theory

Feedback Thought
in Social Science and Systems Theory

Feedback Thought in Social Science and Systems Theory

George P. Richardson

UNIVERSITY OF PENNSYLVANIA PRESS   Philadelphia

The author gratefully acknowledges permission to reprint figures initially published and copyrighted by Agathon Press; Aldine Publishing Company; Ballinger Press; Dover Publications; the *Economic Journal* published by Basil Blackwell; Harper and Row Publishers; Harvard University Press; John Wiley and Sons; Macmillan Press; MIT Press; Random House Incorporated; *Sloan Management Review* published by MIT Press; University of Chicago Press; and the *Coevolution Quarterly* published by the Whole Earth Review. Individual acknowledgments are included in the figure captions.

Copyright © 1991 by the University of Pennsylvania Press
All rights reserved
Printed in the United States of America

Library of Congress Cataloging-in-Publication Data
Richardson, George P.
   Feedback thought in social science and systems theory / George P. Richardson.
     p.   cm.
   Includes bibliographical references and index.
   ISBN 0-8122-3053-1 (cloth). — ISBN 0-8122-1332-7 (pbk.)
   1. Social sciences—Methodology.   2. Feedback control systems.
I. Title.
H61.R487   1991
300'.72—dc20   90-22746
                                                                                CIP

To Gail,
Joshua, and Sara

Contents

Preface   ix

Chapter 1—Introduction

1.1—Problem, Purpose, and Rationale	1
1.2—Concepts and Definitions	5
1.3—Perspective and Methods	9
1.4—Overview	14

Chapter 2—Prehistory and Emergence

2.1—Engineering Servomechanisms and Control Theory	17
2.2—The Loop Concept in Mathematical Models	32
2.3—Homeostasis and the Loop Concept in Biology	46
2.4—Logic Loops	53
2.5—The Loop Concept in Social Science Literature Before 1945	59

Chapter 3—Two Feedback Threads

3.1—Emergence	92
3.2—The Cybernetics Thread	94
3.3—The Servomechanisms Thread	129
3.4—Origins and Prospects	161

Chapter 4—Developments in the Cybernetics Thread

4.1—Introduction	169
4.2—Management Science and the Work of Stafford Beer	170
4.3—The TOTE Unit	189
4.4—Magoroh Maruyama and Deviation-Amplifying Feedback	200
4.5—Political Science: Karl Deutsch and David Easton	204
4.6—Karl Weick and the Study of Organizations	228
4.7—The Psychology of William Powers	240
4.8—Summary and Directions	263

Chapter 5—Developments in the Servomechanisms Thread

5.1—Introduction	271
5.2—Feedback in the Later Work of Herbert Simon	271
5.3—Use of the Feedback Concept in Economics	277
5.4—System Dynamics	296
5.5—Summary and Directions	313

Chapter 6—Issues and Implications

6.1—Introduction	318
6.2—Two Arguments	320
6.3—Themes and Issues	332
6.4—Reflections and Conclusions	341

Bibliography	349
Index	363

Preface

The seeds of this work were planted almost twenty years ago when I first became aware of the feedback concept. I wondered then, and still ponder, how my education could have missed it. A powerful way of thinking—linking concepts of control and self-reinforcement, stability and instability, structure and behavior, mutual causality, interdependence, and uncounted numbers of the deepest ideas in the natural, social and behavioral sciences—yet the concept was never mentioned in my undergraduate years. Only gradually in serendipitous ways did I come to learn of the power of the idea and its rich intellectual history, which this work traces.

It is a story of patterns of thought weaving among people and through time—an important piece of intellectual history. It is an intriguing piece because it shows ideas diverging as well as converging over time, creating separate patterns of thought where one is commonly believed to exist.

But the book is also a testament to the richness of circular thinking in the social sciences. Usually implicit, sometimes explicit, feedback thought is embedded in the foundations of much of social science and systems theory. It is a building block. Part of the purpose of this book is to illuminate the significance of feedback thinking in social science and policy.

The story shows that feedback thought is both venerably old and puzzlingly new. It shows that in the modern era the concept of feedback moved into prominence in the 1940s and 1950s and has since appeared to wane. Some consider it an outmoded idea, a metaphor for the social sciences that has had its day and has been supplanted by new metaphors. Others, perhaps as a consequence, have not encountered it at all. Still others see it not as a metaphor at all but as a natural and crucial property of social systems.

This book assembles an historical record that argues for the importance of feedback thought in social science, systems theory, and social policy—current as well as classical. Moreover, the book provides an explanation for differing points of view on the importance of the feedback concept. Modern feedback thought diverges into at least two distinct threads of scholarly inquiry and application. In one thread, the concept is seen as a metaphor; here, for the most part, it has had its day. In the other thread, the concept remains vital, with important implications for theory and policy. Without reading further, the reader knows in which thread he or she lives. It would be intriguing to know where people stand after they have made their way through the pages that follow.

It was the writings of Jay Forrester that started me in 1973 on the road leading to this point. I am deeply indebted to him. He contributed significantly to this book both by his own works and by steadfast support of the undertaking. I am also especially indebted to Hayward Alker and Donald Schön for their contributions and encouragement. Their enthusiasm for an investigation of feedback thought, their voluminous knowledge of social science literature, and the wise reflections they shared with me were invaluable.

I have also benefited greatly from the support and contributions of other colleagues. Many contributed quotations and sources. Others cheered me on or provided help through the numerous stages leading to this book. I wish to thank in particular Peter Senge, Jim Hines, Mark Paich, Jack Homer, John Morecroft, Barry Richmond, Karl Clauset, Bill Carver, Michelle Marto, Alison Anderson, Jack Gunnell, David Andersen, and John Sterman. To David and John go special thanks for scholarly advice and friendly counsel that helped me greatly with this work and continue to sustain me.

Most of all I thank my wife and friend Gail, who never lost faith. This work is dedicated to her. And to Joshua and Sara. They earned it.

George P. Richardson
Albany, April 1990

Chapter 1
Introduction

> ... action and counteraction, which in the natural and in the political world, from the reciprocal struggle of discordant powers draw out the harmony of the universe.
>
> Edmund Burke, *Reflections on the Revolution in France*

1.1 Problem, Purpose, and Rationale

Within the last forty years, the engineer's concept of feedback has entered the social sciences. The essence of the concept, defined more precisely in section 1.2, is a circle of interactions, a closed loop of action and information. The patterns of behavior of any two variables in such a closed loop are linked, each influencing, and in turn responding to, the behavior of the other. Thus the concept of the feedback loop is intimately linked with the concepts of interdependence and mutual or circular causality, ideas with a rich history in the social sciences. This book is an investigation of the use of the loop concept in social science that lies at the heart of feedback and circular causality.

In some areas of the social and policy sciences the feedback loop, by whatever name it is known, has become a fundamental center of attention, a vital concept in the analysis of societal problems and the construction of theory. In other areas, however, the concept is noted but its applicability and explanatory power are seen to be very limited. In still other corners, the concept is largely unrecognized.

This book investigates how people in the social sciences became aware of the feedback loop concept, what they have accomplished with it, and what they perceive its significance and limitations to be. I will argue that the answers to these questions are linked—that what individuals accomplish with the concept of feedback and what they perceive its strengths and limitations to be depend on the sources of their understandings of it. I will show that ideas from five or six intellectual traditions weave over time into two main lines of development of the feedback idea in the social sciences, which I will refer to as the servomechanisms thread and the cybernetics thread. Practitioners in these two threads have significantly different perspectives on the proper use, problems, and potential of the feedback concept in the social sciences.

I have five reasons for undertaking an investigation of the evolution of feedback thought in the social sciences. Listing them will both lay out some initial presumptions and preview some of what we will find in the following chapters. Together they foreshadow a central claim of this book—that the loop concept underlying feedback and circular causality is one of the most penetrating fundamentals in all social science. The work demonstrates that great social scientists are feedback thinkers, and great social theories are feedback thoughts.

First, loop concepts are central to an emerging view of social reality. For some time the unidirectional view of cause-and-effect has been giving way to the circular, looping perspective of mutual causality. For decades we have applied variations of a research paradigm that strives to derive the causal connection between a "dependent" variable Y and "independent" variables $X_1, X_2, \ldots, X_n$. What characteristics of a teacher correlate with teaching effectiveness? What will consumer demand for a company's product be next year? How does a manager's leadership style affect worker productivity? How do parents socialize children? Increasingly, social scientists are taking the point of view that the dependent-independent variable view is inadequate for some such questions. Teaching effectiveness is apparently the result of an interaction of characteristics of teacher and student. The demand for a company's product is not simply the result of consumer behavior but is partly determined by the company's own behavior in setting price, maintaining quality, being early to market and ship, and setting a balance among such goals. A manager's behavior is not independent of worker characteristics: worker productivity can turn around and affect a manager's leadership style. And children socialize parents. In each of these examples, and in a vast range of others, causality appears to be circular. Social scientists, public policy makers, business people, and ordinary folk have to learn to deal with the emerging fact of circular causality in social systems.

Second, the concept has already played an important role in the social and policy sciences. We shall see how in more detail in the chapters that follow, but a brief name-dropping excursion is appropriate here to make the point. The loop concept lies just under the surface of some of the most significant contributions in the social sciences. It is hidden but discernible in the writings of Malthus, Adam Smith, Marx, Mill, Keynes, and many others. It is almost explicit in Merton's concept of the self-fulfilling prophecy, Festinger's cognitive dissonance, the vicious circles in American race relations illuminated by Myrdal, and all the macroeconometric models descended from Tinbergen's analysis of business cycles and Samuelson's multiplier/accelerator model. It was discovered before its time as "schismogenesis" by the anthropologist Gregory Bateson.

Once the concepts of feedback and homeostasis became known to social scientists, they became central, organizing concepts for a number of important thinkers. Wiener's *Cybernetics* and the cybernetics movement that went with it influenced the thinking of Bateson, Kurt Lewin, Margaret Mead, Karl Deutsch, and hosts of others directly through their own work and indirectly through others' writings. As an alternative to "rationality" and optimization, Herbert Simon suggested the feedback concept as a basis for models for human behavior. Tustin identified servomechanisms as *The Mechanism of Economic Systems* and proposed the mathematical methods of control engineers for the analysis of economic dynamics. The TOTE unit of Miller, Galanter, and Pribram's *Plans and the Structure of Behavior* was explicitly based on a feedback view of psychological mechanisms. Deutsch's *The Nerves of Government* urged the scholarly exploitation of the analogy between communication and feedback in neural networks and their counterparts in governmental networks. Forrester's *Industrial Dynamics* initiated a modeling approach for management science that is explicitly centered on the feedback concept. In James Grier Miller's monumental *Living Systems,* the most complete statement of the general systems theorist's view of the world, the concept appears repeatedly at all levels, from the kidney to the "supranational" system. More recently, the feedback loop concept has played an important role in Weick's *The Social Psychology of Organizing* and provides the foundation for a form of psychological counseling called reality therapy, as described in William Glasser's *Stations of the Mind.* Currently, in works like Gleick's *Chaos* and Prigogine and Stengers' *Order out of Chaos,* feedback loops are emerging at the heart of the phenomena of complex nonlinear dynamics and unpredictable deterministic systems. Thus psychotherapy, psychology, sociology, anthropology, social psychology, economics, political science, and management have all been touched and changed by the discovery of the potential significance of the feedback loop in human affairs. We shall investigate these appearances of feedback and circular causality much more thoroughly in the chapters that follow.

Third, in spite of these developments the concepts of feedback and circular causality are not well understood. Some see several very distinct concepts here—interdependence, mutual causality, circular causality, recursion, self-reference, feedback, knowledge of results, causal loops—while others lump various ones of these together. Some limit the concept of feedback to homeostasis and control, implicitly or explicitly restricting it to refer to negative feedback.[1] Some derive the dynamic behavior of a system from its feedback structure, while others fail to connect feedback to dynamics. Still others try to draw dynamic inferences from feedback structure but produce demonstrably false conclusions. In formulating or critiquing quantitative models involving feedback, some use data in inappropriate ways. Few take adequate account of compensating feedback effects in making policy decisions.[2] And some see problems that others ignore or bypass with the application of the feedback concept. This investigation is a step toward improved understandings.

Fourth, an emerging conclusion from some applications of the feedback perspective holds that the concepts of feedback and circular causality are essential to reliable policy analysis. Experience with dynamic, nonlinear models of feedback systems repeatedly shows that failure to take account of existing feedback effects in the analysis of a policy initiative can cause exactly the wrong conclusions to be reached. More subtly, information feedback can be used to explain the observed tendency of social systems to be "policy resistant," to react more weakly and more perversely to policy shifts than some experts predict. A promising key to understanding and anticipating such behavior is the notion of compensating feedback.

I have a fifth, more personal, set of reasons for investigating the evolution of the feedback perspective in social science. The concept of feedback lies at the core of my field, system dynamics. It is a relatively new field, originating just thirty years ago, using computer simulation to help understand complex social systems. But it reflects traditions and patterns of thinking that are far older and far more widespread than is generally recognized. Exposing the evolution of the feedback concept and its uses in the social sciences will help to map the intellectual context of my field.

Furthermore, the implications of the role of feedback and circular causality in the computer simulation methods of system dynamics are not well understood. The feedback loop is the fundamental building block of system dynamics models, and it is the basic unit of analysis and communication of system behavior. Moreover, it is the feedback notion pressed to an extreme that leads to the endogenous point of view that is perhaps the single most characteristic and significant feature of the field.[3] I believe that an essential step in understanding the potential of the system dynamics approach is the illumination of the deep meaning and significance of the feedback concept within the field and within the social sciences at large.

There are thus three questions which have guided this investigation.
1. How has the loop concept underlying feedback and circular causality evolved in social science?

2. How has it come to pass that different people interpret the feedback concept differently and see different potential in it?
3. Is it possible to evaluate different uses and interpretations of the concept, to identify its most fruitful future directions in social science theory and social policy?

1.2 Concepts and Definitions

A basic premise of this investigation is that there is a unifying loop concept underlying a number of superficially diverse ideas in the social sciences. Servomechanisms engineers, whose work came eventually to influence the social sciences, capture feedback systems in terms of differential equations. Econometricians express interdependencies in an economy using difference equations. Other social scientists use words to paint verbal pictures of circular causal processes: vicious circles, self-fulfilling prophecies, homeostatic processes, and invisible hands. Underlying all these representations, as we shall see, is the concept of a closed loop of causal influences.

Since the concept of a loop is fundamentally visual, we will use visual terms to give an intuitive definition of the loop concept underlying feedback and mutual causality. (See Figure 1.1.) An arrow drawn from A to B will be taken to mean A "causally influences" B. Thus if A influences B, and B in turn influences A, we would have a pair of arrows that form a loop of mutual or circular causality. To this elemental loop we attach an additional idea, which we shall refer to as the *polarity* of the loop. It is the concept of polarity that gives the causal loop its perceived analytic and explanatory power.

The polarity of a circular causal loop reflects the loop's tendency either to reinforce or to counteract a change in any one of its elements. Following the terminology that has become common since the emergence of the feedback concept from engineering into the social sciences, we shall characterize such loops as "positive" or "negative":

- A causal loop that characteristically tends to reinforce or amplify a change in any one of its elements is called a positive loop.

In a positive loop, an increase in an element A feeds around the loop and tends to cause A to increase still further; likewise a decrease in A tends to cause A to decrease still further. Similarly,

- A causal loop that characteristically tends to diminish or counteract a change in any one of its elements is called a negative loop.

In a negative loop, an increase in A feeds around the loop and tends to cause A to slow or reverse its increase; likewise a decrease in A tends to cause A to slow or reverse its decrease.

The motivation for the positive and negative labels comes from the way loop polarities can be obtained from the polarities of the individual causal links that combine to form the loop. The "arithmetic" of causal links parallels the arithmetic of multiplying signed numbers. To establish the parallelism, define a causal influence from A to B to be positive if a change in A tends to produce a change in B in the same direction: an increase in A tends to produce an increase in B, a decrease in A tends to produce a decrease in B. Similarly, define a causal influence from A to B to be negative if a change in A tends to produce a change in B in the opposite direction: an increase in A tends to produce a decrease in B, a decrease in A tends to produce an increase in B. It is then easy to argue that the polarity of a causal loop is the product of the polarities of its causal links. We have the following unambiguous definition:

- A causal loop is positive if it contains an even number of negative links, and is negative if it contains an odd number of negative links.

An example best illustrates this arithmetic of link and loop polarities. Consider the circular causal loop shown in Figure 1.1, which happens to be abstracted from two classic works on the study of discrimination in the United States (Myrdal 1944, Merton 1948). It contains two positive links and two negative links. The causal link from Prejudice to Discrimination is labeled positive to capture the assumption, in this causal argument, that an increase in prejudice in a majority group tends to produce an increase in discrimination, or, in the more hopeful direction, a decrease in prejudice tends to produce a decrease in discrimination. A change in prejudice is thus assumed in this diagram to tend to cause a change in discrimination in the same direction. Hence, the link is positive.

The causal link from Discrimination to the Achievements of the minority is labeled negative to indicate an inverse sort of relationship. A change in discriminatory practices by the majority is assumed to lead to an opposite change in minority achievements in standard of living, education, health, and so on. An increase in discrimination tends to decrease potential achievement; a decrease in discrimination tends to increase potential achievement. Thus the link is negative by our definition. The other links in the diagram can be similarly explained.

The positive polarity of the closed causal loop that these influences form together can be obtained two ways. Blindly applying the "multiplication rule" for loop polarities, we see that the loop should be called a positive loop because it contains an even number of negative links, and the product of an even number of negatives is a positive. More insightfully, we see that the loop describes a self-reinforcing process in which prejudice feeds on the fruits of its own discriminatory tendencies. An increase in prejudice, traced around the loop, tends to increase discrimination, which tends to lower minority achievements, which in turn increases the majority's perception of the inferiority, providing additional support for the majority's prejudice. More beneficially, a decrease in prejudice traced around the loop tends to lessen the majority's perception of the inferiority of the minority and leads in the direction of further reductions in prejudice.

The circular causal process described in Figure 1.1 thus fits our characterization of a positive loop as a self-reinforcing circular causal process. It does so precisely because the two negative causal links behave in effect like one positive link: the net effect of an increase in discrimination is an increase in majority prejudice. Thus the signs on the causal links operate together like the multiplication of signed numbers. The product of two negatives is positive; analogously, the polarity of a causal loop with an even number of negative causal links is positive. A similar development is easy to construct for negative loops, containing an odd number of negative links, as self-correcting or deviation-counteracting processes.

Thus our focus in this investigation is on signed causal loops—mutual or circular closed-loop causal processes characterized as either positive or negative. Throughout, we shall refer to a mutual or circular causal loop with its associated loop polarity as a feedback loop. This study is thus an investigation of the evolution of the implicit and explicit use of positive and negative feedback loops in social science.

I have deliberately glossed over a number of subtleties in the preceding development. First, there are a host of questions about *causality* in

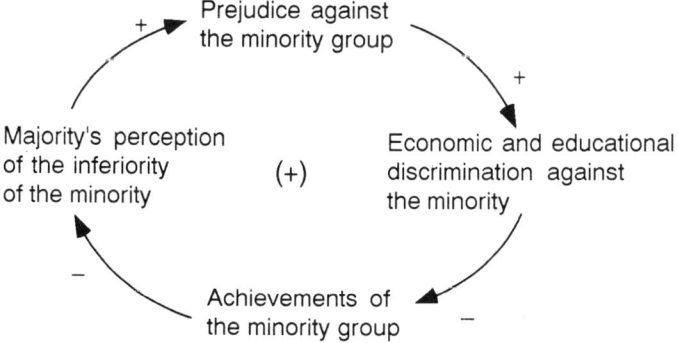

FIGURE 1.1: Causal loop used in the text to illustrate the determination of link and loop polarities (abstracted from Myrdal 1944 and Merton 1948).

social science, including whether the concept has any scientific meaning at all. The debate could be said to have been sparked by Aristotle, who asserted that men of art are wiser than men of experience, for men of art know the *why* of a phenomenon. "Clearly," he asserted, "wisdom is a science of certain causes and principles" (Apostle 1966, p. 14). In stark contrast, David Hume held that there is no basis for the concept of cause-and-effect because we only experience "one object following another" (Hume 1739; Book I, part I, section VII). Bertrand Russell concluded that "The word 'cause' is so inextricably bound up with misleading associations as to make its complete extrusion from the philosophical vocabulary desirable." He added that the only reason some philosophers persist in thinking in terms of causes rather than functional laws is that they do not know enough mathematics (Russell 1913). Still, John Stuart Mill built his *System of Logic* (1846) on a "law of universal causation" that "every consequent has an invariable antecedent." Closer to the present day, MacIver (1942) made a thorough attempt to defend and restore the place of causality in social science, identifying no fewer than eight kinds of *why*'s social scientists are concerned with.

A summary of the arguments appears in Lessnoff (1974). The reader interested in the issues is invited to peruse these references, as well as Bunge (1963), Craik (1943/1952), Meehan (1968), Stinchcombe (1968), von Wright (1971), and particularly Lerner (1965). The latter contains an acclaimed paper by Herbert Simon, whose work figures prominently in this book, which argues, in essence, that even if we do not know what causality means in social reality we can be precise about what it means in models of reality. It is an interesting idea, one that absolves causal thinkers from any residual guilt, but we shall not address these issues. I choose simply to presume that the concept of cause in the social and policy sciences has meaning, from which we can derive a meaningful idea of closed loops of circular causality. Or perhaps more guardedly, I choose to begin, as MacIver did, concerned not with the validity of the causal concept but only with its universality—in particular, its widespread use underlying the notions of feedback and circular causality.

Second, the characterizations of positive and negative links in feedback loops given above are a bit too loose to cover without ambiguity causal links that represent additions and subtractions from accumulations in a feedback loop (see Richardson and Pugh 1981, pp. 25–28). Without going into detail, we note that some positive and negative signs in causal-loop diagrams may represent not proportional change but actual addition and subtraction. The presence of additive and subtractive links does not alter the characterization of positive and negative loops given above, so further detail here is not really necessary for our purposes.

Third, there are a number of people in the evolution we are about to trace who would take offense at the use of the phrase "feedback loop" to stand for all signed circular causal processes. Some, for example, tend to use the term "feedback" for one link in a conversational or dialectical process, as in "giving feedback" to someone. Others who take a loop view of feedback prefer to reserve the concept for the explicit, deliberate "feeding back" of information in a control process. Some justifiably see useful distinctions among the ideas of interdependence, mutual causality, circular causality, and feedback. At this introductory point, we are not in a position to discuss the pros and cons, and we need a reasonably short name for the concept on which we are focusing. I will use the term "feedback loop" to represent the common loop concept that I perceive to underlie all signed circular or mutual causal processes. I will defend the choice at this point only by observing that in at least one line of thinking in the social sciences the notion of the "feedback loop" covers all closed-loop processes that can be represented in signed diagrams more or less similar to Figure 1.1. Starting anywhere in any such loop, something (presumably information) "feeds around" the loop and eventually "feeds back" to its point of origin. I should also note that there is probably a normative component to my use of a single term to cover these myriad ideas: I see a value to using a single label, so that what is in common to these ideas becomes the focus.

1.3 Perspective and Methods

Value Positions

Tracing the evolution of an idea in the social sciences is an investigation in intellectual history. This work is both something more and something less. The central characteristic that distinguishes it from pure intellectual history is that its author is not a dispassionate chronicler. As a practitioner in a field that emphasizes the feedback concept in formulating and analyzing simulation models for social policy analysis, I am part of the evolution I am tracing. A late comer, to be sure, but nonetheless I am a participant observer.

That fact forces caution on my part and the reader's. Throughout this work I have tried not to let my own perspective cloud or distort my presentations and interpretations of other perspectives. That effort has resulted in a number of cautions I have had to try to impose upon myself. There are, no doubt, places where I have not been successful.

Tracing the evolution of an idea soon convinces one that "priors" are very powerful.[4] My priors—my background and previous understandings of ideas related to feedback and circular causality—have no doubt colored my perceptions in ways I have not been able to detect. Part of the reader's caution therefore must be to exercise enough care to uncover my unjustified biases when they slip through. The obverse of this caution applies as well: the informed reader will also bring priors to this investigation. Perhaps we can hope that the misperceptions that come despite our mutual best efforts will balance out.

The effort to be unbiased, however, must not be allowed to prevent us from trying to perceive strengths and weaknesses in uses of the feedback concept that we shall encounter. The narrative contains a normative component, which is reflected periodically throughout and brought together in a discussion of issues and implications in Chapter 6.

Bibliographic Methods

The feedback-related works investigated in detail in Chapters 3, 4, and 5 represent only a fraction of what has been written employing the feedback concept, and only a part of what I have uncovered. They form a selected history. Although all histories are selective, it is fair to ask on what basis the selections were made.

The most reliable way to uncover feedback works that are considered significant by other scholars is to work backward in time—from recent works back toward books, articles, and authors cited. Feedback authors and works that repeatedly appear in such an investigation of citations clearly deserve to be analyzed for the nature and scope of their contributions to the evolution of the feedback concept. Part of the approach followed in this investigation involved working backward from known works to frequently cited authors and works. That process was particularly helpful in exposing lines of thinking that contributed to the origins of the cybernetics and servomechanisms threads. It was also somewhat helpful in tracing out the directions of the two threads.

A second way is to work the other way in time, from a given author's work forward to those citing or making use of it. This task is much more troublesome, particularly for frequently cited works. According to the Social Sciences Citation Index, a typical work discussed in Chapters 4 and 5, for example, appears in more than 800 citations from 1960 through 1982. It is clearly not possible to investigate all such citations, even for one work. And it is impossible to tell how many and which of such citations would be relevant to a study of the use of the feedback concept. A more promising approach is to look for intersections: a set of works that cite two or more earlier works. The resulting lists are more manageable. This approach uncovered, among other things, the work of William Powers (see section 4.7).

In this investigation of the evolution of the feedback concept, however, the intersection approach tended to produce anthologies or review articles, particularly if one looked for sets of works that cite three or more significant feedback writers. Although initially a discouraging result that made it difficult to ferret out more recent important feedback works, that fact is significant. It contributes to the conclusion that there are distinct threads of feedback thinking in the social sciences. There are, for example, only nine authors that reference all three of Wiener, Ashby, and Forrester (see sections 3.2, 3.3, and 5.4). Only two of these involve actual applications, one to accounting information systems and the other to ecosystems. The others are survey discussions of the systems literature that do not, in themselves, make use of the feedback concept.

A third approach, distinct from citation searches, would be to search the titles of books and articles for occurrences of the word "feedback" or its various synonyms or associated concepts. This is a fruitless approach and one that is apt to be profoundly misleading. The most important social science works employing the concept of feedback and circular causality that I encountered had no reference to such terms in their titles. Furthermore, many of those that do contain the word "feedback" use it to refer to something akin to "constructive criticism," not the concept of circular causality with positive or negative polarity that is the subject of this investigation.

A number of these approaches could be facilitated by using computerized databases of social science literature. Garfield (1964) and Price (1965) have pioneered in such methods. Such techniques are particularly useful for finding titles containing given words or phrases, searching out intersection sets of citations common to two or more authors or works, and identifying networks of authors. I did not make use of computerized databases in this investigation, however. Investigations of the Social Sciences Citation Index by hand and eye, such as those described briefly above, did not yield enough new information to suggest that the results would be worth the expense. Furthermore, the fact that such databases are computerized only back to the early 1970s made them useless for much of the searching required in this investigation.

Of these feedback-related works in the social sciences that surfaced in the course of these various citation searches and readings, I have selected a subset to describe in detail and discuss. Many others are mentioned briefly in the course of the narrative, and still more are contained in the bibliography. The selections were made with four aims in mind: first, to show the breadth of the significant uses of feedback and circular causality in the social sciences; second, to establish the presence of a split in the use of these concepts into two separate threads of feedback thinking; third, to show counterexamples to the split; and fourth, to display, through the authors' own words, issues and implications of this separation of feedback threads. It is interesting to contemplate that this work is itself a small part of the evolution of the feedback concept in the social sciences.

Conceptual Threads

Throughout the investigation I have found the metaphor of "conceptual threads" to be helpful. We shall see that since the emergence of the feedback concept in the social sciences in the 1940s there have been significant differences in the use of the concept among various feedback thinkers. The differences lump together initially into two lines of thinking in the social sciences, which I will refer to as "conceptual threads." The distinctions between the two feedback threads will be developed in sections 3.1 through 3.3. Here it is appropriate to motivate and define the general idea of a conceptual thread.

We are tracing the evolution of a concept in the social sciences. Its meaning and uses move through time, perhaps changing during one period and holding relatively constant during another. The concept is passed from individual to individual, usually in writings but also in direct conversation. The language can vary from words to mathematics or even pictures. Each individual to whom the concept is passed selects some of what is communicated, emphasizing some aspects and de-emphasizing or passing over other aspects. Furthermore, each individual brings to the concept notions and nuances from his or her own background. Thus as it moves through time the concept gathers together different ideas into one conceptual thread, much the way a wool thread is spun from diverse strands frequently contributed by different sheep.

In Chapter 2 I shall argue that modern interpretations of the concept of feedback in the social sciences are blends of ideas from a number of intellectual traditions: classical ideas in the social sciences about self-reinforcing and self-correcting phenomena; the notion of homeostasis from biology; loop concepts from formal logic; understanding derived from mathematical models of dynamic systems in biology, economics, and econometrics; and, of course, the term and concept of feedback itself from engineering. The "thread" metaphor is thus quite appropriate: feedback thinking in the social sciences is the result of the weaving together of ideas from five or six conceptual strands. I shall argue in Chapter 3 that subsets of these six strands weave together to form two rather distinct threads of feedback thinking in the social sciences.

The thread metaphor is also suggestive as a concept evolves over time. Threads can unravel into separate strands. A concept can move in different directions spawning somewhat different lines of thinking. Evidence that we shall encounter in Chapter 4 suggests that one of the two main feedback threads in the social sciences splits into two or three directions as it evolves through time, with one of its strands heading off in the direction of the other main thread of feedback thinking in the social sciences, the servomechanisms thread.

A conceptual thread, as I intend the phrase, is an interaction of people and ideas. People contribute to a line of thinking, and people select from it as they encounter it over time. People should appear in our pictures of the threads of feedback thinking in the social sciences. But a conceptual thread is a way of thinking, not a collection of people.

There are two ways people become connected with a conceptual thread. Usually, people are linked to a line of thinking by direct communication through the scholarly literature. People read about feedback, and what they carry away with them depends upon what they read and also what background they bring to the encounter. Thus one way that people become linked to a feedback thread is through the network of social scientists using the concept and writing about their work. One can find direct evidence of this sort of linking in the citations in published papers. One could call this being connected sociologically to a conceptual thread.

It is conceivable, however—and indeed it happens in this investigation—that some thinkers arrive at much the same patterns of thought, not by selecting from the same literature but by developing the same concepts and methods independently. People who are in completely different scholarly networks can be linked to the same feedback thread because they make use of the concept in the same ways. Thus the notion of a conceptual thread has a methodological dimension as well as a sociological dimension. We can objectively document sociological connections using citations evidence, but methodological similarities are more debatable and require detailed investigations of what people are actually doing with the feedback concept. For that reason, we shall make extensive use of analysis of direct quotations from feedback thinkers, in addition to citation evidence.

The purpose of pursuing the idea of different feedback threads in the social sciences is not simply to split hairs or threads. There is a serious significance to the existence of two separate lines of thinking

that share elements of a common concept. The significance derives from two simple propositions about the history and ecology of ideas.
1. People connected with the same conceptual thread will tend to think in similar ways, and their similarities can be predicted from knowledge of the characteristics of the thread.
2. People connected with a given conceptual thread perceive related ideas as if they were an established part of their own thread.

The first proposition claims that ideas are something of a mixture of inertia and grease. Once a scholar sets an idea in motion, it tends to move in the same direction in the hands of others, and they find it easier to push. Newton used a far more elegant metaphor with similar meaning when he asserted that if he had seen further than others it was because he had "stood on the shoulders of giants." The second proposition suggests a dark side to the evolution of ideas, and it stems, I think, from the fact that everybody knows the first proposition. People looking outward from one line of thinking may be blinded to advances in another, closely related line of thinking, precisely because of the tendency to interpret new developments as old familiar ideas. Newton's image could be turned upside down—a scholar can fail to rise to new heights because of the weight of giants on his shoulders.[5]

If there are significant but subtle differences in the use of the feedback concept in the social sciences—if there are different conceptual threads that are unrecognized as such—there would be tendencies for people from those different conceptual threads to misperceive, in terms of their own understandings, work they encounter in the other threads. Strengths attributed to the feedback concept in one thread would be attributed to the other, perhaps without foundation.

Similarly, weaknesses or failures linked to the concept in one thread would be attributed to work in the other, again perhaps without foundation. To enable people to do the most with the feedback idea, we would first have to perceive the differences in usage and perceived potential, in order to move toward a more enlightened, presumably more powerful synthesis of feedback views.

Thus the real goal of illuminating different conceptual threads in the evolution of the feedback concept is the strengthening of the feedback perspective in the social sciences.

1.4 Overview

Chapter 2 lays out elements of six intellectual traditions, some of which can be traced to the ancient Greeks, which contribute ideas to modern understandings of the meaning and significance of the feedback concept in the social sciences. The content of this chapter is historical background for the developments traced in Chapters 3, 4, and 5. It is brought together in section 3.1, and retrospectively interpreted in section 3.4 in light of modern understandings of the feedback concept, to show its links to the evolution of the feedback concept.

Chapter 3 describes in detail the beginnings of the two distinct threads of feedback thinking in the social sciences that emerge in the 1940s and 1950s. The characteristics that make these threads distinct are explored. The development makes considerable use of direct quotations and citation evidence to establish the existence and nature of the separation of feedback threads. In section 3.4 an effort is made to account for their differences by linking them to different aspects of the six intellectual traditions described in Chapter 2.

Chapters 4 and 5 explore developments in the two feedback threads in the social sciences, spanning roughly 1960 to 1980. Here the implications of the separation between the two threads begin to be felt. We shall find that the lack of recognition of the differences between these two feedback threads has muddled the concept of feedback and blended disparate ideas. Chapter 6 reflects on the preceding chapters and discusses a series of issues raised throughout the investigation. Of particular interest will be the following questions.

- For what purposes is the feedback concept employed in the social sciences? Where does it appear to be most promising?
- What explanatory functions does the concept of the feedback loop serve?
- What are other characteristics associated with a given feedback perspective? In particular, in what terms is behavior described? In what terms is behavior analyzed?
- What characterizes the causal views of different feedback perspectives?
- What problems is a feedback perspective particularly suited to address?
- In short, what are the elements of the "world views" of different feedback perspectives?

Notes

1. Negative feedback loops are circular causal processes that tend to counteract deviations from an equilibrium condition. See section 1.2 for definitions of terms and concepts.

2. Compensating feedback is a phenomenon in circular causal systems in which an external intervention produces natural feedback effects within the system that counteract the intended effect of the intervention. See section 5.4.

3. The endogenous point of view looks inside a complex system for the causes of its own significant behavior patterns. It can be contrasted to an

exogenous point of view, in which problems are seen to be caused by forces external to the system. See sections 3.3 and 5.4.

4. See Meadows (1980) for discussion of the significance of priors.

5. A famous social science principle akin to these ideas is Mannheim's paradox: All knowledge "is dependent upon the subjective standpoint and the social situation of the knower". Karl Mannheim, *Ideology and Utopia*, quoted in Landau (1972), pp. 34–42. The concept of a scientific paradigm may apply here (Kuhn 1962/1970).

Chapter 2
Prehistory and Emergence

> For one good deed leads to another good deed, and one transgression leads to another transgression.
>
> *Pirke About (The Wisdom of the Fathers)*

The concept of the feedback loop in twentieth-century social science is a blend of intuitions and ideas from at least six intellectual traditions: engineering, economics, biology, mathematical models of biological and social systems, formal logic, and classical social science literature itself. To understand the evolution of the feedback idea in the social sciences, we must investigate the development and use of loop concepts underlying feedback and circular causality in these other areas. We shall find that ideas from all six traditions are influential, but not equally so across all areas and authors in the social sciences. One feedback thread is more directly linked, both sociologically and methodologically, with engineering servomechanisms and mathematical models in biology (sections 2.1 and 2.2). The other is more influenced by homeostatic mechanisms in biology and ideas from formal logic (sections 2.3 and 2.4). Both threads select from the engineering and the social science literature, but in subtly different ways. At the close of this chapter, we will be in a position to see the beginnings of these different feedback threads in the social sciences.

2.1 Engineering Servomechanisms and Control Theory

The engineers' contribution to our understanding of feedback is often thought to originate with the dramatic proliferation of feedback devices in the late eighteenth century. The classic example, frequently cited as the first conscious use of a feedback device, is James Watt's centrifugal governor for the steam engine (1788) (Figure 2.1). The negative loop nature of this controller is clear from the description in the figure. A change in the speed of the engine produces forces in the governor that counteract the change and return the engine to its normal operating speed.

18 Prehistory and Emergence

But feedback devices by no means originated with Watt. His centrifugal governor was, in fact, a variant of the "centrifugal pendulum" patented in 1787 by Thomas Mead, who applied the device to regulating the speed of windmills by automatically furling and unfurling the windmill's sails (Mayr 1970, pp. 102–104). Moreover, as Figure 2.2 shows, both Mead and Watt came rather late into the feedback business.

The Ancient History of Feedback

Mayr (1970) traces the history of feedback all the way back to the golden age of Greece. He credits Ktesibios (250 B.C.) with the first

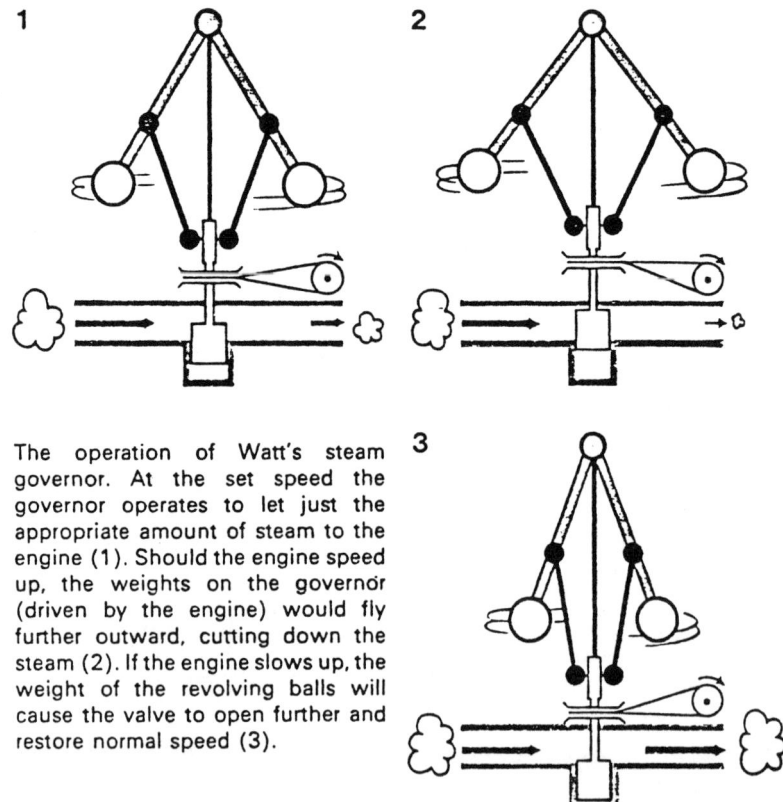

The operation of Watt's steam governor. At the set speed the governor operates to let just the appropriate amount of steam to the engine (1). Should the engine speed up, the weights on the governor (driven by the engine) would fly further outward, cutting down the steam (2). If the engine slows up, the weight of the revolving balls will cause the valve to open further and restore normal speed (3).

FIGURE 2.1: James Watt's governor for controlling the speed of a steam engine. Source: Beer (1967).

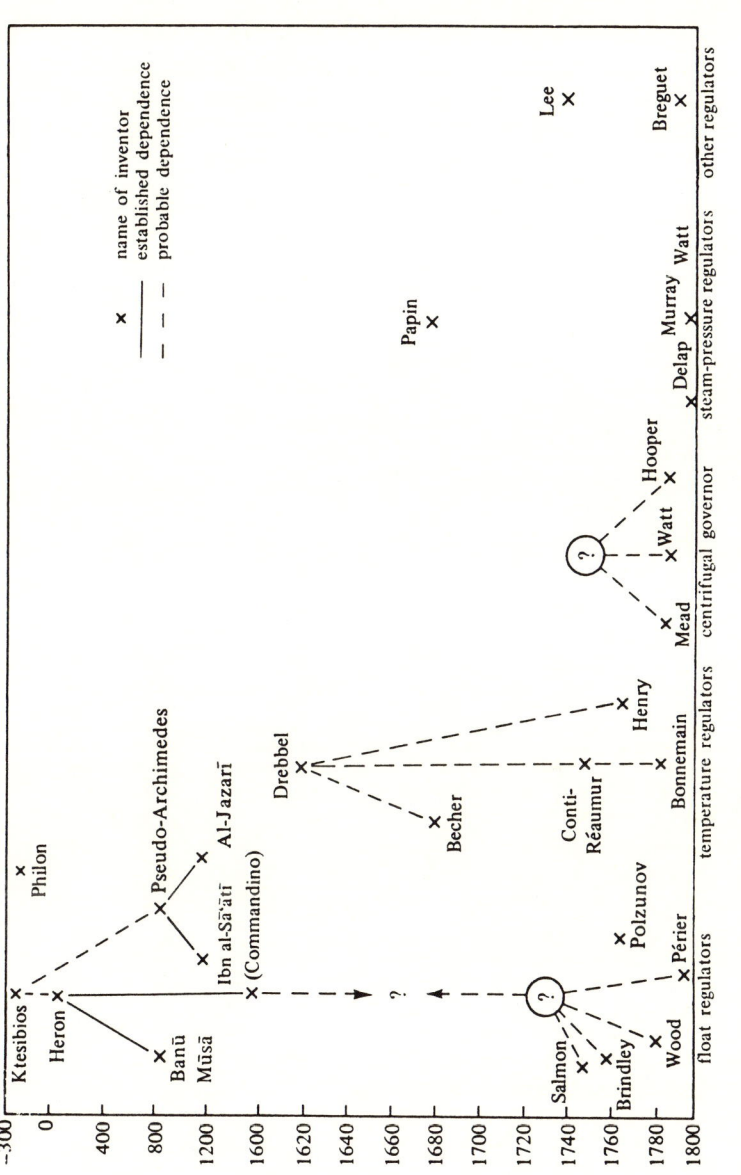

FIGURE 2.2: Chronology of feedback devices. Source: Mayr (1970, p. 126).

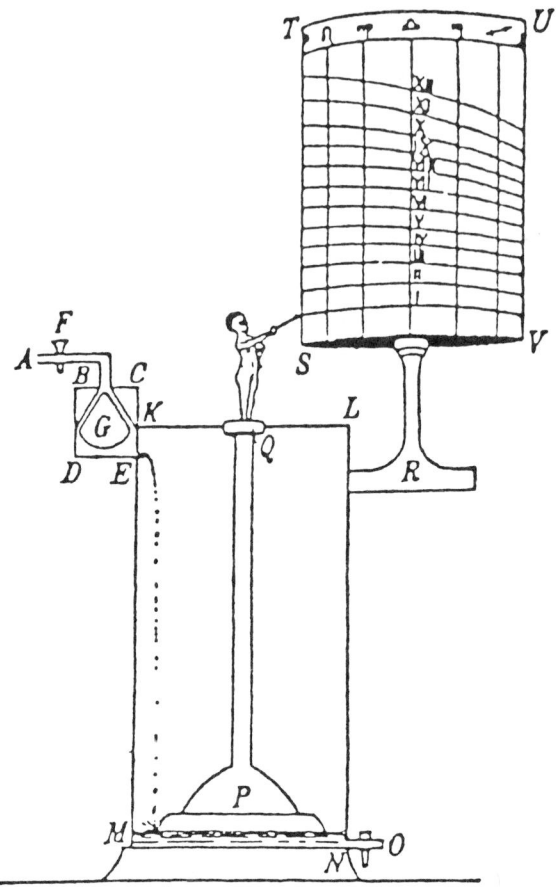

FIGURE 2.3: The first known feedback device—the float valve of Ktesibios (250 B.C.) used to create an accurate water clock. Source: Mayr (1970, p. 12).

known feedback device—a float valve used to create a steady drip of water into a cylindrical vessel, enabling the construction of an accurate water clock. Before Ktesibios, the rate of water dripping from a holding tank into the clock depended upon the water pressure in the holding tank. That pressure, in turn, depended upon the height of the water in the holding tank. As the water dripped out of the holding tank into the clock, the level of water in the clock would rise less and less rapidly. With a linear scale on the water clock, it would appear as if time were slowing down. Ktesibios interposed a float valve between the holding tank and the clock, as shown in Figure 2.3. The rate of water flow into the clock then depended upon the height of the water in the valve, not the height of water in the holding tank.

Ktesibios' float valve is a true self-regulating device that operates on the feedback principle. Figure 2.4 sketches the feedback loops involved. As the water pressure in the holding tank drops over time, the inflow to the valve tends to lessen, causing the float to start to drop, which in turn opens the inlet further and increases the inflow. Thus the tendency for the inflow to decrease over time is automatically compensated for by the action of the valve. The height of the water in the valve tends to remain constant, and the resulting water clock is able linearly to trace the flow of time.

As early as 60 A.D. there existed a book on fluid control mechanisms, Heron's *Pneumatics*. Besides illustrating several variations on Ktesibios's water clock, it contained remarkably clever devices for (among other things) maintaining the level of oil in a lamp and keeping constant the level of wine in a bowl at a party. The latter device now finds almost universal use in modern toilets.

Feedback Devices in China

Needham has claimed that the Chinese south-pointing chariot was the first homeostatic machine in human history (Needham 1954, vol. 4.2,

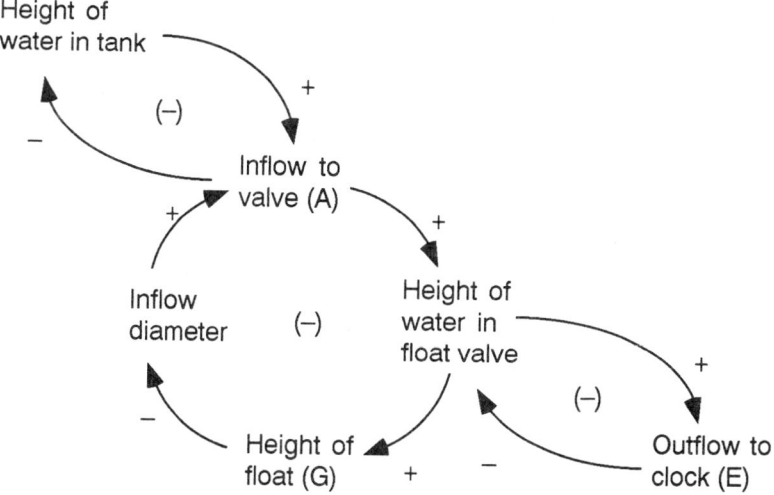

FIGURE 2.4: The feedback loop structure Ktesibios's float valve.

pp. 286–303). Since it dates, as near as is known, somewhere around the twelfth century A.D., it was certainly not the first. Furthermore, it was not a true closed-up feedback mechanism. The purpose of the device was to keep a chariot headed in a constant direction in a trackless wilderness. But the driver initiated the steering corrections, so, as Needham himself pointed out, the driver had to be included in the description of the negative feedback control loop. Mayr concludes, quite appropriately:

> Any human action can be interpreted as a case of feedback control, in which the human performs the functions of sensing and corrective action. There is no difference in principle between crossing a desert with the help of the south-pointing device, and simply walking up to a tree. The south-pointing chariot thus is no feedback device by our definition (Mayr 1970, p. 50).

Another device potentially involving a feedback mechanism was a monumental water clock built by Su Sung around 1090 A.D. Unfortunately, the descriptions of the clock are not sufficiently detailed to show for certain whether some sort of float valve similar to that of Ktesibios was employed to maintain a constant flow of water. In this case, both Needham and Mayr conclude that the clock is probably not an example of an early feedback device (Mayr 1970, p. 52). Mayr also rules out another water clock attributed to Wang P'u (c. 1135 A.D.). Somewhat surprisingly then, we must conclude that early feedback control devices are essentially western inventions. Similar developments apparently do not occur in the Orient until after the rise of feedback devices in Europe. Certainly, the difficulty of finding evidence of such devices guarantees that if any did exist they had no effect on the views of twentieth-century social scientists.

European Feedback Devices

As indicated in Figure 2.2, feedback devices were apparently limited to fluid control mechanisms up to the seventeenth century. The first departure from fluid devices, and the first feedback system to be invented in modern Europe, was the thermostatic furnace of Cornelius Drebbel (1572–1633) (Mayr 1970). "Furnace" must be interpreted a bit broadly, for one of the applications of Drebbel's temperature regulator was to maintain the proper warmth for the hatching of chickens. The idea of the device was to control the flow of air to the heating flame, increasing the airflow when the system was too cold, and decreasing it when too hot. The trick was to convert information about the temperature of the system automatically into the mechanical action that varied the airflow to the flame.

Drebbel put alcohol in a closed vessel connected to a U-shaped tube containing mercury. As the alcohol heated up, pressure in the vessel would build, forcing the level of mercury in the free end of the U-tube to rise, which in turn pushed a lever that constricted the air supply to the flame and cooled the vessel down. The system thus tended to seek and maintain a particular temperature, vapor pressure, and vapor volume in the alcohol vessel, enabling it to be used to keep constant the heat of a furnace or the temperature of an incubator. The goal-seeking, negative feedback loop character of the structure is clear: too cool, more air to the flame, more heat; too hot, less air, less heat; just right, no change. With luck and repeated design attempts, seventeenth-century engineers were able to build such regulators that could zero in on any reasonable desired temperature apparently without repeatedly overshooting and undershooting the set point.

Figure 2.2 shows that there was an explosion of activity in the invention and use of feedback devices in Europe toward the end of the eighteenth century. It was the time of the industrial revolution, there was an explosion in the invention and use of all sorts of machinery. Nonetheless, Mayr does not believe that this activity fully accounts for the burgeoning use of feedback devices:

Neither purely technical factors nor events in the economic environment have been able to account for the increased use of the concept of feedback in the 18th century. Since the inventions involved are most varied in external form, and since their inventors for the greater part worked independently of each other, one suspects that the true causes for this upswing are to be found in the intellectual currents of the era. Between the 17th and 18th centuries the attitudes toward the problem of regulation underwent a radical change (Mayr 1970, p. 128).

Was automatic regulation simply "in the air" of late eighteenth-century Europe? The question has interesting implications for our investigation of the feedback concept in the social sciences of that period. (See section 2.5.)

James Clerk Maxwell and the Analysis of Feedback Devices

Although feedback devices were used with increasing frequency from the time of Ktesibios, and although formal analysis of their structure and behavior began as early as 1868 with Maxwell's paper "On Governors," the loop concept underlying them remained completely unrecognized until the twentieth century (Mayr 1970, p. 131). Before Maxwell, regulators were constructed with little theorizing and much experimenting. The tendency of a regulator to overshoot and undershoot its set point, known as the problem of "hunting," was solved largely by tinkering.

Maxwell (1868) opened up an entirely new approach to the design of feedback regulators. He showed that it was possible to represent the dynamic characteristics of a feedback system in the form of differential equations. The first result of his approach was to expose the reason why some regulator mechanisms, Watt's governor among them, failed to hold a rotating system at a particular desired speed of rotation. Suppose, he said, that the governor is arranged to change the driving power on the axis of the machine by a quantity proportional to the gap between actual and desired velocity:

$$F \cdot \left(\frac{dx}{dt} - V\right) = \text{the force that the governor applies whenever } \frac{dx}{dt} \neq V,$$

where

$\frac{dx}{dt}$ = the actual angular velocity of the system,

V = the desired angular velocity,

F = a force factor.

Then the differential equation for the motion of the rotating system with this regulator is

$$\frac{d}{dt}\left(M \frac{dx}{dt}\right) = P - R - F \cdot \left(\frac{dx}{dt} - V\right),$$

where

M = the mass of the system,

P = the driving power on the rotating axis,

R = the resistance (friction) of the rotating system.

(The left side of this equation is the rate of change of momentum, while the right is the net force acting on the axis of the machine. The equation is simply Newton's statement, force equals mass times acceleration, applied to this rotating system.)

Maxwell observed that when this type of system comes to a steady state, the left side of this equation is zero (velocity is constant, acceleration is zero), so the steady state velocity is given by $dx/dt = V +$

(P − R)/F. Thus the system's actual velocity dx/dt is different from its desired velocity V by a quantity depending upon the driving power P and resistance R of the system and the correcting force F from the regulator. In modern terms, Maxwell showed that "proportional control" leaves a "steady state error" that can be lessened but never totally eliminated by increasing the power of the controller (represented by the factor F). Maxwell suggested that the word "governor" should be reserved for regulators that showed no steady state error. Thus Watt's governor was to Maxwell merely a "moderator." He devoted just one sentence to it. Not only is Watt's governor far from being the first feedback device, and not the first use of the "centrifugal pendulum," but it is not even a true governor.

Maxwell went on in his paper to show the conditions required to create true governors. The goal of a governor is to maintain "a uniform motion" at a set velocity. But, as those who tried to eliminate the steady state error in Watt's "moderator" found out, designing a system to hold to a set speed often resulted in endless oscillatory "hunting" around the set point. Maxwell's most significant contribution was the demonstration that control without sustained oscillatory tendencies was

> mathematically equivalent to the condition that all the possible roots, and all the possible parts of the impossible roots, of a certain equation shall be negative (Maxwell 1868, p. 271).

In modern terms the "certain equation" is the characteristic equation of the system. The "possible roots" are what we now call real roots; the "impossible roots" are imaginary or complex numbers involving the square root of -1. Maxwell's conditon thus requires that the real parts of all the roots of the system's characteristic equation must be negative.

> The actual motions corresponding to these impossible roots are not generally taken notice of by the inventors of such machines, who naturally confine their attention to the way in which it is designed to act; and this is generally expressed by the real root of the equation. If, by altering the adjustments of the machine, its governing power is continually increased, there is generally a limit at which the disturbance, instead of subsiding more rapidly, becomes an oscillating and jerking motion, increasing in violence till it reaches the limit of the action of the governor. This takes place when the possible part of one of the impossible roots becomes positive. The mathematical investigation of the motion may be rendered practically useful by pointing out the remedy for these disturbances (p. 272).

He applied his result to various regulators of his day and showed the design characteristics they required for smooth control.

In the history of feedback in engineering, Maxwell's paper is enormously significant. It opened the way for the use of mathematical methods in the design of control systems. Tinkerers now had help. Most significantly, Maxwell's paper exposed the role of complex numbers in the analysis of the behavior of feedback systems. That development would eventually lead to the use of operator notation, Laplace transforms, and transfer functions. Perhaps surprisingly, it is these high level mathematical notions, coupled with the use of feedback in electrical circuits, that revealed the simple loop concept in feedback control.

The Loop Concept in Twentieth-Century Feedback Engineering

Maxwell's work led directly to extensive use of differential equations in the design and analysis of feedback control devices. The idea of a closed loop of interactions is largely hidden in such equations, however. To have seen a feedback loop in the equation cited above from Maxwell, one would have to shift from left-to-right mathematics to a picture, something like that shown in Figure 2.5. To make such a shift in perspective requires either serendipity or a strong need for a schematic pictorial representation. But the feedback devices of the nineteenth century were physical contrivances; pictures of them—design sketches—captured their mechanical components, not an abstract schematic structure. And the mathematical analyses of the nineteenth century, the more famous of which include Maxwell (1868), Routh

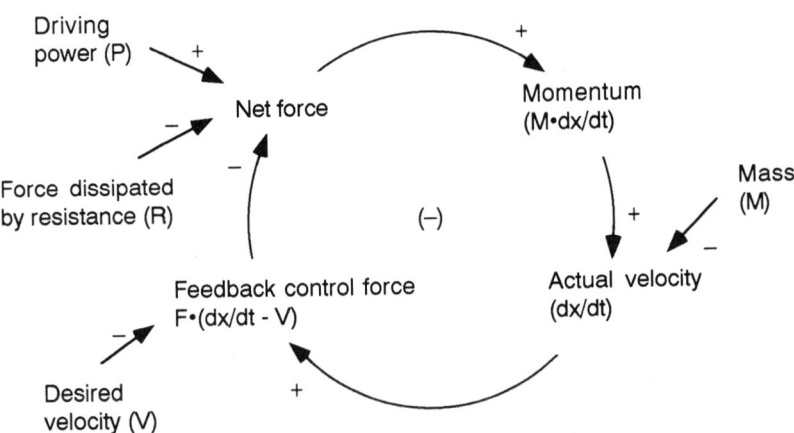

FIGURE 2.5: Implicit feedback loop hidden in Maxwell's differential equation for a moderator.

(1877), Lyapunov (1892), and Hurwitz (1895), apparently had no need for pictures. The focus was on the dynamic implications of differential equations. The mathematics itself was the schematic representation.

I do not know the first use of a loop diagram or the first reference to feedback control as a closed-loop phenomenon in engineering devices. It seems likely, however, that electrical networks were the stimulus for pictorial representations that exposed the loop nature of feedback control devices. Figure 2.6, for example, was used by Bode (1960) to describe Harold Black's 1927 invention of the negative feedback amplifier. Although now modified out of all recognition, it is still the germ of the idea behind modern high fidelity. Figure 2.6a captures Bode's basic idea. A signal coming in from the left travels along two paths. One path gets amplified by μ (in the triangle), and in the process the signal acquires unwanted noise. To get rid of the noise, the amplified signal is reduced back down to its original strength by θ ($= 1/\mu$), and the original signal is subtracted from it. Emerging from that subtraction is just the noise acquired by the amplification. Black's idea was to amplify that noise (in the other triangle) back up to the strength of the amplified signal and subtract. The system would then send out to the right the original input signal, amplified, but containing no amplification noise.

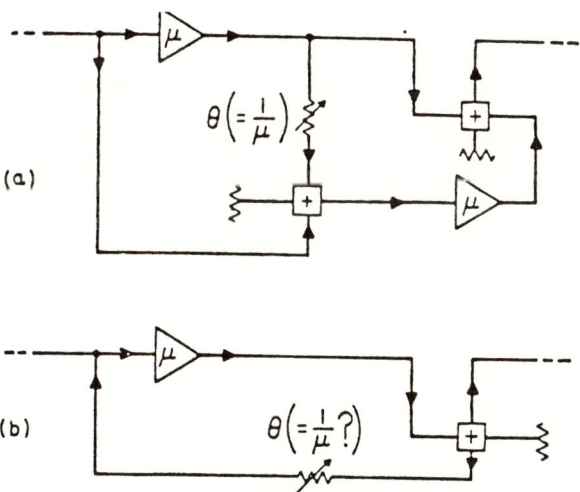

FIGURE 2.6: Schematic representations of the idea behind Harold Black's 1927 low noise amplifier. Source: Bode (1960).

As Bode recounts:

It is tantalizing to observe that we need a whole second amplifier identical with the primary one just to provide the compensation. One would like to think that it should somehow be possible to use the same amplifier twice. A naive approach to this problem is indicated in the schematic of [Figure 2.6b] (Bode 1960, p. 6).

The "naive approach" results in a negative feedback loop. Unfortunately, the idea didn't work: the noise does not get subtracted out once and for all, but instead a signal goes round and round the circuit. Nonetheless Black was able to take this naive approach and refine it to where it did produce the desired result.

The point here for the development of the feedback concept is that Black's schematic is vividly a loop. Indeed, the loop idea is the key to Black's design of a low noise amplifier, and to improvements later developed by Bode and Macmillan (Bode 1960, pp. 10–16). Furthermore, this explicit representation of a feedback circuit as a loop was reinforced by the use of operator notation for differential equations and the development of Laplace transform techniques for circuit design. (Bode 1960, p. 9, traces the latter to L.A. MacColl in the 1920s and early 1930s.) Those mathematical tools apparently led naturally to the block diagram, in which the loop idea is blatant.

Put simply, the engineer's block diagram rests on a translation of calculus concepts into the language of algebra. The operation of differentiation d( )/dt can be represented by an "operator" D. The operation of integration

$$\int ( )dt,$$

being the inverse of differentiation, can be represented by 1/D. Thus, if

$$\frac{d}{dt} x(t) = u(t)$$

so that

$$x(t) = \int u(t) dt,$$

then we can write

$$D\, x(t) = u(t) \quad \text{and} \quad x(t) = \frac{1}{D} u(t).$$

The operators D and 1/D have all the nice multiplicative properties that ordinary algebraic quantities have. Most importantly for our investigation, the operators D and 1/D push differential equations toward diagrammatic representations that expose their loop character. Consider exponential smoothing, for example. Its differential equation is

$$\frac{d}{dt} x(t) = \frac{u(t) - x(t)}{T}.$$

In operator notation this becomes

$$D x = \frac{(u - x)}{T}$$

or

$$x = \frac{1}{D} \frac{(u - x)}{T}.$$

There is an x on both sides of the equation. If one thinks of u as the input to the system, the equation says "x is subtracted from u, the difference is then divided by T and then operated on by 1/D, and the result is x." There is a loop in the sentence and therefore in the equation; engineers were led to represent this loop in a diagram such as the one in Figure 2.7. The blocks in the diagram represent multiplicative operations—multiplication or division, differentiation (D), or integration (1/D). The circle represents a sum or difference. The result is a visual representation of a differential equation. In it we can see

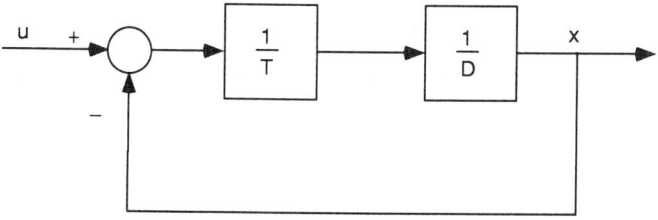

FIGURE 2.7: Block diagram of the differential equation for exponential averaging or smoothing, exposing the loop nature of the system: x is subtracted from u; the result is divided by T and then operated on by 1/D, to represent the equation $D x = (u - x)/T$.

the loop nature that Maxwell and his nineteenth-century colleagues missed.

Block diagrams gained considerable importance with the discovery of the significance of Laplace transforms for control theory analysis (Ogata 1970; Takahashi, Rabins, and Auslander 1970). It just so happens that these more sophisticated mathematical devices have the same symbolic characteristics as the simpler differential operators: the Laplace transform of a derivative is symbolized by multiplying by the Laplace variable s; integration by 1/s. We need no understanding of the details, however, to appreciate the significance of these engineering developments in the evolution of the feedback concept. In block diagrams, which were (and still are) used extensively in feedback control engineering, the loop nature of feedback systems symbolized mathematically is at last unmistakable.

By the 1930s, the concept of closed loops in amplifiers and engineering control mechanisms was apparently well established. By the 1940s the idea was so basic that Gordon Brown, founder of the servomechanisms laboratory at the Massachusetts Institute of Technology, began his early text with the statement, "This book is about closed-loop control systems" (Brown 1948). He pictured the basic idea as shown in Figure 2.8.

Brown's book is of interest to us for two other reasons, which, though somewhat ahead of our story, may be mentioned now to show linkages and directions. First, one of the important figures in the evolution of the feedback concept in the social sciences is Jay Forrester, who came to MIT's servomechanisms lab in 1940 and who was profoundly influ-

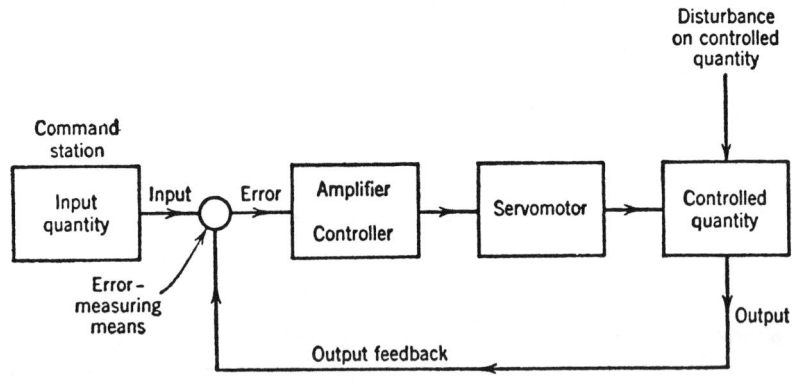

FIGURE 2.8: Closed-loop control system as pictured in Brown (1948).

enced by Gordon Brown and the activity there. We shall hear phrases like the following echoed in various forms in Forrester's and others' applications of the feedback concept to social systems:

> The specific objective of this book is to set forth in fairly simple terms a procedure for understanding the complexities involved in the system behavior and a technique for the direct approach to system design as a substitute for the approach through experimentation. An aim of the book is to establish in the mind of the designer a concept of synthesis as contrasted with mere analysis.... In [closed-loop control], the control acts in accordance with the dictates of an arbitrary quantity and in accordance with what happened as a result of the control operation.... A closed-loop control system is thus an error-sensitive system and, being such, it acquires certain peculiarities and idiosyncrasies which, in large measure, are the reasons for this book (Brown 1948, pp. 2,3).

Second, Brown's book contains historical references and comments, unusual in such a text, that together with some remarks of Bode will help us to understand one of the possible origins of the split we shall observe in the use of the feedback concept in the social sciences. There was not one kind of feedback engineer, but at least two. Servomechanisms people and electronic amplifier people made use of the feedback concept in somewhat different ways. The details of the differences will be more significant to us after investigating some uses of the feedback concept in the social sciences, so we shall return to these ideas in section 3.4.

The Impact of Control Theory

Lest it be forgotten in our tracings of the concept of the feedback loop, it should be emphasized that the concern of feedback engineers was control. What emanated most strongly to the social sciences from engineering control theory was that control, or management, requires the feeding back of information. In particular, the necessary information is the "error signal"—the extent to which the actual state of the system differs from the desired state. Is the temperature of Drebbel's furnace, the speed of Watt's steam engine, or the size of a car dealer's inventory equal to its desired value? If not, adjust. For the control theory engineer, the goal is to adjust automatically, as promptly but as smoothly as is feasible. The goal of a manager, or a governor or legislature, might be stated similarly.

The control engineer's concept of feedback made it easier to see a particular characteristic of human systems. What is the difference between a washing machine, for example, and a primitive woman washing clothes in a river by beating them clean with a stick? Before awareness of feedback, many people would have asserted that the difference is cleaner clothes from the washing machine. Enlightened by the feedback view, however, one has the chance to observe that the machine stops without ever "knowing" whether the clothes have reached a desired state of cleanliness. The "primitive" woman is not as primitive as the machine; she continues cleaning until she is satisfied with the results. Think how difficult it would be to design a washing machine with that talent. Such efforts at automatic control led to our awareness of the role of feedback in the functioning of individuals, groups, and societies.

2.2 The Loop Concept in Mathematical Models

Though it remained hidden until the 1900s, the loop idea is present in all the differential and difference equations used to represent dynamic phenomena in the natural and social sciences. The uncovering of that loop structure in the mathematical models of control engineers was sketched in section 2.1. But there has been a totally separate tradition in the use of mathematical models in the biological and social sciences. Eventually coming together with engineering control theory developments, the mathematical modeling tradition in the social sciences contributes something of its own point of view to our modern conception of the feedback loop.

The Population Equation of Verhulst

The logistic population equation of P.F. Verhulst (1838) provides an instructive example of the loop concept embedded in differential equations applied to societal phenomena:[1]

$$\frac{dP}{dt} = aP - bP^2.$$

The term aP in this equation represents the tendency of population to grow at a rate proportional to the size of the population, the standard assumption of exponential growth. Verhulst intended the term involving the square of population to represent conflict and stress arising from contacts between people, which he assumed would be roughly proportional to P times P. The result is the familiar nonlinear logistic equation, which exhibits sigmoid growth.

Figure 2.9 shows a representation of this equation as a pair of feedback loops. The positive loop corresponds to the tendency of the population to grow at a rate proportional to itself. The negative loop corresponds to the growth-limiting effects Verhulst envisioned in conflict and stress. For low levels of population the growth effects predominate, and the population would appear to exhibit essentially unrestricted exponential growth. The term bP^2 representing conflict and stress would grow more rapidly than aP, however. Eventually, it would overtake aP and bring population growth to a halt, as shown in Figure 2.10. Thus over time the system changes its own growth tendencies. In feedback terms, the system shows a gradual shift in *loop dominance*. When population is low, the positive loop dominates and increasingly rapid growth ensues. When population gets high enough, the negative loop begins to dominate and the growth rate begins to decelerate, becoming zero when the population equals $a/b.^2$

The logistic differential equation and its closed-form solution are very familar. What the feedback view reveals here is an underlying loop structure to the system that is not frequently recognized. (There is no evidence, for example, that Verhulst had in mind loops like those shown in Figure 2.9, nor that he made use of any loop notions to formulate his equation.) Yet the loop structure is even more basic than the logistic equation itself. Many systems consisting of a positive loop and a negative loop will generate S-shaped behavior over time, differing in only minor ways from the precise logistic pattern. The primitive

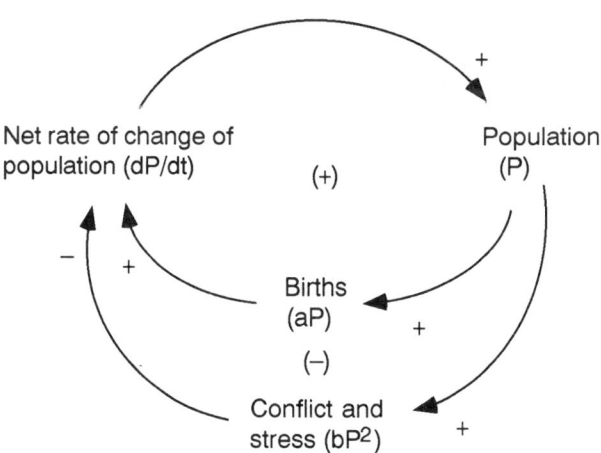

FIGURE 2.9: Loop structure of the Verhulst population equation.

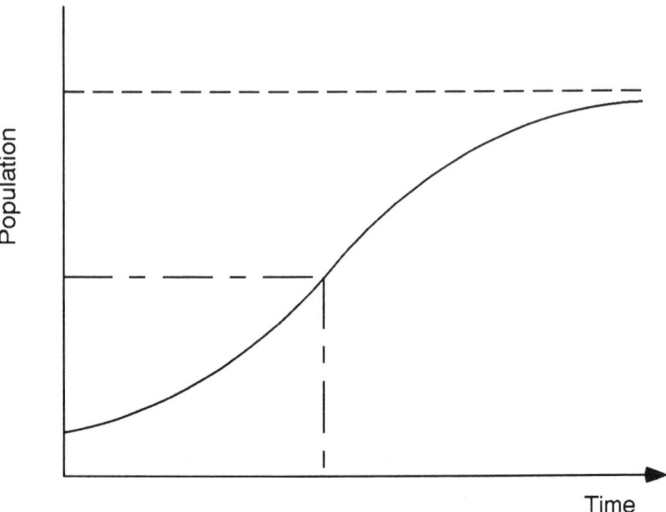

FIGURE 2.10:
Behavior of the Verhulst population equation, showing the effect of the shift in loop dominance from positive to negative when the population reaches half its maximum.

concept here is not the logistic equation, but the coupled positive and negative loops it exemplifies.

A second insight can be derived from Verhulst's equation. We have observed that it displays a shift in dominance from the positive to the negative loop as population grows, and that this shift in dominance is generated by the equation itself. No outside influence causes the change. In mathematical terms the shift is a consequence of the nonlinear structure of the differential equation. Initially, the term aP is much larger than bP^2. But as P grows, P^2 grows more rapidly than P. Eventually, bP^2 will overtake aP, no matter how great a head start aP was given. The shift in loop dominance is a consequence of the nonlinear form of the equation. If the rate of change of population had been written as a linear function, as $aP - bP$ rather than $aP - bP^2$, the shift in dominance between the positive and negative terms would not have taken place. Moreover, if the rate of change had been written as a cubic, $aP - bP^3$, the negative term again would eventually overtake the positive one and shift the dominance between loops.

Nonlinear Models and Shifting Loop Dominance

The ability to shift loop dominance is a characteristic of nonlinear mathematical models. (One might also argue that it is a characteristic of reality.) Nonlinearity, in turn, is frequently a characteristic of the more famous mathematical models employed in biology that have found their way into the collective consciousness of the natural and social sciences. Thus many of these models acquire their characteristic patterns of behavior from shifts in loop dominance. A classic example is the common epidemic model (Kermack and McKendrick 1927) that has been used in varying disguises to study the diffusion of information:

$$\frac{dS}{dt} = -aSI,$$

$$\frac{dI}{dt} = aSI - bI,$$

where

S = the number of susceptible people,
I = the number of infectious people,
a = proportionality parameter for the rate of infection,
b = proportion of people removed from the infectious pool per unit time (by cure, quarantine, death, or other causes).

The loop structure of this model, shown in Figure 2.11, is most clearly revealed when the term aSI common to both dS/dt and dI/dt is recognized as the infection rate. The diagram shows that a single negative loop affects the susceptible population, while two loops, one positive and one negative, influence the infectious population. As in the Verhulst model, whether the infectious population grows or not

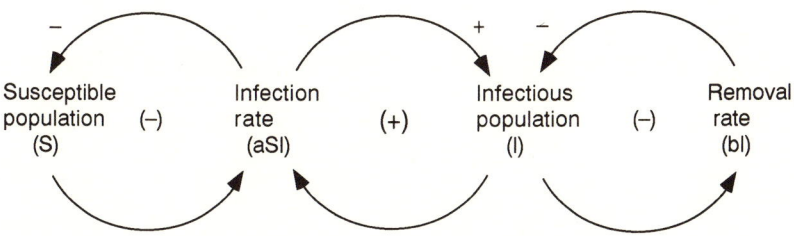

FIGURE 2.11: Loop structure of an epidemic model.

depends upon which of these two loops dominates. And again, the dominance shifts in the course of epidemic because of a nonlinear term. (Here the nonlinearity comes from the product SI, analogous to the product P^2 in the Verhulst model.) As long as there is a sufficient number of susceptibles around, the positive loop dominates and the epidemic grows in severity. When S declines sufficiently, the negative loop comes to dominate, and I then declines, eventually to zero. The behavior of the susceptible population is usually S-shaped, something like a mirror image of the pattern of population in Verhulst's logistic model. The similarity of the behavior patterns in the two models is a consequence of their similar coupled positive/negative feedback loop structures.

It should be emphasized again that the loop structure of these differential equation systems is a modern observation. There is no evidence that these early authors viewed their equations as circular-causal structures or used a loop perspective to guide their formulations. Loop dominance is a modern concept that emerged when the engineer's notion of feedback came together with nonlinear models of the sort described here. One of the principal sources of such models for social scientists was the work of the mathematical biologist Alfred Lotka.[3]

The Work of Alfred Lotka

Lotka's *Elements of Physical Biology* appeared in 1925 and was reprinted again in 1956. The focus of the work was pattern in the dynamics of living organisms, from the sweep of evolution down to the five-day life cycle of a bacterial colony. Here we have "cycles" appearing in the form of the circulation patterns of water, carbon dioxide, nitrogen, and phosphorus in the biosphere (Lotka 1925, pp. 209–258). Hints of feedback loops appear again in Lotka's discussion of "receptors" and "effectors"—sensorimotor capabilities in living organisms that receive information from the environment and act or react upon it (pp. 339–340). Later in the same book Lotka also considered adaptation in evolutionary processes, although not in loop terms. The explicit linking of the loop idea with adaptive mechanisms would come later, after the emergence of the engineers' concept of feedback into the social sciences. As a part of that adaptive evolutionary process, Lotka attempted to account for the origin and development of consciousness. But unquestionably his most quoted contribution is the model of a closed ecosystem attributed jointly to him and to Volterra (1931).

The Lotka-Volterra equations are a model of a predator-prey system that exhibits sustained oscillations:

$$\frac{dx}{dt} = ax - bxy,$$

$$\frac{dy}{dt} = cxy - dy,$$

where

x = the prey population,
y = the predator population,

and a, b, c, and d are parameters relating to births, frequency of contacts between prey and predators, and deaths. The equations reflect the assumptions that
- in the absence of predators, prey grow without limit (y = 0 implies dx/dt = ax);
- in the absence of prey, predators die out (x = 0 implies dy/dt = −dy);
- consumption of prey by predators is proportional to the number of contacts between prey and predator (−bxy);
- births of predators are directly related to the health of the predator population and are assumed to be proportional to the consumption of prey (cxy).

Once again this is a nonlinear system of differential equations with an implicit causal loop structure, shown in Figure 2.12. Neither Lotka nor Volterra observed this loop structure, however, as they focused on solving the system of nonlinear equations in closed form.

The oscillatory nature of this predator-prey system can be viewed as a consequence of a regular pattern of shifts in dominance among the loops in Figure 2.12. Each population is affected by a positive growth loop and a negative decay loop. A major negative loop operates between the two populations to shift the dominance between these minor positive and negative loops. When both populations are small, contacts between prey and predator are few and far between. The positive feedback loop affecting the prey dominates, and the prey grow almost exponentially, while the predator population, dominated by its negative loop, declines even further. Once the prey population has grown sufficiently, the dominance in the predator population shifts to its positive loop, and predators begin to flourish. Eventually, that increase in predators brings about another shift in loop dominance: the nega-

38 Prehistory and Emergence

tive loop in the prey population comes to overpower the positive growth loop, and the prey begin to decline. These continuing shifts in loop dominance result eventually in the dominance of the negative loops in both populations, reducing them down to small numbers, at which point the cycle repeats. Although not a rigorous argument, this heuristic analysis in terms of loop dominance makes the existence of sustained cycles in the Lotka-Volterra equations very plausible.

Dynamic systems like these predator-prey equations eventually contributed to the modern concept of feedback and mutual causality in social systems in two ways. First, such a system of equations has a loop structure that eventually became recognized as synonymous with the concepts of feedback and circular causality. Second, such a system derives its dynamic patterns from its own internal structure. It is an endogenous theory of a pattern of behavior. No external forces are needed to intervene to create the turning points in the Lotka-Volterra equations. That endogenous view also eventually becomes explicit and gets linked with feedback (see section 3.3).

The Arms Race Models of L.F. Richardson

The closest to a feedback view of systems of equations in the social sciences prior to the 1940s appears in the work of Lewis Richardson

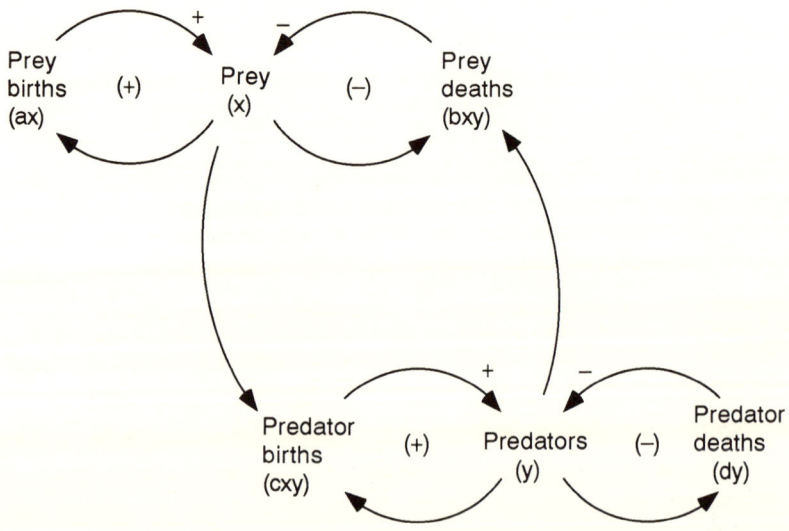

FIGURE 2.12: Loop structure of the Lotka-Volterra predator-prey equations.

(1919, 1935, 1938, 1939),[4] who developed a sequence of mathematical models of arms races. Richardson's early work appeared after the first World War and was applied retrospectively to data before the war. His "Generalized Foreign Politics" (1939) used current data to look ahead to a possible second World War. The models are of interest because they attracted attention in the social sciences and they focused on the potential for societies to generate and then try to control a deadly positive feedback loop.

Richardson presented his models as

> merely a description of what people would do if they did not stop to think ... what would occur if instinct and tradition were allowed to act uncontrolled (Richardson 1960, p. 12).

He attempted to capture, in equations simple enough to be solvable, the tendencies of nations to resist defiance, suspect defense to be concealed aggression, find difficulties in paying for armaments, perceive insults or threats in the behavior of other nations, and so on. His concerns were fear and rivalry among nations, what these motives lead nations to do, and what processes can act to control them. He was aware that some would find his mathematical models inappropriate for such investigations or too simple to be useful, but he perceived the mathematics to be indispensable:

> What the result of all these motives may be is not at all evident when they are stated in words. It is here that mathematics can give a powerful aid (1960, p. 13).

There are two themes here that we shall see repeated (by some) as the feedback perspective is applied later in the social sciences. First, mathematical models are seen to be necessary to clarify complex situations and allow accurate deductions from hypotheses. Second, the models formulated are seen to be predictive in only a very limited sense: they reflect what would occur if traditional patterns of interaction do not change.

Richardson postulated that arms races develop as nations expand their own arms production to counter or defend against the stockpile of arms in an opposing country. Mathematically,

$$\frac{dx}{dt} = my,$$

$$\frac{dy}{dt} = nx,$$

where

x = stockpile of arms in nation I,
y = stockpile of arms in nation II,
m, n = positive "defense" coefficients.

The structure is a single positive feedback loop: the stockpile of arms in country I leads to an increase in the arms of country II, which in turn leads to still more arms in country I. The only stable condition in such a situation is for the arms in both countries to be zero—otherwise, a never-ending arms buildup, presumably ending in war.

But Richardson observed that this formulation does not take into account the willingness and ability of the countries to produce arms. The models became:

$$\frac{dx}{dt} = my - ax,$$
$$\frac{dy}{dt} = nx - by,$$

where a and b represent positive "fatigue and expense" coefficients. The model has now acquired two negative feedback loops that can diminish the tendency to stockpile arms. Missing in this more complete model, however, is any reflection of a tendency in either country to build up its arms independent of the actions of the other.

Thus Richardson postulated the following as his basic arms race model:

$$\frac{dx}{dt} = my - ax + g,$$
$$\frac{dy}{dt} = nx - by + h,$$

where g and h represent "grievances" or "aggressive intentions." The loop structure of these equations is shown in Figure 2.13.

Two other models involving the difference $(y - x)$ between the stockpiles of the two countries were developed and analyzed. One of these was used to analyze "rivalry" (Richardson 1919/1935), while the other dealt with the potential for "submissiveness" in the face of large inequities in arms (Richardson 1960, p. 56). The latter was Richardson's most complete model, including as it did the above models as special cases. It was the only nonlinear structure he presented.

It is clear that the mathematics available to Richardson limited his choice of the form of his equations (1960, p. 56). Given more powerful analytical tools, his investigations would have been somewhat different. He was naturally criticized for the high aggregation of his models and the feeling that little of significance could come from such simple hypotheses, however deeply they were developed. He himself would have been skeptical were it not for a remarkable fit between his theories and data for arms expenditures leading to the first World War (p. 33). Data in the 1930s convinced him that the world was again on an unstable path, and he pushed to have "Generalized Foreign Politics" published in an effort to awaken antiwar forces. Nonetheless, his work in the long run probably reinforced the feelings of many that mathematical analyses of this sort do not significantly help solve political problems.

For us the significance of Richardson's work is its nearness to an explicit treatment of feedback loops in the dynamics of a social system. Richardson knew he was dealing with a powerful self-reinforcing tendency in international relations. In *Arms and Insecurity* (1947, 1960), the book that reiterated and expanded on the ideas of "Generalized Foreign Politics," Richardson associated some of his ideas with work of Ashby (1945) and with Bateson's concept of "schismogenesis" (Bateson 1935, 1936), both of which figure prominently in the explicit development of the concept of the feedback loop in the field of cybernetics. (See sections 2.5 and 3.2.)

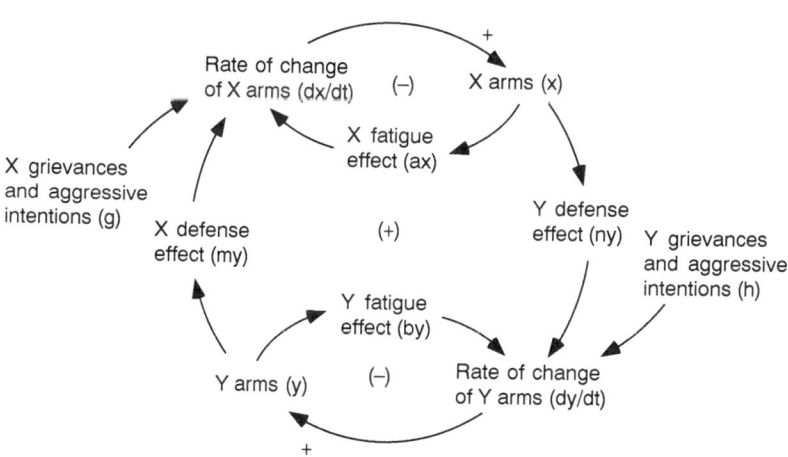

FIGURE 2.13: Loop structure of Richardson's linear model of an arms race.

Feedback Loops in Economic Models

No discussion of the loop concept in mathematical models influencing the social sciences is complete without reference to economic models. Efforts to use differential equations, difference equations, and mixed difference/differential systems to capture economic dynamics flourished, beginning particularly in the 1930s. (See Frisch 1933, Kalecki 1935, Harrod 1939 for examples.) The loop concept is implicit in all of them. Perhaps the most significant illustration is the classic multiplier/accelerator model of Paul Samuelson (1939).

Samuelson's model, a blend of two previously existing models, was expressed in terms of finite difference equations:

$$Y_t = C_t + I_t + G_t,$$
$$C_t = aY_{t-1},$$
$$I_t = b(C_t - C_{t-1}) = ab(Y_{t-1} - Y_{t-2}),$$
$$G_t = \text{an exogenous quantity},$$

where

Y_t = national output (= income) at time t,
C_t = consumer demand (= consumption) at time t,
I_t = estimated investment demand (= investment) at time t,
G_t = government spending at time t,
a = the marginal and average propensity to consume,
b = the "acceleration coefficient."[5]

The subscript t−1 denotes the value of a quantity at a previous point in time. The period between t and t−1 was unspecified in Samuelson's original paper, but must have something to do with the time over which aggregate consumption and investment decisions are planned and carried out.

That causal loops are present in this model is evident from the equations that show mutual dependence. The first two equations state that output and consumer demand depend on each other. The first and third equations show output and investment depend upon each other. Figure 2.14 shows the resulting feedback loop structure of the model.[6] The multiplier is a positive loop, while the accelerator consists of both a positive and a negative loop. By expressing the system as a single second-order difference equation, Samuelson showed the structure has the potential to oscillate, to asymptotically approach an equilibrium, or to grow without bound, depending upon the values of the parameters a and b. The tendency to oscillate comes from the negative loop in the accelerator structure. In addition to multiplying the effect of exogenous government spending, the multiplier loop tends to stretch out the pattern of behavior the rest of the structure produces.[7]

The Econometric Tradition

Samuelson's analysis focused on the dynamic patterns the multiplier/accelerator structure could produce. He found that the behavior of the system of difference equations depends significantly on the values of the parameters chosen. That conclusion reinforced well the statistical estimation efforts in the economics of the time, exemplified by the

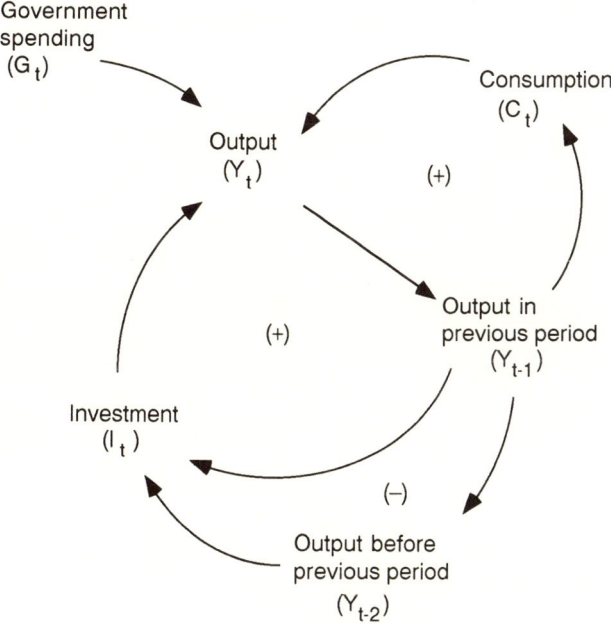

FIGURE 2.14: Feedback loop structure of Samuelson's multiplier/accelerator model.

work of Koopmans (1936) and Tinbergen (1937, 1939). In the econometric tradition established by these authors and further developed by Klein (1950) and many others, loops of information feedback and mutual causality play a major, although largely unacknowledged, role.

Tinbergen defined "econometric business cycle research" as "a synthesis of statistical business cycle research and quantitative economic theory" (Tinbergen 1939, I, p. 11). His *Business Cycles in the United States of America, 1919–1932* (1939, II) was one of the earliest attempts to construct a complete economic system on a statistical basis. It is the progenitor of all modern econometric models. Composed of finite difference equations, similar to those in Samuelson's multiplier/accelerator, Tinbergen's econometric model for U.S. business cycles is replete with feedback loops. He came about as close to recognizing them as such as one could without actually doing so.

Tinbergen's equations for investment and profits and his comments about them provide striking examples. He presented the equations and regression results in an unusual but rather revealing graphic form, reproduced in Figure 2.15. In the top pair of curves in each plot, the solid line represents the actual time series to be "explained" by the regression. The dotted line gives the estimated time series. The curves below give the time series of the components of the estimate in the least squares regression, as weighted by their regression coefficients. The shaded graph at the bottom gives the pattern of deviations between the actual and the estimated time series.

Tinbergen observed a loop implied by these equations for profits (Z) and investment demand (V). He noted that V is the principal explanatory variable in the regression for Z in one graph, while in the other a component of Z ($Z^C_{-1/3}$) similarly "explains" a component of V (v'). As he viewed it,

Taking the fall in general investment from 1929 to 1930—which contributed considerably, according to graph [2.15b], to the fall in profits in 1930—we find from graph [2.15a] that profits one half-year before were the chief explanatory series. Here we meet a very important feature. It would seem as if this were a circular reasoning: profits fell because investment fell, and investment fell because profits fell (1939, II, p. 127).

The equations suggest a positive feedback loop involving profits and investment. To Tinbergen, however, their circularity was not a structural characteristic but a potential source of internal contradiction that needed to be explained away. He went on to say,

This is, however, an inexact statement. Profits in period t fell because investment in period t fell, but the latter fell because of a fall in profits in period t − ½; and owing to this time lag there is no danger of circular reasoning.

The lag around the loop prevents "circular reasoning." If the lag were not there, Tinbergen explained, one could still avoid the contradictions inherent in such circularities by assuming that both profits and investment were both determined by other variables (presumably the same set of other variables). It is fascinating to observe that what Tinbergen saw as a potential logical flaw, others in the social sciences would later see as a fundamental structural characteristic of social systems.[8]

From the time of Samuelson and Tinbergen on to the present, loops of information feedback and circular causality have characterized econometric models.[9] Klein's contribution (1950) came at a time when he conceivably could have drawn connections between the structure of econometric models and the emerging feedback literature in the social sciences. There is, however, no mention of the word feedback in his work, nor any apparent tendency to make use of the concept in the formulation or analysis of the models he discusses.

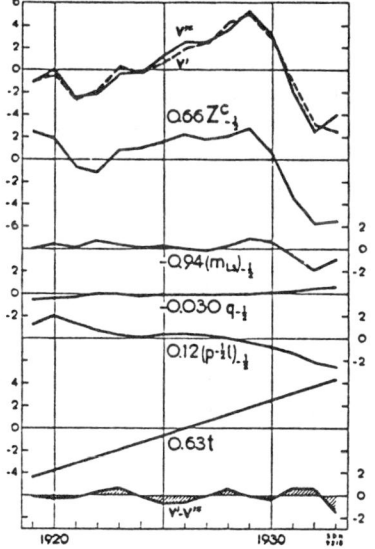

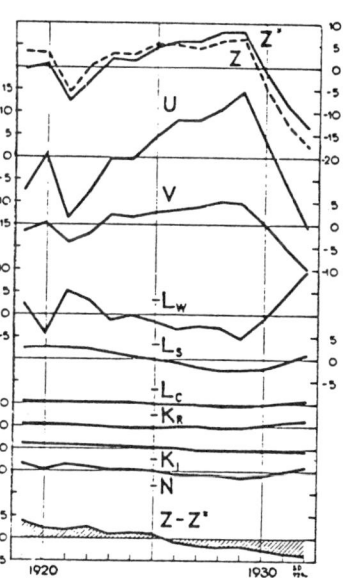

FIGURE 2.15: Graphical representations of regression equations for Demand for Durable Producers' Goods (v') and Profits (Z) in Tinbergen (1939, II, pp. 47, 124).

We may conclude that although the early econometric authors recognized, in fact clearly intended, the mutual dependencies in their models, they did not picture their systems in loops nor associate patterns of model behavior with loop structure and polarities. Before 1940 the feedback concept had not yet emerged from engineering, and the other loop notions in the social sciences (that we shall encounter in section 2.5) were not associated with this quantitative, statistical line of thinking.

Summary

Two distinct mathematical modeling traditions implicitly embodying feedback and mutual causality are evident here. One, emerging largely from biology, emphasized nonlinear models and systems of differential equations. The other, in economics, focused almost exclusively on linear difference equations. In the former, efforts were directed at deriving the dynamic behavior of an aggregate system from a simple, clear structure of interrelationships. In the latter, the major effort was the accurate estimation of parameters to fit a set of linear equations to economic data. In the former, parameters played a relatively minor role; the nonlinear structure of equations (with the shifts in loop dominance it engendered) was largely responsible for the dynamic patterns of the system. In the linear econometric models, parameters were found to be the determining factors in whether a system would exhibit the behavior of real economic variables. In neither of these mathematical modeling traditions was the loop concept recognized before it emerged from engineering servomechanisms and control theory.

2.3 Homeostasis and the Loop Concept in Biology

In addition to the loop idea implicit in mathematical models of ecosystems, biology contributed the concepts of kinesthesia, proprioception, and homeostasis, each of which became identified with the feedback concept once it emerged from engineering. The first two ideas are closely related. The third, homeostasis, had a dramatic impact on the evolution of the feedback concept in the social sciences.

Kinesthesia and Proprioception

Kinesthesia, a word constructed from the Greek words for motion (*kinēma*) and perception (*aisthēsis*), states its own meaning—the perception of motion. The perception, however, is not a conscious act but an autonomic process within the body. Kinesthesia is the internal sense by which one knows without looking how some part of one's body is moving. It is the sense that enables a person to pass a common sobriety test: spread your arms wide, close your eyes, and then touch your index finger to your nose in one smooth motion. The central nervous system sends out messages to appropriate muscles to carry out the action. In turn, sensory nerves continuously transmit back to the brain the position and motion of the tip of the finger, the hand, the elbow, and so on, allowing the sober person to bring finger to nose without difficulty. Presumably, the presence of enough alcohol in the blood stream interferes with the sensors or transmitters sufficiently to cause a person to miss the mark.

In this description of information transmitted and received within the nervous system, a closed loop clearly exists. Signals originating in the brain direct the muscles to act, and sensors in the muscles report back. Such loops were known or hypothesized to exist in the early 1800s (Bell 1826). One component of such loops are the sensory receptors, the receivers and perceivers of stimuli arising within the body. These sensory receptors are called proprioceptors, and the act of receiving stimuli is known as proprioception.

Perhaps the earliest nearly explicit description of a loop process linking the senses, muscles, nerves, and brain was given by Descartes in his *Traité d l'Homme*. The Treatise was apparently written between 1629 and 1632 and was paraphrased in his *Discourse on Method* (1636) but not published until twelve years after his death (Descartes/Hall 1972). In one section, Descartes described the flow of blood and the operation of the nervous system in the human body. He likened both to aspects of the fountains in the king's gardens, which contained statues that could move in various ways as people walked through the gardens and stepped on stones that activated the statues' automated, pneumatically powered mechanisms. One can perceive the shadow of a negative feedback loop in Descartes's description of how the *rational soul* in the brain can respond to external disturbances:

External objects which merely by their presence act on the organs of sense and by this means force them to move in several different ways, depending on how the parts of the brain are arranged, are like strangers who, entering some of the grottoes of these fountains, unwittingly cause the movements that then occur, since they cannot enter without stepping on certain tiles so arranged that, for example, if they approach a Diana bathing they will cause her to hide in the reeds; . . . And finally when there shall be a rational soul in this machine, it will have its chief seat in the brain and will there reside like the turncock [*le fontenier*] who must be in the main to which all the tubes of these machines repair when he wishes to excite, prevent, or in some manner alter their movement (Descartes/Hall 1972, p. 22).

In this passage, disturbances are sensed, and messages about them are communicated to the rational soul in the brain [*le fontenier*] which can respond in various ways. If the rational soul were to respond to *counteract* a disturbance, then Descartes is describing here a process of negative feedback control. Elster (1975, p. 57) finds this passage to be the first use of the concept of negative feedback to describe human behavior. The loop is implied in Descartes's image of the "tubes" in the body which emanate from the brain and return to it, carrying information and motive force between the brain and the muscles. It should be noted, however, that without an intelligent control mechanism the processes Descartes describes, in both the body and the king's gardens, are open-loop machines that merely respond to stimuli but do not necessarily act to eliminate discrepancies between some desired state and the perceived disturbed state.

Once the feedback concept became well known in the social sciences, it was linked to these previously existing psychobiological ideas. To illustrate, the entry for "feedback" in the 1958 edition of *A Comprehensive Dictionary of Psychological and Psycho-Analytical Terms* ends with the following editorial comment:

It is only recently that the feedback principle has been isolated in mechanics and extensively made use of: in organisms, the corresponding principle has long been recognized (see knowledge of results, coenesthesia, proprioception, kinesthesia); it is only the term which is new (English and English 1958, p. 204).

Homeostasis

Recognizing the feedback loop concept in proprioception and kinesthesia largely amounted to applying a new name to old ideas. Little additional insight was gained in the recognition. But the linking of the feedback concept with the biological notion of homeostasis was of a wholly different order of significance. To some in the social sciences, the feedback concept became identified, virtually synonymous, with homeostasis. As we shall see, that identification eliminates completely the consideration of positive feedback loops.

Homeostasis refers to the remarkable capacity of higher organisms to maintain physiological stability in the face of dramatically varying external and internal conditions. At the heart of the idea is the notion that living organisms can apparently react automatically to counter disturbances from a preferred or normal status quo. The term for the phenomenon was coined by Walter Cannon (1926, 1929, 1932), although he himself traced elements of the concept back as far as the ancient Greeks. Hippocrates (460–377 B.C.), for example, had apparently supported the idea that disease can be cured by natural powers within the body. Cannon observed that such a *vis medicatrix naturae* "implies the existence of agencies which are ready to operate correctively when the normal state of the organism is upset" (1932, p. 26).

As early as 1912 Cannon and his colleagues at the physiological laboratory at Harvard University had published research about self-equilibrating mechanisms in the body. They investigated, among others, the phenomena of hunger and thirst, blood clotting, traumatic shock, adrenal secretion, and blood pressure dynamics, each of which involve processes that work toward the maintenance of stable body conditions. Cannon concluded that the word "equilibrium" did not do justice to such self-regulating phenomena:

The coordinated physiological processes which maintain most of the steady states in the organism are so complex and so peculiar to living beings—involving, as they may, the brain and nerves, the heart, lungs, kidneys, and spleen, all working cooperatively—that I have suggested a special designation for these states, homeostasis. The word does not imply something set and immobile, a stagnation. It means a condition—a condition, which may vary, but which is relatively constant (Cannon 1932, p. 24).

Precursors of the Concept of Homeostasis

Darwin had argued for the possibility of the evolutionary adaptation of organisms to their environment. L.J. Henderson (1913) addressed what might be thought of as the flip side of the evolutionary story. In *The Fitness of the Environment* he explored the numerous and highly improbable ways the physical and chemical properties of the biosphere are precisely right for the support of life. He argued that "fitness is a mutual or reciprocal relationship between the organism and the environment" (Henderson 1913, p. 267). He structured his argument around three characteristics of organisms which he considered to be essential for life, one of which is, in essence, the notion of homeostasis:

Living things, still more the community of living things, are durable. But complexity and durability of mechanism are only possible if internal and external conditions are stable. Hence, automatic regulation of the environment and the possibility of regulation of conditions within the organism are essential to life (Henderson 1913, p. 31).

Although the idea of homeostatic regulation was described quite explicitly by Henderson (1913, 1928, 1935) in both physical and social contexts, he did not originate the idea. Cannon noted that by the 1800s a number of physiologists had begun to observe and speculate about homeostatic mechanisms in living organisms. Among the earliest and most important was Claude Bernard (1878–79). Bernard had suggested that the surprising stability of specific internal operations of a living organism was the result of a relatively constant general "milieu interne," a concept that Cannon later associated with what he called the "fluid matrix" of the body (the blood and lymph systems):

In the main, stable states for all parts of the organism are achieved by keeping uniform the natural surroundings of these parts, their internal environment or fluid matrix. This 'milieu interne,' as Claude Bernard pointed out, is the product of the organism itself. So long as it is kept uniform, a large number of special devices for maintaining constancy in the workings of the various organs of the body are unnecessary (Cannon 1932, p. 269).

Bernard's idea seems rather distant from the feedback concept. But other early physiologists came quite close to a closed-loop view of such self-maintained steady states. Cannon (1926, p. 26) cites several examples: Pflüger in 1877 asserted that "the cause of every need of a living being is also the cause of the satisfaction of the need." Fredericq in 1885 observed that "the living being is an agency of such sort that each disturbing influence induces by itself the calling forth of compensatory activity to neutralize or repair the disturbance." By 1900 Richet had concluded that the key to such bodily stability was what he called its "slight instability": "By an apparent contradiction, [the body] maintains its stability only if it is excitable and capable of modifying itself according to external stimuli and adjusting its response to the stimulation." Each of these quotations is tantalizing close to a description of the generic negative feedback loop, in which any discrepancy between the actual state of a system and its desired state continuously generates pressure for action that brings the actual closer to the desired.

The Wisdom of the Body

The same implicit loop structure is present in all of Cannon's descriptions in *The Wisdom of the Body* (1932). To illustrate, some systems in the body maintain relatively steady states by an overflow process:

If the hydrogen-ion concentration in the blood is altered ever so slightly towards the acid direction, the especially sensitized part of the nervous system which controls breathing is at once made active and by increased ventilation of the lungs carbonic acid is pumped out until the normal state is restored (Cannon 1932, p. 270).

The sentence itself traces a closed loop starting and ending with the hydrogen-ion concentration. Cannon observed that other homeostatic mechanisms control not the quantity or concentration of some material but the extent of the operation of a process. The closed-loop nature of some mechanisms is still vivid:

If the blood pressure falls and the necessary oxygen supply is jeopardized, delicate nerve endings in the carotid sinus send messages to the vasomotor center and the pressure is raised (p. 270).

Some complex body systems maintain homeostasis by controlling both processes and materials:

A combination of the use of reserves and the use of the altered rate of processes is found in the complex of mechanisms which operate to assure uniformity of the oxygen tension. It will be recalled that besides the faster return of blood to the heart, the faster heart beat, and the high head of arterial pressure which results in a faster blood flow—all accelerations of continuous processes—there is a setting free of the concentrated red corpuscles from the store in the spleen. These corpuscles now join those already in the hurried current and help them to meet the exigency which confronts the laboring cells (p. 278).

The wise body indeed! In these descriptions we clearly see discrepancies between actual and desired or normal states, sensors that continuously monitor the actual state, and information passing from sensors to other parts of the body that carry out corrective action, altering the state of the system.

It was clear to Cannon that such homeostatic mechanisms have to exist whenever a condition in a living organism is held relatively constant. Even when the information transmitters, the sensors, or the activating agents had not been determined, Cannon asserted they had to exist. He noted, for example, that changes in the concentration of blood proteins and blood calcium bring about "alarming disturbances" in the organism. He concluded that calcium concentration must be tightly controlled, even though he did not know how the body accomplished it:

In all probability a slight movement in the direction of change is signalled, just as in other instances which we have considered above, and then the tendency is corrected. But what sends in the signal and how the signal sends orders to the organs which make the correction, must remain a mystery until further physiological research has disclosed the facts (1932, p. 271).

Cannon's General Principles on Homeostasis

Cannon pushed his understandings of homeostatic mechanisms toward generalizations about control in living organisms. The following were among a list he first published in 1926:
- In an open system, such as our bodies represent, compounded of

unstable material and subjected continually to disturbing conditions, constancy is in itself evidence that agencies are acting, or ready to act, to maintain this constancy.
- If a state remains steady it does so because any tendency towards change is automatically met by increased effectiveness of the factor or factors which resist the change.
- The regulating system which determines a homeostatic state may comprise a number of cooperating factors brought into action at the same time or successively.
- When a factor is known which can shift a homeostatic state in one direction it is reasonable to look for automatic control of that factor, or for a factor or factors having an opposing effect (1932, pp. 281–282).

He repeatedly emphasized the second principle, in which change is met by "increased effectiveness" of the factors that resist the change. It is tantamount to a modern definition of a negative feedback loop. The first principle thus essentially asserts that stability in a dynamic system necessarily implies the existence of one or more negative feedback loops. Cannon himself, however, did not expose the loop nature inherent in the regulatory processes he investigated. The linking of homeostasis with the feedback loop came about only later, through the collaboration of Cannon's colleague Auturo Rosenbleuth with Norbert Wiener (see section 2.5).

Such generalizations suggested to Cannon that the concept of homeostasis had application beyond the physiological organization of single living organisms and could be applied meaningfully to societies. He ended *The Wisdom of the Body* with a chapter on "social homeostasis." Central to his thinking was the principle that a degree of constancy in a system was evidence of the necessary existence of homeostatic mechanisms striving to maintain that constancy. Society showed the beginnings of such tendencies:

A display of conservatism excites a radical revolt and that in turn is followed by a return to conservatism. Loose government and its consequences bring the reformers into power, but their tight reins soon provoke restiveness and the desire for release. The noble enthusiasms and sacrifices of war are succeeded by more apathy and orgies of self-indulgence. Hardly any strong tendency in a nation continues to the stage of disaster; before that extreme is reached corrective forces arise which check the tendency and they commonly prevail to such an excessive degree as themselves to cause a reaction. A study of the nature of these social swings and their reversal might lead to valuable understanding and possibly to means of more narrowly limiting the disturbances. At this point, however, we merely note that the disturbances are roughly limited, and that this limitation suggests, perhaps, the early stages of social homeostasis (1932, pp. 293–294).

Cannon's descriptions in this paragraph also strongly call to mind the problem of oscillations in engineering feedback control systems that Maxwell had addressed some sixty years before. Yet, as noted above, Cannon did not make the link. Instead, he focused on the still primitive nature of societal homeostasis. He concluded that the mechanisms necessary to maintain the stability of society's "fluid matrix" or "milieu interne" have not yet developed sufficiently. To him it was a matter of societal evolution. The higher an organism is on the evolutionary ladder, the more developed its homeostatic capabilities, and the freer its individual cells and organs are to specialize. Cannon suggested that society would follow a similar pattern. The evolutionary goal for society was the development of automatic societal stabilization mechanisms, ultimately operating below the level of conscious control (pp. 302–306). To Cannon it was an exciting goal, full of potential for releasing the "highest activities of the nervous system for adventure and achievement" (p. 305).

The Impact of the Notion of Homeostasis

The central theme of the idea of homeostasis is control. In feedback terms, it is control through the operation of negative feedback loops. Self-reinforcing positive loops have no role in homeostatic mechanisms. The clear link between Cannon's development of the concept of homeostasis and the work of control engineers led some to overemphasize negative feedback and control in socio-economic applications of the feedback idea. As we shall see, Cannon's observations about mechanisms of social homeostasis also reinforced the urge to add on societal feedback control mechanisms, rather than to strive to perceive those feedback structures already contributing to observed societal dynamics.

2.4 Logic Loops

The modern concept of a feedback loop is also influenced by loop notions arising from formal logic and computing. Three kinds of logic loops have been important, in different ways, in shaping the way various modern authors work with the feedback concept: the vicious circle, the self-referring argument, and the modern DO-loop from computing.[10] The former is a very old notion that is completely compatible with the engineer's concept of feedback. The two ideas eventually merge smoothly, with the vicious circle being seen as merely another name for a positive feedback loop. The self-referring argument and the iterative DO-loop, however, are quite distinct ideas. They are exceptionally powerful concepts with important implications for the social sciences, but they do not derive their power from the concept of positive and negative feedback. Nonetheless, in the writings of some social scientists they interweave with the concept of feedback, with mixed results.

The Vicious Circle

Today we interpret the phrase "vicious circle" to refer to a bad situation that leads to its own worsening. My children, for example, are sometimes "so tired that they can't get to sleep." But the phrase and the concept actually have their origins in formal logic.

In Elizabethan times, the word "vicious" meant, among other things, "flawed" or "faulty" (OED 1933). Any fallacious logical argument was "vicious." One particular form of vicious argument well known in the 1600s (and much earlier) was circular reasoning—basing an argument on the very proposition to be proved. Such an argument naturally came to be known as a vicious circle (*Encyclopædia Brittannica* 1792, cited in OED 1933), meaning simply a process of reasoning that is faulty because it is circular.

But "vicious" also meant evil, harmful, and threatening. By the mid-1800s the concept of the vicious circle had evolved from its narrow meaning of flawed logic to a more general notion of circular causality in which bad leads to worse. The following is an early example from the French:

> I'd need rest to refresh my brain, and to get rest, it's necessary to travel, and, to travel, one must have money, and, in order to get money you have to work, create, etc.: I am in a vicious circle [*cercle vicieux*], from which it is impossible to escape (Balzac 1850, p. 32).

Here the circle of propositions describes a closed sequence of causes and effects, not a logical fallacy. The result is an explicitly circular process, perceived as characteristically self-perpetuating and self-reinforcing—a positive feedback loop. A sketch of the loop underlying Balzac's statements is shown in Figure 2.16.

The concept of a vicious circle in this sense is now so universal that it would be hard to trace the extent of its application or to document its very first appearance. We shall see it developed into a serious working tool for the social sciences in the work of Gunnar Myrdal (see section 2.5). We should note, however, that such a process never occurs by itself in reality. There are always constraints that prevent a self-reinforcing process from expanding itself beyond all bounds. And since such processes dominate at some times and not at others, there must be influences that shift loop dominance between positive and negative loop processes. That is, there must be structural changes, nonlinearities, or external influences that affect the interplay between self-reinforcing and self-opposing loops. However, in the early appearances of self-reinforcing vicious circles, such as Balzac's statement above, only the positive loop is described.

Self-Referring Statements

The second logic loop idea eventually to influence some modern scholars' views of feedback is associated with classic paradoxes. Perhaps the earliest version is the so-called paradox of Epimenides, or the "liar's paradox." Epimenides is supposed to have said,

> All Cretans are liars.

Not so bad by itself, but a disturbing paradox appears with the realization that Epimenides himself was a Cretan. His statement is akin to saying "This statement is false." The paradox is that we cannot decide whether "This statement is false" is true or false. Either option leads to its own contradiction. If the statement is true, then it must be false, and if it is false then it must be true. In Epimenides's version, if it is true that all Cretans are liars, then Epimenides is a liar and is presumably lying

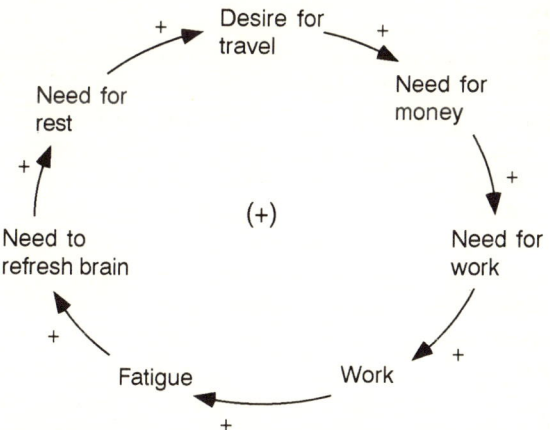

FIGURE 2.16: The vicious circle contained in Balzac's (1850) statement, represented as a positive feedback loop.

when he made his statement, so his statement would be false—a contradiction of what we supposed in the first place.

The problem in such statements is that they refer to themselves. Because of that self-reference, Hofstadter (1979) has called such contradictory statements and arguments "strange loops." Their strangeness results from the fact that they form loops of assertions that close back upon themselves and in the process contradict themselves. In a sense, the paradox of Epimenides is the opposite of a vicious circle. A vicious circle argument supports itself and therefore can be self-consistent and yet be completely false. The Epimenides paradox undermines itself. It is completely self-inconsistent and can be neither true nor false.

Since the time of Epimenides there have been many variations of his paradox, all deriving their paradoxical nature from the fact that they refer to themselves. "The barber of Seville," it is said, "shaves everyone in Seville who doesn't shave himself. Who shaves the barber of Seville?" Such logical conundrums would probably have remained at the level of children's puzzles were it not for the development of set theory in mathematics and a logical paradox of deep significance posed within it by Bertrand Russell.

Russell's paradox grows out of an exquisite argument used by the mathematician Georg Cantor (c. 1871) to prove that no set can be put into a one-to-one correspondence with all of its subsets. The conclusion from Cantor's theorem is that there are somehow "more" subsets of integers than there are integers themselves. There are "more" intervals on a line than there are points on it. There are, in short, different "orders" of infinity—a conclusion that is deeply disturbing to many. The method of proof was almost as disturbing as the conclusion itself. It was an indirect argument in which the contradiction that proved the theorem took the form "statement p is true if and only if statement p is not true." The mathematical world was stunned by the theorem and its proof, and many argued that Cantor's contradiction argument was somehow seriously flawed. Some argued that the fault lay in Cantor's concept of "set" itself, which he had characterized neatly as "a multitude conceived of as a one." Perhaps, some said, mathematicians should not be allowed to conceive of such a thing.

Russell's Paradox

Russell took Cantor's idea and turned it into a paradox casting doubt on the very foundations of mathematical thought. Russell merely said, consider the set R defined to be "the set of all sets in the universe that are not members of themselves." Now it is hard to imagine a set that *is* a member of itself, so R ought to be a fine, healthy set with a lot of perfectly sensible members. Russell then asked, "Is R a member of itself?" It is, as the reader can check, if and only if it is not, and it is not if and only if it is—a thorough paradox.

Notwithstanding the deep difficulties Russell exposed with this paradox, Cantor's argument as he used it is not flawed. There is no internal contradiction in different orders of infinity in the sense of one-to-one correspondences that Cantor developed. Cantor's conception is a sound mathematical creation, and has had fruitful consequences for later mathematicians. Russell's is a genuine paradox, however, and it acquires its paradoxical nature from self-reference.

It was enormously influential. It forced a reconsideration within mathematics of the concept of a "set" and how it can legitimately be used. Outside of mathematics it led to reconsiderations of what are allowable kinds of statements. More than Epimenides or the barber of Seville, Russell's paradox caused people to question the limits of logic and language. Russell's own solution to the dilemmas revealed by the paradox was the theory of "logical types," that in essence barred the use of self-referential statements in logic. Without self-reference, the paradoxical settings of Epimenides, the barber of Seville, and Russell's own paradox can not even be stated, so the paradoxes can not appear.

The grandest and most astonishing result in this line of reasoning about self-referential statements is a theorem and proof in formal logic by Kurt Gödel (1931). Gödel proved that it is not possible to construct an arithmetic proof that shows that arithmetic itself is logically consistent (Nagel and Newman 1956). The result actually applies to any logical system that is at least as logically complex as arithmetic. It is equivalent to saying that if the logical system is rich enough, it will admit true statements that it cannot prove and false statements that it cannot disprove. While enormously significant in mathematical logic, Gödel's conclusion is less important to us than his method of proof. The proof is based on the rigorous construction of a self-referential logical statement that asserts, in essence, that it itself is not provable.[11]

Implications

These self-referential arguments acquire both their contradictory tendencies and their potential logical power from the meanings attributed to their component parts. They form linguistic loops. They connect to lines of thinking in the social sciences through the notion of the dialectic, which is also a linguistic, closed-loop process focusing on meaning—a generalized conversation. Although it may be hard to see in the mathematical formalisms, there are elements of Russell and Gödel in Hegel and Marx.

Furthermore, there are direct links from these self-referential linguistic processes to modern day notions of computer programs that can create copies of themselves or rewrite their own structure—examples of so-called "self-reproducing automata." These are self-referring structures that perceive, act on, and even rewrite their own structure and meanings. Seeing that societies can apparently do these things, some social scientists place the highest significance on the linguistic loop notions of self-reference, self-reflection, self-transformation, and even self-creation.[12]

Such self-referential systems are perilously close to our concept of feedback, particularly if one views feedback, as Wiener phrased it, as "the transmission and return of information." Yet it is important to realize that they are not the same as the positive and negative feedback loops that are the central focus of this study. Circular causal, feedback processes as we have characterized them do not have the potential to be self-contradictory. Loops whose elements are statements or messages can be paradoxical in that fashion, but the primary focus of this investigation is on loops containing variables that can be interpreted as quantities that increase or decrease over time. If an increase in a quantity in such a loop feeds back eventually to produce a decrease in that quantity, we have a negative loop, but not a self-contradictory loop.

We shall see, however, that self-referring, self-contradictory "message loops" influenced how some social scientists interpreted and worked with the feedback concept. In the evolution of the feedback concept traced in this book, I shall take note of that influence when it occurs, but such "message loops" must remain something of a diversion from my main focus on causal loops with positive and negative polarity. In fact, in the evolution of the concept of positive and negative circular causal processes, such potentially contradictory message loops *are* a diversion. As we shall soon see, the idea draws some feedback thinkers away from self-reinforcing and self-correcting loop structures and leads them into discussions of meta-social structure analogous to the meta-language structures that follow from Russell's theory of types. I do not mean to denigrate these ideas. This sort of "diversion" in our investigation may turn out to be the main stream for subsequent social science. While a good treatment of the significance of self-referring linguistic loops is beyond the scope of this study, interested readers could well start with Hofstadter (1979).

2.5 The Loop Concept in Social Science Literature Before 1945

The social sciences themselves contribute two patterns of thinking to the modern conception of the feedback loop. In one pattern, containing relatives of biological homeostasis and goal-seeking negative feedback, the loop idea is only implicit. In the other, positive loops play the major role, and there is frequent explicit reference to "circular processes" and loops of "mutual causality." The two patterns of thinking appear to have remained largely separated until the 1940s when social scientists became aware of the engineer's concept of feedback.

The Loop Concept Implicit in Social Science Literature

It is hard to resist drawing parallels between the self-equilibrating economy of Adam Smith's *Wealth of Nations* (1776) and the burgeoning diversity of feedback devices of the late 1700s. Indeed, Otto Mayr, an authority on the early history of the feedback concept, has investigated the connections in detail and concludes that Smith's work "implies a conception of the closed causal loop that is in principle the same as the feedback loop" (Mayr 1970, p. 129).

Before investigating this claim, we should note that the notion of a self-regulating system appeared even earlier in the social sciences in the writings of David Hume, one of Smith's closest friends. In his essay "On the Balance of Trade" (1752), Hume argued that there is a basic law operating to keep international trade in equilibrium. His argument was phrased in the following thought experiment, surprisingly similar to an engineer's test of a step input to a control system:

Suppose four-fifths of all the money of Great Britain to be annihilated in one night, and the nation reduced to the same condition, with regard to specie, as in the reigns of the Harry's and Edward's, what would be the consequence? Must not the price of all labour and commodities sink in proportion, and everything be sold as cheap as they were in those ages? What nation could then dispute with us in any foreign market, or pretend to navigate or to sell manufactures at the same price, which to us would afford sufficient profit? In how little time, therefore, must this bring back the money which we had lost, and raise us to the level of all the neighbouring nations? Where, after we have arrived, we immediately lose the advantage of the cheapness of labour and commodities; and the farther flowing in of money is stopped by our fullness and repletion (Hume 1752, cited in Mayr 1971).

The negative feedback loop is this balancing argument is vivid to a modern reader; see Figure 2.17. Hume, however, made use of a physical analogy to explain the self-regulation. "All water," he said, "wherever it communicates, remains always at a level ... [W]ere it to be raised in any one place, the superior gravity of that part not being balanced, must depress it, till it meet a counterpoise." He therefore concluded that "it is impossible to heap up money, more than any fluid, beyond its proper level" (Hume 1752, cited in Mayr 1971).

The Wealth of Nations

It is not clear whether Adam Smith thought in terms of the same fluid balancing metaphor or whether he had some other self-regulating mechanism in mind. Yet in his hands the concept of a self-regulating socioeconomic system became a real working tool, applied to a wide variety of situations. Mayr (1971) cites three particularly outstanding examples, each of which can be easily translated into a negative feedback loop structure. First is Smith's famous argument that the contributions and rewards of all occupations in an economy must be the same:

> The whole of the advantages and disadvantages of the different employments of labour and stock must, in the same neighborhood, be either perfectly equal or continually tending to equality. If in the same neighborhood, there was any employment evidently either more or less advantageous than the rest, so many people would crowd into it in the one case, and so many would desert it in the other, that its advantages would soon return to the level of other employments. This at least would be the case in a society where things were left to follow their natural course, where there was perfect liberty, and where every man was perfectly free both to chuse what occupation he thought proper, and to change

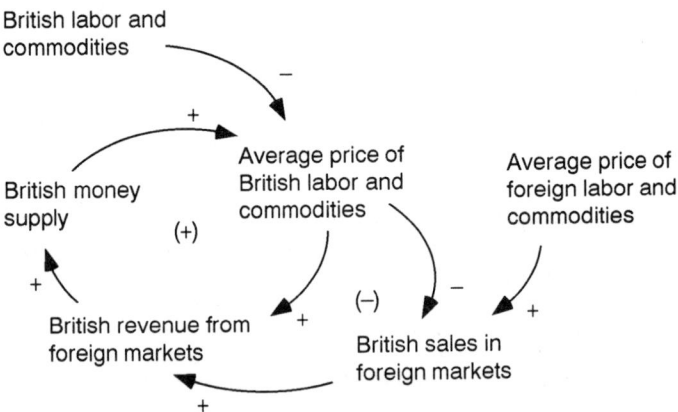

FIGURE 2.17: Negative feedback loop implicit in David Hume's "On the Balance of Trade" (1752).

it as often as he thought proper. Every man's interest would prompt him to seek the advantageous, and to shun the disadvantageous employment (Smith 1776, p. 99).

Smith thus argued that the "relative attractiveness" of all occupations ought to tend to be the same. (The term is from Forrester's *Urban Dynamics* (1969), appearing almost 200 years after *The Wealth of Nations* and applied to land areas not occupations, but the concept is the same.) Smith intended the attractiveness of a given occupation—the "whole of its advantages and disadvantages"—to be defined broadly, including risk, cost of training, and spiritual as well as material rewards. The result is an expression of the tendency of supply and demand for labor in a given occupation to be in equilibrium. The underlying loop structure of the argument is shown in Figure 2.18.

Smith gave several other examples of the self-equilibrating tendency of supply and demand, and then stated an abstract, general argument that places the market price at the center of the equilibrating mechanism:

When the quantity of any commodity which is brought to market falls short of the effectual demand, all those who are willing to pay the whole value . . . cannot be supplied with the quantity which they want. Rather than want it altogether, some of them will be willing to give more. A competition will immediately begin among them, and the market price will rise more or less above the natural price (Smith 1776, p. 56).

Smith argued similarly in the opposition situation: when supply exceeds demand, the price must go down. Thus the market price is a

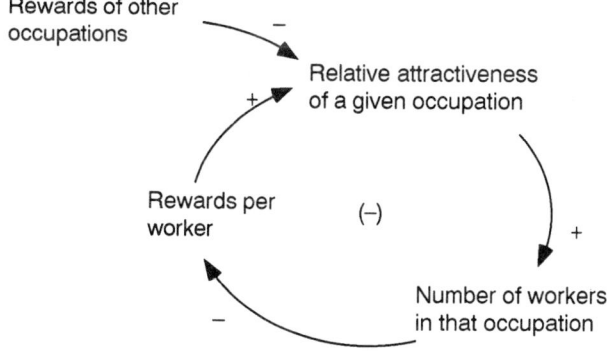

FIGURE 2.18: Negative feedback loop implicit in Adam Smith's argument that the rewards in all occupations must tend to be the same.

function of supply and demand. One could formulate this functional relationship in terms of a negative feedback loop, but the most obvious loop implicit in Smith's argument emerges from the effect of price on production:

> If . . . the quantity brought to market should at any time fall short of the effectual demand, some of the component parts of its price must rise above their natural rate. If it is rent, the interest of all other landlords will naturally prompt them to prepare more land for the raising of this commodity; if it is wages or profits, the interest of all other labourers and dealers will soon prompt them to employ more labour and stock in preparing and bringing it to market. The quantity brought thither will soon be sufficient to supply the effectual demand. All the different parts of its price will soon sink to their natural rate, and whole price to its natural price (Smith 1776, p. 57).

Similarly, he argued that production would decrease if supply exceeded demand. The result is the negative loop shown in Figure 2.19: any persistent discrepancy between the current market price and the "natural price" would generate pressures that would push production in the direction that would decrease the discrepancy. Smith concluded that in the absence of disturbances the system should equilibrate with the market price equal to the natural price of the commodity:

> The natural price, therefore, is, as it were, the central price to which the prices of all commodities are continually gravitating. Different accidents may sometimes keep them suspended a good deal above it, and sometimes force them down even somewhat below it. But whatever may be the obstacles which hinder them from settling in this center of repose and continuance, they are constantly tending towards it (Smith 1776, p. 58).

Smith then observed that disturbances could combine with the equilibrating tendency of the system to produce "occasional and temporary fluctuations in the market price of any commodity" (Smith 1776, p. 59). It is very tempting to believe that this conclusion came from similarities Smith observed between the behavior of prices and the oscillatory "hunting" behavior of some of the regulated machines of his day. Smith may have perceived intuitively that both are in some similar sense controlled, and that controls can produce oscillations. (Maxwell, as we have seen, later shows mathematically why such oscillatory behavior can arise in a negative feedback system.)

One of Smith's applications of his general theory of the self-regulating nature of supply and demand concerns the size of the working population itself.

The demand for men, like that for any other commodity, necessarily regulates the production of men; quickens when it goes on too slowly, and stops when it advances too fast (Smith 1776, p. 80).

The mechanism of population adjustment for Smith was not the voluntary control of birth rates, but rather involuntary control through infant mortality. If the supply of workers exceeds demand, wages would be low, he argued. Poor living conditions would cause infant mortality to rise and reduce the growth rate of the working class. Eventually—it would presumably take a generation or two—the supply of workers would be reduced to the demand. On the other hand, if workers were in high demand, wages would tend to remain high. Conditions would favor the survivability of infants and children, and the growth rate of the worker population would increase. Some twenty years later, Thomas Malthus (1798) published his famous pessimistic essay addressing the same subject from a very similar point of view (see below). Whether the argument is considered right or wrong is less interesting for us than its clear relationship to the feedback concept.

The feedback loop is thus implicitly but strongly present in Smith's thinking. It is natural to wonder to what extent he was influenced by the growing diversity of mechanical control devices that flourished toward the end of the eighteenth century. Certainly, he saw some connections between human systems and machines. In an early philosophical work, for example, he had produced the following delightful definition of a "system" as an "imaginary machine":

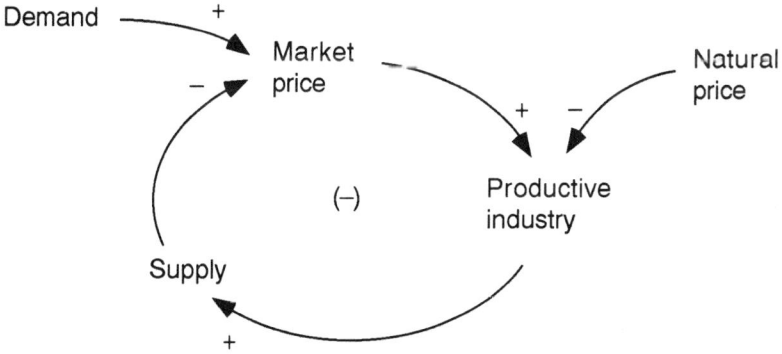

FIGURE 2.19: Negative feedback loop structure of Adam Smith's general theory of the equilibration of supply to demand.

Systems in many respects resemble machines. A machine is a little system, created to perform, as well as to connect together, in reality, those different movements and effects which the artist has occasion for. A system is an imaginary machine invented to connect together in the fancy those different movements and effects which are already in reality performed (Smith, cited in Mayr 1971, p. 17).

Furthermore, the notions of "living machines" and "living automata" trace back at least as far as the "Monadology" of Leibniz (1714).

However, to my knowledge, in *The Wealth of Nations* Smith never used an explicit analogy between his mechanisms of socioeconomic self-regulation and the automatic control devices of his day. Comparing the spread of the ideas of economic liberalism and feedback control of machines, Mayr concludes:

It would be wrong to interpret one of these developments as a direct consequence of the other. A theoretician like Adam Smith may have observed feedback devices in operation, and this may have made his analysis sharper and his formulations more concrete, but the beginnings of economic liberalism lie in the early 18th century, antedating the breakthrough of the feedback concept in technology. On the other hand it is also highly unlikely that 18th-century inventors should have obtained their conceptions by materializing abstract theories of political economy (Mayr 1971, p. 129).

The most that I can say is that self-regulation appears to have been simply a part of the spirit of the times in late-eighteenth-century Britain.

Self-regulation in the Federalist papers

Some support for that conclusion can be seen in other writings of the time. Platt (1966) argues persuasively that the "checks and balances" assiduously written into the Constitution of the United States were a conscious effort to design a system of "stabilization feedbacks." Evidence for the claim can be found in the *Federalist* papers, a series of 85 newspaper articles written by Alexander Hamilton, John Jay, and James Madison to persuade the public to favor the new Constitution.

One of the concerns in these papers, for example, is instability. Madison argued that pure democracy of the Athenian sort is prone to turbulence:

A common passion or interest will, in almost every case, be felt by a majority of the whole; a communication and concert results from the form of government itself; and there is nothing to check the inducements to sacrifice the weaker party or an obnoxious individual. Hence it is that such democracies have ever been spectacles of turbulence and contention; have ever been found incompatible with personal security or the rights of property; and have in general been as short in their lives as they have been violent in their deaths (*The Federalist*, cited in Platt 1966, p. 112).

"Nothing to check the inducements to sacrifice the weaker party or an obnoxious individual" sounds like an uncontrolled positive feedback loop. The solution was not to abandon the goal of democratic government, but to design it so that its tendencies toward instability were countered by the form of government itself. The key to that design was the self-interest of the participants:

The great security against a gradual concentration of the several powers in the same department, consists in giving to those who administer each department the necessary constitutional means and personal motives to resist the encroachments of the others. . . . Ambition must be made to counteract ambition. The interest of the man must be connected with the constitutional rights of the place (*The Federalist*, cited in Platt 1966, pp. 112–113).

Smith found socioeconomic self-regulation in the dictum that "Every man's interest would prompt him to seek the advantageous, and to shun the disadvantageous." The writers of the United States Constitution sought to use similar, powerful self-interests to assure the self-regulation of government. Again the argument sounds vaguely loop-like: governments exist to control the self-interests of people, while the same sort of personal self-interests control government.

Maxwell (1868) showed, and others before him knew intuitively, that controls do not guarantee stability and, in fact, can generate their own instabilities. Platt (1966, pp. 114–115) observes that the Federalists had a sophisticated view of the requirements for control with stability. In modern terminology, they emphasized different "time constants" of response to different sorts of disturbances. Executive actions required "energy and dispatch." The House of Representatives, with its two-year terms, was designed for moderately rapid response to the changing will of the people and assigned duties accordingly, such as the initiation of taxation and spending legislation. The Senate, with its six-year terms, was deliberately focused on longer term issues and adjustment. The Supreme Court, with its lifetime appointments, was designed for very long-term stability and independence from disturbance. Finally, the amendment process itself had to be designed to avoid "that extreme facility which would render the Constitution too mutable; and that extreme difficulty which might perpetuate its discovered faults."

The drive for a governmental structure that naturally regulates itself is evident in the *Federalist* papers. Yet the implicit feedback-loop nature of Adam Smith's arguments is even less apparent here. Regulation may have been in the spirit of the times, but the concept of the feedback loop remained hidden in the spirit's shadow.

Thomas Malthus

In his essays on population, the gloomy parson Thomas Malthus (1798) came remarkably close to an explicit feedback view of population dynamics. It may appear dangerous to say so, for Malthus is generally regarded as having been "disproved" by subsequent history. Nonetheless, in the structure of his arguments Malthus correctly illuminated the potential for exponential growth contained in what we would now call a positive feedback loop. In addition, he correctly identified several plausible negative loops that strive ultimately to control that potentially runaway loop. Whatever predictions he is considered to have made that have been judged false, his implicit feedback view is fundamentally sound.

Malthus was addressing himself to a philosophical discussion of his day, on the "perfectability of man and of society," and he saw a problem.

Population, when unchecked, increases in a geometrical ratio. Subsistence increases only in an arithmetical ratio. . . . By that law of our nature which makes food necessary to the life of man, the effects of these two unequal powers must be kept equal. This implies a strong and constantly operating check on population from the difficulty of subsistence. This difficulty must fall somewhere; and must necessarily be severely felt by a large portion of mankind (Malthus 1798, p. 14).

The specter of this difficulty made it hard for Malthus to accept promises of utopias that some in his time were describing, "where all narrow luxuries would be contemned; where [people] would be employed only in collecting the necessities of life; and where, consequently, each man's share of labour would be light, and his portion of leisure ample" (1798, p. 11).

Why does population grow geometrically, or, as we might say, exponentially? The mechanism, which Malthus thought too obvious or too graphic to state blatantly, is that babies tend to become parents:

Population, could it be supplied with food, would go on with unexhausted vigour, and the increase of one period would furnish the power of a greater increase the next, and this without any limit (p. 107).

Prehistory and Emergence   67

The positive feedback loop from births to population and back to births is almost explicit. Births add to population and "furnish the power" for even more births and more population, potentially ad infinitum. Food production does not have this self-reinforcing property, however. Acreage can be added into the land cultivated for food production, but the increase of one period does not "furnish the power of a greater increase the next." Indeed, there would be less remaining land available, and probably of lesser productivity. This is the basis of Malthus's claim that food production tends to grow "arithmetically," while population grows "geometrically." An additional acre is just a one-time increase in potential food production. An additional person increases the potential for future increases in population. Malthus is completely correct in these observations.

If population growth has this self-reinforcing character, why has it not long since surpassed the ability of the land to support it? Malthus proposed two types of mechanisms, which act continually to control population growth:

a foresight of the difficulties attending the rearing of a family, acts as a preventative check; and the actual distress of some of the lower classes, by which they are disabled from giving the proper food and attention to their children, acts as a positive check, to the natural increase of population (pp. 62–63).

Preventative checks are societal responses, voluntary in some sense, even if imposed by culture and tradition; positive checks are all those "vices" and "miseries" that shorten life. Preventative checks therefore work on the birth rate; positive checks work on the death rate. These checks regulate population, keeping it in line with the food production. It is tempting to add "much like Watt's governor regulated the speed of a steam engine," but Malthus, like Adam Smith, did not make such a connection explicit.

It is also tempting to sketch the feedback structure of Malthus's argument, as shown in Figure 2.20. We must be careful about rewriting history here, however. Malthus himself did not mention the closed-loop nature of his argument. He came close to an explicit statement of a feedback loop in his description of the cause of exponential population growth, cited above, but the closed causal loops in his preventative and positive checks were hidden in the language of control and regulation common to his time.

Malthus argued that the structure he described operated continually to hold down the potentially explosive growth rate of population. He suggested that the preventative and positive checks imposed should result in oscillatory ebbs and flows of population growth:

The constant effort towards population, which is found to act even in the most vicious societies, increases the number of people before the means of subsistence are increased. The food therefore which before supported seven millions, must now be divided among seven millions and a half or eight millions. The poor consequently must live much worse, and many of them be reduced to severe distress. The number of labourers also being above the proportion of the work in the market, the price of labour must tend toward a decrease; while the price of provisions would at the same time tend to rise. The labourer therefore must work harder to earn the same as he did before. During this season of distress, the discouragements to marriage, and the difficulty of rearing a family are so great, that population is at a stand. In the mean time the cheapness of labour, the plenty of labourers, and the necessity of an increased industry amongst them, encourage cultivators to employ more labour upon their land; to turn up fresh soil, and to manure and improve more completely what is already in tillage; till ultimately the means of subsistence become in the same proportion to the population as at the period from which we set out. The situation of the labourer being then again tolerably comfortable, the restraints to population are in some degree loosened; and the same retrograde and progressive movements with respect to happiness are repeated (pp. 63–64).

Two things about this argument are interesting from our point of view. First, it is a feedback argument: it begins and ends with population and contains in between a number of closed causal loops, as shown in Figure 2.21. Second, it argues for the potential for oscillations to arise from the controlling influences on population. It calls to mind the phenomenon of "hunting" that troubled eighteenth-century designers

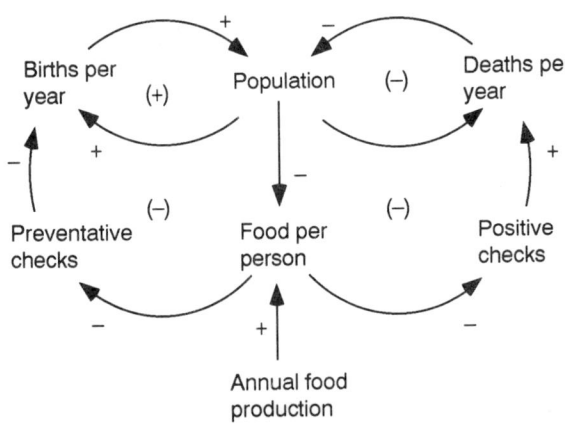

FIGURE 2.20: Implicit feedback structure of Malthus's theory of population growth.

of governors for steam engines. Although it seems a very likely connection, we can, however, again only speculate on whether Malthus was influenced by the feedback devices of his day.

As noted, the arguments of Malthus reflect some of the thinking of Adam Smith. Both held that low wages would suppress population growth and bring it more in line with subsistence. Both, in fact, pessimistically assumed that control would tend to come predominantly through positive checks, such as rising infant mortality, rather than voluntary preventative checks such as late marriages or birth control. Malthus carried the argument further, however, by exposing the engine of population growth, the self-reinforcing positive feedback loop that exists in the nature of the parenting process. Smith focused on the interrelationships adjusting the supply of labor to demand. His theories contain only negative loops. Malthus focused on the mechanisms that could conceivably control runaway growth caused by an unavoidable positive loop. Both argued from the same fundamental philosophical view, however: socioeconomic systems are, by their nature, self-regulated in something approaching an automatic sense, with or without the conscious action or acquiescence of people. In the work of both Smith and Malthus, the closed-loop nature of their arguments and their relationship to automatic control in machines remained implicit.

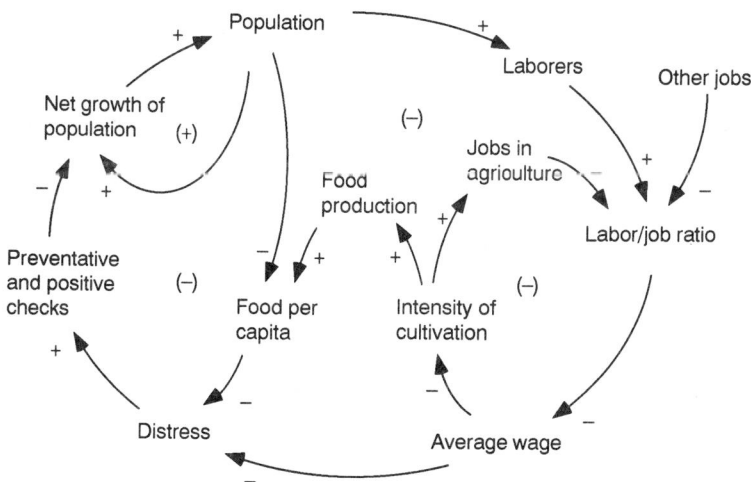

FIGURE 2.21: Closed-loop structure of Malthus's argument suggesting oscillatory tendencies in population growth.

The first unequivocal recognition, that I know of, of a connection between automatic control in engineering devices and self-regulation in living systems is contained in a famous essay sent to Darwin by Alfred Russel Wallace, the "other man" most responsible for the theory of evolution. In it, Wallace writes:

> The action of this principle [the struggle for existence] is exactly like that of the steam engine, which checks and corrects any irregularities almost before they become evident; and in like manner no unbalanced deficiency in the animal kingdom can ever reach any conspicuous magnitude, because it would make itself felt at the very first step, by rendering existence difficult and extinction almost sure to follow (Wallace, cited in Bateson 1972, p. 428).

Wallace's analogy here is something like the discovery of the Americas by Norsemen: only much later did other discoverers open the way for widespread exploitation of the idea.

By the 1900s, analogies between the regulation of machines and socioeconomic systems were made more frequently, and there were attempts to derive substantive implications. Albert Aftalion (1909, 1927), for example, suggested that business cycles stemmed from actions of entrepreneurs that were analogous to attempts to maintain manually the temperature of a furnace. When the temperature is very low, there is a tendency to put in more coal than is necessary in steady state, thereby causing the temperature to rise too high. High temperatures, on the other hand, tend to cause people to wait too long to add more coal, and the temperature drops too far before the need for more coal is perceived. Similar under- and over-reactions in production and ordering decisions, Aftalion argued, could cause economic cycles (Goodwin 1951a; see section 3.3).

The Wealth of Nations was instrumental in changing the way people thought about economic phenomena. It helped smooth the way for "the system of natural liberty" (modern capitalism), and it dramatically advanced economics as a field of scholarly study. Most importantly for us, subsequent economic theoreticians adopted the style of Smith's analyses, including arguments phrased implicitly or explicitly in terms of causal loops. Mayr (1971) observes that the work of David Ricardo, for example, contains numerous loops and is easily translated into the mathematics of feedback systems. (We shall see other examples in the work of other economists in section 5.3.) As in Smith's work, the feedback loops implicitly deal with regulation and are almost always negative.

Other implicit examples of the feedback loop concept

In the generic negative feedback loop, a discrepancy between the actual state of a system and some desired condition results in action designed to reduce the discrepancy. In one form or another this pattern has appeared throughout the history of the social sciences. If one sets out to use the feedback perspective as a kind of lens through which to view older theories, one can invariably see them as implicit feedback theories. I will briefly sketch some additional examples and then comment on what we can learn from such an exercise.

The dialectic of Hegel, and Marx's variation on that theme, contain "discrepancies between desired and actual conditions." The contradictions between thesis and antithesis set up pressures that eventually force a new state of affairs, the synthesis. The common mental picture of the phenomenon, shown in Figure 2.22a, contains no loops. The closed-loop version shown in Figure 2.22b is an attempt to capture the idea that the synthesis emerges from a restructuring of thesis and antithesis. The "desired condition" is synthesis, the elimination of contradiction and conflict between thesis and antithesis. The loops shown in Figure 2.22 are both negative: conflict between thesis and antithesis bring about a restructuring that reduces or eliminates (negates) the conflict.

The loop nature implicit in Marx's views has been pointed out more rigorously by Stinchcombe (1968). In his study of patterns of explanation in the social sciences, Stinchcombe argued that two related kinds of

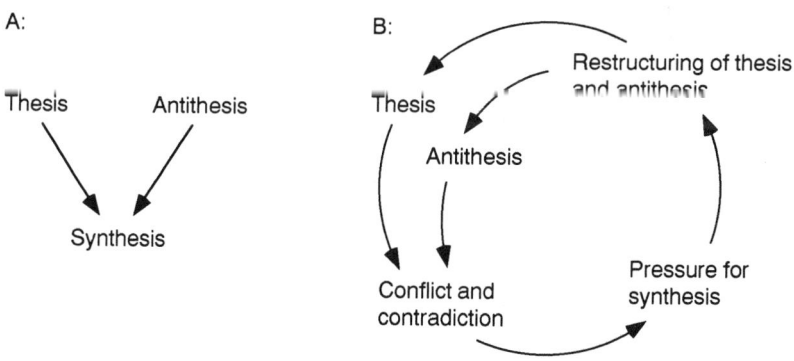

FIGURE 2.22: Dialectic of Hegel and Marx viewed as a feedback structure. a) open-loop view; b) closed-loop view.

causal imagery having a closed-loop character have frequently been employed in the social sciences. In the writings of Marx, the functional anthropologists such as Malinowski and Radcliffe-Brown, and the work of Robert King Merton (among others), he found a circular causal view that he labeled functional causal imagery.

Stinchcombe defined a "functional explanation" to be one in which

the consequences of some behavior or social arrangement are essential elements of the causes of that behavior (Stinchcombe 1968, p. 80).

He sketched the general causal loop structure of such a functional explanation as shown in Figure 2.23a, and interpreted a particular argument of Marx in those terms with the diagram shown in Figure 2.23b.

In Figure 2.23a, the symbols are defined as

H: the homeostatic variable, defined as "the consequence or end which tends to be maintained, which in turn functions indirectly as a cause of the behavior or structure to be explained";

S: social structure or behavior that has a causal impact on H;

T: tensions or difficulties or other causal forces that tend to disturb or prevent H.

Applied to Marx's argument about power in "bourgeois democracy," the symbols in Figure 2.23b become

S: parliamentary republicanism as a form of government (Marx's "bourgeois democracy");

Hn, Hp, Hw: the consequences of parliamentary democracy for the nobles, the bourgeoisie, and the proletariat, respectively;

Pn, Pb, Pw: the power of the nobles, the bourgeoisie, and the proletariat, respectively.

Figure 2.23b tries to capture the Marxist view that parliamentary republicanism has positive effects on the bourgeoisie and negative effects on the nobles and the proletariat. Each group therefore applies its power either for or against the establishment and maintenance of that form of government. Their efforts determine the evolution of governmental structure, which is in turn a further cause of their continuing efforts. The closed-loop structure of the argument is clear. Marx expected that the growing power of the proletariat would lead to a workers' revolution, altering the structure of government and the future interplay of power among groups.

From this point of view, then, Marx becomes something of a feedback thinker. Stinchcombe refrained from using the word, however, and may in fact have seen some differences between the closed causal loops of functional explanations and the feedback writings he would have been familiar with in 1968.

Other modern authors have noted that several classical theories of motivation in psychology have a similar negative-loop character. In the drive-reduction theory of C.S. Hull and E.C. Tolman, drive can be thought of as the pressure to move an organism from its current state toward a preferred state. A person is hungry. The hunger drive leads to finding food and eating, which reduces the hunger drive, at least until digestive and metabolic activity exhaust the food consumed. Less physical processes were phrased in similar terms in the theory of tension reduction appearing in psychoanalysis and Gestalt psychology. The loop nature of these processes is clearly discernible, as drives and tensions both affect and are affected by the state of the organism. The generic negative feedback loop is evident from the tendency of such systems to counteract and try to eliminate disturbances from some goal state.

In the conceptual scheme for motivation developed by John Dewey and George H. Mead, stages in the process were labeled impulse,

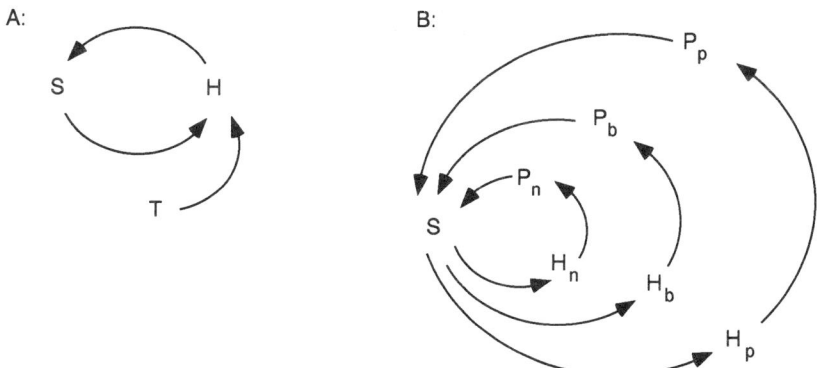

FIGURE 2.23: Circular causal views of functionalist explanations. a) elementary causal structure of a complete functional explanation; b) example of the causal loop structure of Marxian functionalism. Source: Stinchcombe (1968, pp. 89, 94). See text for explanation of symbols.

perception, manipulation, and consummation. Looking in retrospect on their work, one author observed:

> An impulse is a disturbance, any lack of adjustment between an organism and its milieu—pique over an imagined slight, hunger pangs, concern over the whereabouts of a friend who is overdue. An organism is set into motion by a disruption of its steady state; any discomfort leads to attempts to eliminate it. The key principle is that an act, once under way, tends to persist until the discomfort is removed (Shibutani 1968, p. 332).

Again the generic negative loop structure is apparent. A disturbance from steady state leads to impulses, which are perceived by the individual and which lead to manipulating the situation until "consummation," that is, until the impulses signaling the disturbance are no longer perceived. The "manipulations" are what we can observe. Mead emphasized that manipulations give rise to new impulses and new perceptions, which serve to guide the continuing action. It is this emphasis that makes Mead's conception very much a feedback theory of motivation (Shibutani 1968, p. 333).

Finally, in the stimulus-response-reinforcement theories as they developed from Pavlov to Skinner and beyond, modern authors have found implicit suggestions of loop processes. Reinforcement—what follows the response to a stimulus—affects future responses. However, there is a good deal of confusion, some of it semantic and some substantive, about whether positive reinforcement reflects positive or negative feedback. Kramer (1968), for example, considers three rather similar, classic reinforcement situations—a cat repeatedly put in a box and repeatedly scrabbling to get out, Pavlov's dog learning to salivate at the sound of a buzzer, and a rat learning to press a bar to get food pellets. He concludes that the cat's behavior is an example of negative feedback, and the rat's reflects positive feedback up to the point of satiation, while the dog's behavior does not involve any feedback loop at all. I consider the attempt to label an entire complex behavior pattern as exemplifying either positive or negative feedback to be a misguided use of the feedback concept, but nonetheless, Kramer's comments here are a serious attempt to interpret reinforcement theory in feedback terms.

These observations of feedback-like concepts in the social sciences have been repeatedly made by scholars since the emergence of the engineer's concept of feedback in the 1940s. The observations tell us two things. First, they suggest that the feedback concept in one form or another has been implicitly present in the thinking of some of our most respected and influential social scientists, and it underlies a number of the most significant social science concepts and theories. Second, they imply that little is accomplished if all we do is relabel theories with the feedback stamp of approval.

However, even in the most pessimistic view, relabeling social science ideas as feedback theories serves to expose a common and powerful pattern of thinking. I side with Stinchcombe who observed that

Explicit knowledge accessible to intelligent beginners is obviously more efficient for a science than knowledge perceived by the intuition of its geniuses (Stinchcombe 1968, p. 148).

If loop thinking characterizes the insights of great social scientists, then exposing that loop structure for others to adopt is reason enough to relabel theories as feedback views. In Chapters 3, 4, and 5 we shall be concerned with what people perceive the additional promise of the feedback concept is in the social sciences, once it has been made explicit and well-understood.

Feedback loops in the thinking of John Dewey

One final precursor of the feedback notion in the social sciences falls midway between the implicit loop notions of this section and the explicit loops of the following. In a remarkable ten-page essay published in 1896, John Dewey displayed a conception of the feedback loop in psychological processes that was far ahead of its time (Dewey 1896; Slack 1955). The essay has also been credited with foreshadowing the functionalist school in sociology and anthropology, the Gestalt point of view in psychology, and modern criticisms of behaviorism (Dennis 1948, p. 355). More than sixty years after its publication it contributed significantly to the development in psychology of the feedback idea called the TOTE unit (Miller, Galanter, and Pribram 1960; see section 4.3).

Dewey was disturbed by the rising influence of the reflex arc in psychology. The concept originated in neurophysiology and refers to the neurological path linking a stimulus to a response: stimulus → receptor → afferent nerve → connective fibers → efferent nerve → effector → response (Miller, Galanter and Pribram 1960, p. 22). In Dewey's time, the reflex arc was being groomed as the organizing principle for all psychological phenomena. His primary criticism was that the reflex arc did not solve the old dualism between sensation and idea:

The sensory stimulus is one thing, the central activity, standing for the idea, is another thing, and the motor discharge, standing for the act proper, is a third. As a result, the reflex arc is not a comprehensive, or organic unity, but a patchwork of disjoined parts (Dewey 1896, in Dennis 1948, pp. 355–356).

What Dewey desired for the organizing principle for psychology was the reality that tied the disjoined parts of the reflex arc together. Taking a cue from physical activity, he termed that reality a "coordination." He envisioned stimulus and response as phases of a coordination. Attention is stimulated by "a conflict within the coordination," an uncertainty about how to complete it. A stimulus is that phase "requiring attention." Motion, as response, is "whatever will serve to complete the disintegrating coordination" (p. 363). The "conflict in the coordination" sounds suspiciously like the feedback engineer's "error signal," or the discrepancy between desired and actual conditions in a negative feedback loop. Indeed, Dewey explicitly argued for a reinterpretation of the reflex arc as a circuit.

To exemplify his arguments, Dewey analyzed two classic situations, William James's example of a child reaching for a candle flame and Baldwin's discussion of a person hearing a loud unexpected sound and reacting by running away. He concluded that the beginning of the candle sequence is not the sensation of light, but the act of seeing, which is an optical-ocular, sensorimotor activity in which movement is the primary element and sensation secondary (p. 356). Then followed a distinctly feedback-like description of the connection between seeing the flame and reaching for it:

> Now if this act, the seeing, stimulates another act, the reaching, it is because both of these acts fall within a larger coordination; because seeing and grasping have been so often bound together to reinforce each other, to help each other out, that each may be considered practically a subordinate member of a bigger coordination. More specifically, the ability of the hand to do its work will depend, either directly or indirectly, upon its control, as well as its stimulation, by the act of vision. If the sight did not inhibit as well as excite the reaching, the latter would be purely indeterminate, it would be for anything or nothing, not for the particular object seen. The reaching, in turn, must both stimulate and control the seeing. The eye must be kept upon the candle if the arm is to do its work; let it wander and the arm takes up another task (p. 356; emphasis added).

The eye stimulates and controls the hand, while the hand stimulates and controls the eye—a clear conception of a closed-loop process involving mutual causality and information feedback.

In analyzing the situation of a person hearing a loud, unexpected sound and running from it, Dewey exposed the circuit he believed to be present.

> Just as the "response" is necessary to constitute the stimulus, to determine it as sound and as this kind of sound, of wild beast or robber, so the sound experience must persist as a value in the running, to keep it up, to control it. The motor reaction involved in the running is, once more, into, not merely to, the

sound. It occurs to change the sound, to get rid of it. The resulting quale, whatever it may be, has its meaning wholly determined by reference to the hearing of the sound. . . . What we have is a circuit, not an arc or broken segment of a circle. This circuit is more truly termed organic than reflex, because the motor response determines the stimulus, just as truly as sensory stimulus determines movement (p. 359).

The concept of the reflex arc does not capture this mediating circuit. Consequently, to Dewey it was an inadequate organizing principle for psychology. It is clear that what he wanted was a closed-loop concept closely analogous, if not identical, to the feedback loop. Moreover, he advocated a continuous interplay between elements in the loop. One of his criticisms of the notions of stimulus and response in the reflex arc was their discrete, discontinuous character:

In its failure to see that the arc of which it talks is virtually a circuit, a continual reconstitution, [the reflex arc] breaks continuity and leaves us nothing but a series of jerks, the origin of each jerk to be sought outside the process of experience itself, in either an external pressure of "environment," or else in an unaccountable spontaneous variation from within the "soul" or the "organism" (pp. 357–358).

In Dewey's view stimulus and response do not follow discretely one after the other in time, but are contemporaneous phases of one and the same coordination (p. 365). They are elements in a closed-loop process in which behavior is continuously controlled by conditions and conditions are in turn continuously altered by behavior.

Dewey thus concluded that the appropriate foundation for psychology was to be found in a "circular coordination" that we would now recognize as the feedback loop:

It is the coordination which unifies that which the reflex arc concept gives us only in disjointed fragments. It is the circuit within which fall distinctions of stimulus and response as functional phases of its own mediation or completion (p. 365).

Not until 1960 were these suggestions pursued seriously by psychologists (Miller, Galanter, and Pribram 1960; Powers, Clark, and McFarland 1960; see sections 4.3 and 4.7).

Circular Processes and Loops of Mutual Causality

The previous discussion focused on a set of ideas in the social sciences in which the feedback concept is largely implicit. In contrast, another pattern of thinking in the social sciences that contributes to the modern concept of a feedback system makes explicit reference to "circular processes" and loops of "mutual causality." Explicit loop-like processes appeared in the social sciences as early as the 1840s, as exemplified by the following vivid description of speculation by John Stuart Mill:

> When there is a general impression that the price of some commodity is likely to rise, from an extra demand, a short crop, obstructions to importation, or any other cause, there is a disposition among dealers to increase their stocks, in order to profit by the expected rise. This disposition tends in itself to produce the effect which it looks forward to, a rise of price: and if the rise is considerable and progressive, other speculators are attracted, who, so long as the price has not begun to fall, are willing to believe that it will continue rising. These, by further purchases, produce a further advance: and thus a rise of price for which there were originally some rational grounds, is often heightened by merely speculative purchases, until it greatly exceeds what the original grounds will justify. After a time this begins to be perceived; the price ceases to rise, and the holders, thinking it time to realize their gains, are anxious to sell. Then the price begins to decline: the holders rush into the market to avoid a still greater loss, and, few being willing to buy in a falling market, the price falls much more suddenly than it rose (Mill 1848).

Speculation, as Mill saw it, is a self-reinforcing process—a positive feedback loop (see Figure 2.24). A tendency for the price to rise feeds back to produce a still greater tendency for the price to rise. Mill's description of it is particularly striking because it indicates that he was clearly aware of the closed-loop nature of the phenomenon. In addi-

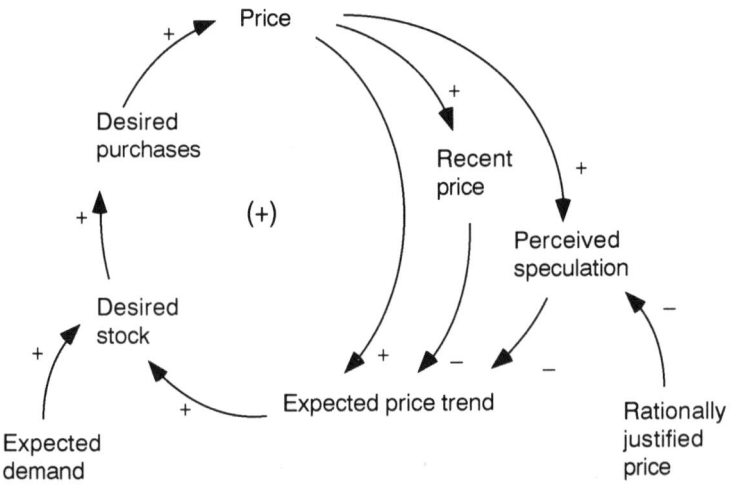

FIGURE 2.24: Mutual causal loops in John Stuart Mill's (1848) description of speculation.

tion, he traces the implications of the closed causal loop in both the rise and the fall of a speculative market. The same self-reinforcing loop operating through trends and expectations tends to exacerbate both rising and falling prices. Finally, Mill's description even contains a hint of the feedback structure that could act to halt the rise of speculative behavior and start its collapse. When the price greatly exceeds the rationally justified price, the lack of support for the high price begins to be perceived. Speculators come to think the price will stop rising, so they start to sell, and indeed the price stops rising and starts to fall. This assumption is captured in the negative loop in Figure 2.24 passing through "perceived speculation."

The vicious circle

By the 1900s, observations of circular processes exhibiting loops of mutual or circular causality had become commonplace in the social sciences. There is the vicious circle, for example, in which a bad situation leads to its own worsening. As we have seen (section 2.4), the term actually had its origins in formal logic. Starting from the notion of flawed, circular reasoning, the concept has come to represent an explicitly circular causal process, perceived as characteristically self-perpetuating and self-reinforcing—a positive feedback loop. In that form it has entered the folklore of common conversation.

Myrdal's "principle of cumulation"

In the social sciences, the concept of the vicious circle undoubtedly reached its greatest development in the work of Gunnar Myrdal (1939, 1944, 1957). He re-elevated it from the level of folklore to serious social science and gave it a new name: the "principle of circular and cumulative causation," or more simply, the "principle of cumulation" (Myrdal 1944, p. 75; 1957, p. 23). He preferred the new names because, as he repeatedly observed, the phenomenon can work two ways—in a beneficial sense, as well as in a harmful sense. He recognized that vicious circles could become virtuous. Indeed, the goal of his analyses of American race relations (Myrdal 1944) and the persistent gap between rich and poor nations (Myrdal 1957) was to determine how vicious circles could be turned around.

In the course of the 1,500 pages of his monumental *An American Dilemma* (1944), Myrdal made use of his loop "principle" in no fewer than twenty distinct contexts. At the outset, he stated its central role in his thinking:

A deeper reason for the unity of the Negro problem will be apparent when we now try to formulate our hypothesis concerning its dynamic causation. The mechanism that operates here is the "principle of cumulation," also called the "vicious circle." This principle has a much wider application in social relations. It is, or should be developed into, a main theoretical tool in studying social change (Myrdal 1944, p. 75).

He devoted a section of a chapter and an entire appendix to the development of the idea. Each application of it expressed some aspect of the following general pattern:

Throughout this inquiry, we shall assume a general interdependence between all the factors in the Negro problem. White prejudice and discrimination keep the Negro low in standards of living, health, education, manners and morals. This, in turn, gives support to white prejudice. White prejudice and Negro standards thus mutually "cause" each other (pp. 75–76).

Anticipating later developments, he connected the closed-loop nature of this mutual causality to dynamic behavior (change over time), and in the process came within a hair of inventing on his own the word "feedback":

If, for example, we assume that for some reason white prejudice could be decreased and discrimination mitigated, this is likely to cause a rise in Negro standards, which may decrease white prejudice still a little more, which would again allow Negro standards to rise, and so on through mutual interaction. If, instead, discrimination should become intensified, we should see the vicious circle spiralling downward. The original change can as easily be a change of Negro standards upward or downward. The effects would, in a similar manner, run back and forth in the interlocking system of interdependent causation. In any case, the initial change would be supported by consecutive waves of back-effects from the reactions of the other factor (p. 76).

Myrdal outlined an approach to societal problems that strikingly foreshadowed the work of Forrester and others following in the field that became known as system dynamics. I will quote him at length, because I view his words to be most significant in the evolution of the feedback concept:

It was during this study [*An American Dilemma*] that I first came to realise the inadequacy of the equilibrium approach, and to understand that the essence of a social problem is that it concerns a complex of interlocking, circular and cumulative changes (Myrdal 1957, p. 26).

The main scientific task is, however, to analyse the causal interrelations within the system itself as it moves under the influence of outside pushes and pulls and the momentum of its own internal processes. . . . The scientific ideal is not only to split the factors into their elements and to arrange them in this way, but to give for each of the elements quantitative measures of its ability to influence each of the others, and to be influenced itself by changes in other elements within the system or by changes in exogenous forces (1957, p. 30; also 1944, p. 1068).

Ideally the scientific solution of a problem like the Negro problem should thus be postulated in the form of an interconnected set of quantitative equations, describing the movement—and the internal changes—of the system studied under the various influences which are at work. That this complete, quantitative and truly scientific formulation is far beyond the horizon does not need to be pointed out; but in principle it could be made, and I submit that the working out of such a complete and quantitative solution should be the aim of our research endeavors even when they have to stop far away from the ideal (1957, p. 31; also 1944, p. 1069).

It was not as far off as Myrdal thought. In 1944, when he originally made these observations in *An American Dilemma*, the prospects looked bleak indeed. But he reiterated the comments, slightly revised in the form shown here, in *Rich Lands and Poor* (1957). Just one year later, Forrester's "Industrial Dynamics: a Major Breakthrough for Decision Makers" appeared (see section 3.3), demonstrating the practicality of an approach very much akin to the one Myrdal outlined. In two rather separated parts of the social sciences, and unaware of each other's work, these two authors apparently echoed almost precisely each other's thoughts. Myrdal could easily be considered the grandfather, or perhaps stepfather, of system dynamics, as the following further generalizations about his principle of cumulation suggest:

It is useless to look for one predominant factor, a "basic factor" such as the "economic factor" . . . as everything is cause to everything else in an interlocking circular manner (1957, p. 31; also 1944, p. 1069).

If the hypothesis of cumulative causation is justified, an upward movement of the entire system can be effected by measures applied to one or the other of several points in the system; but this certainly does not imply that from a practical and political point of view it is a matter of indifference where and how a development problem is tackled. The more we know about the way in which the different factors are inter-related . . . the better we shall be able to establish how to maximise the effects of a given policy effort designed to move and change the social system.

Nevertheless, it is unlikely that a rational policy will work by changing only one factor. Thus, though this theoretical approach is bound to suggest the impossibility, in the practical sphere, of all panaceas, it is, on the other hand, equally bound to encourage the reformer. The principle of cumulation—insofar as it holds true—promises final effects of very much greater magnitude than the efforts and costs of the reforms themselves (1957, p. 32).

These remarks are significant in our story for several reasons. First, they show a concept, the vicious circle, that had long since passed from formal logic into folk wisdom, transformed into a serious and powerful analytical tool for the social sciences. Second, they place the concept of the feedback loop at the foundation of social system dynamics. Third, they urge a connection, which we have seen prepared in sections 2.1 and 2.2, between the loop concept of mutual causality and the formulation of mathematical models. They reveal Myrdal's perception of the need for formal models embodying the feedback point of view in policy analysis. Fourth, they advocate a dynamic view, not only of the movement of a system but also its "internal changes." As we have seen in section 2.2, such internal changes can be captured in formal models by nonlinearities.

It seems reasonable that Myrdal, as an economist, was thinking of the econometric tools being developed in the 1930s and 1940s, and he was proposing their use in more general social science policy analyses. If so, then he had clearly made the connection between the loop concept of mutual causality and the structure of mathematical models of the sort described in section 2.2, an important step apparently taken independently of developments of engineering ideas in the social sciences in the 1940s. Presumably, he thought of capturing the "internal changes" of a system with exogenous influences common in econometrics, rather than the nonlinearities appearing in biological models. Yet his prescriptions describe Forrester's independent efforts very closely, as we shall see in section 3.3, particularly if nonlinearities are substituted for exogenous influences. One wonders if Myrdal really had in mind endogenous structural change. The impossibility of solving nonlinear problems at the time of *An American Dilemma* could account for his belief that quantitative models of phenomena like discrimination were "far beyond the horizon." At the very least, the similarity of ideas here from different corners of the social sciences suggest that intellectual developments are as much the product of general currents as they are the genius of any one individual. Zeitgeist again.

The bandwagon effect

Another positive loop concept embedded in the folklore of the social sciences is the "bandwagon effect," meaning the tendency of a movement to gain supporters simply because of its growing popularity. Originally, a bandwagon was the first vehicle in a circus parade, trumpeting the arrival of Barnum and company. The concept came to be used figuratively as a conveyance for a "band" of successful political leaders: "When I once became sure of one majority they tumbled over each other to get aboard the band wagon" (T. Roosevelt 1899, in OED 1972).

Eventually, the bandwagon effect was adopted seriously as an economic concept, defined as "the extent to which the demand for a commodity is increased due to the fact that others are also consuming the same commodity" (Leibenstein 1950, p. 189). As with the vicious circle, it is the self-reinforcing, closed-loop character of the bandwagon effect that gives the concept its appeal. That is to say, its perceived significance is directly due to the fact that it is a generic positive feedback loop.

The self-fulfilling prophecy

In the 1900s before engineering control concepts surfaced in the social sciences, the positive feedback loop continued to be rediscovered in a variety of guises. Still close to the level of social science folk wisdom is the notion of the "self-fulfilling prophecy." Robert King Merton (1936, 1948) is responsible for labeling the idea, but as he pointed out it has a long history. He traced it to Marx and Freud and a host of others, singling out particularly the sociologist W.I. Thomas and a famous theorem attributed to him: "If men define situations as real, they are real in their consequences" (Merton 1948). Merton translated the principle concisely as "social belief fathers social reality." In these expres-

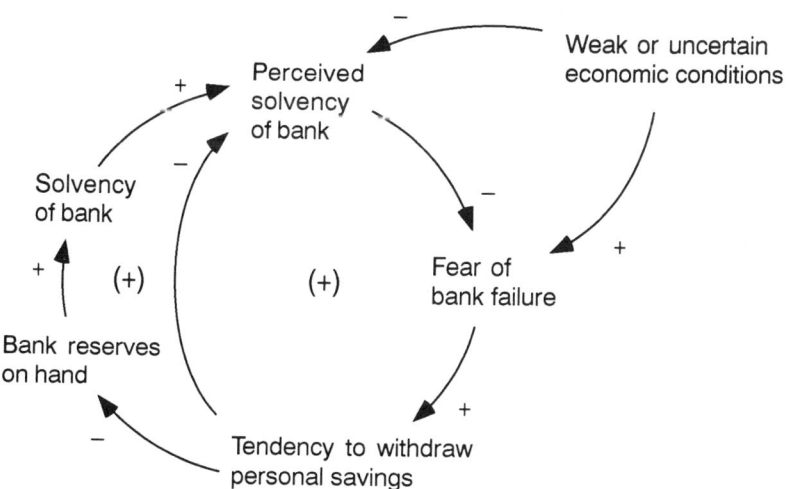

FIGURE 2.25: The self-fulfilling prophecy of a run on a bank in the depression, viewed as a positive feedback loop.

sions, however, it is perhaps hard to see the closed-loop nature of the principle.

As Merton defined it, a self-fulfilling prophecy is an initially false perception of a situation that evokes new behavior that makes the originally false conception come true (Merton 1948). Fearing a collapse of a bank during the depression, people rush to draw out their savings and thereby bring about the very thing they feared. A student, worried sick about his inability to take exams, devotes more time to worrying than studying and consequently fails, just as he knew he would. Two nations, each fearing the war-like moves of the other, stockpile armaments to the point that, as Merton said, "the anticipation of war helps to create the reality." Any of these situations can be viewed as a closed causal loop with a distinctly self-reinforcing character. Figure 2.25 shows the case of a run on a bank as an example. The self-reinforcing property of self-fulfilling prophecies makes the polarity of all such feedback loops positive.

The idea for the self-fulfilling prophecy had occurred to Merton much earlier and appeared briefly in an article addressing the intriguing topic "The Unanticipated Consequences of Purposive Social Action" (Merton 1936). At several points in this essay, Merton thinks in loops. For example, one of the reasons he cites for unanticipated consequences of an action are its ramifications, which feed back upon the system:

Action in accordance with a dominant set of values tends to be focussed upon that particular value-area. But with the complex interaction which constitutes society, action ramifies, its consequences are not restricted to the specific area in which they were initially intended to center, they occur in interrelated fields explicitly ignored at the time of action. Yet it is because these fields are in fact interrelated that the futher consequences in adjacent areas tend to react upon the fundamental value-system (Merton 1936, p. 903).

Merton was generalizing here from Weber's conclusion about the Protestant ethic, namely that "active asceticism paradoxically leads to its own decline through the accumulation of wealth and possessions." The negative feedback loop in Weber's conclusion and Merton's generalization are obvious, even if implicit.

The self-fulfilling prophecy made its brief appearance in this article as the alter ego of the "suicidal prophecy," a phrase Merton attributed to John Venn (1888), of Venn diagram fame. Marx, for example, had predicted that capitalism would progressively concentrate wealth and increase the misery of the masses. Merton observed that Marx's own prediction, by stimulating the spread of the organization of labor, had a hand in slowing up or eliminating the phenomenon he predicted

(Merton 1936, p. 904). Merton linked the phenomenon to a circumstance that he said was "peculiar to human conduct":

Public predictions of future social developments are frequently not sustained precisely because the prediction has become a new element in the concrete situation, thus tending to change the initial course of developments (1936, p. 904).

That is, as we would say, human systems are feedback systems! Merton's thoughts in this essay, emphasizing "self-negative prophecies," eventually led him to the positive loop version of the phenomenon.

Merton (1948) used the self-fulfilling prophecy primarily as a framework in which to view prejudice and discrimination. His brief but perceptive article put a different label on the same self-reinforcing circular causal structure used extensively by Myrdal (1944). Throughout, Merton emphasized that the insidiousness of the phenomenon was that the "facts" of a self-fulfilling prophecy indeed support the circular logic. The results are vicious circles in both of the Elizabethan senses of the phrase: they are harmful, and they are logically flawed.

When the gentleman from Mississippi (a state which spends five times as much on the average white pupil as on the average Negro pupil) proclaims the essential inferiority of the Negro by pointing to the per capita ratio of physicians among Negroes as less than one-fourth that of whites, we are impressed more by his scrambled logic than by his profound prejudices (1948, p. 185).

The mutual causal process such logic sets up, however, contains no contradiction. Instead, its circular, self-reinforcing character creates a very real spiral of discrimination, poverty, and prejudice.

What breaks the vicious circle of an ongoing self-fulfilling prophecy? Merton's answer was "controls":

The self-fulfilling prophecy, whereby fears are translated into reality, operates only in the absence of deliberate institutional controls (Merton 1948, p. 194).

Had he linked the control idea to negative feedback, he would have brought the two patterns of loop thinking in the social sciences together. As it was, even though he wrote "The Self-Fulfilling Prophecy" in 1948, several years after the emergence of the feedback concept in the social sciences, he did not make the connection.

Gregory Bateson's "schismogenesis"

Among the most interesting loop concepts in the social sciences prior to the emergence of feedback is Gregory Bateson's anthropological notion of "schismogenesis," literally, the generation of schism (Bateson 1935, 1936, 1949). One of the reasons it is interesting is that Bateson, through the cybernetics movement, eventually came to a deep understanding of the connections between his thinking and the engineer's concept of feedback. He is a transitional figure who wholeheartedly embraced the new cybernetic notions in the social sciences.

Bateson arrived at his concept of schismogenesis by thinking about possible consequences of two different cultures coming in direct contact with each other. They could fuse together over time into one single culture; one of the cultures (or perhaps both) could be completely destroyed; or both groups could come to persist over time as one large community in a kind of dynamic equilibrium (1935, pp. 64–67). Bateson argued that anthropologists could learn the most from studying this third category. What maintains the cultural differences between nation states in Europe or clans, social classes, castes, or age groups in a society? For that matter, what continually maintains "cultural" differences between the sexes?

Bateson's answer was schismogenesis, which he defined as "progressive differentiation" between cultural groups (1935, p. 68, 1936, p. 175). The word "progressive" is used here in the sense of "self-reinforcing," in much the same way that Myrdal (1944) used the word "cumulative":

If, for example, one of the patterns of cultural behaviour, considered appropriate for individual A, is culturally labeled as an assertive pattern, while B is expected to reply to this with what is culturally regarded as submission, it is likely that this submission will encourage a further assertion, and that this assertion will demand still further submission. We have thus a progressive state of affairs, and unless other factors are present to restrain the excesses of assertive and submissive behaviour, A must necessarily become more and more assertive, while B will become more and more submissive (Bateson 1936, p. 176).

Schismogenesis is thus a sophisticated label for a class of positive feedback loops.

Bateson identified two main types of schismogenesis, "complementary" and "symmetric." Complimentary schismogenesis is exemplified in the previous quotation linking assertive and submissive behavior in a positive loop.

But there is another pattern of relationships between individuals or groups of individuals which equally contains the germs of progressive change. If, for example, we find boasting as the cultural pattern of behaviour in one group, and that the other group replies to this with boasting, a competitive situation may develop in which boasting leads to more boasting, and so on. This type of progressive change we may call symmetrical schismogenesis (1936, pp. 176–177).

If one grants that submissive behavior is essentially the opposite of assertive behavior, then the symmetric situation can be thought of as a feedback loop composed of two positive links, while the complementary case contains two negatives (see Figure 2.26). In the latter case, an increase in assertive behavior in one actor produces a decrease in assertive behavior in the other actor (i.e., an increase in submissive behavior), which in turn reinforces the first to exhibit still more assertive behavior. In both cases, of course, the loop polarity is positive, and the process is self-reinforcing and disequilibrating.

Bateson was originally led to the concept of schismogenesis by reflecting on the profound contrast he uncovered between the ethos of men and women of the Iatmul tribe in New Guinea (1936). Yet he found examples of the phenomenon in everything from suburban marriages to politics. One of the common pathologies in modern marriages in Western culture, he noted, is the tendency of one of the partners increasingly to cast the other in the role of parent. The relationship between mother and child is initially completely complementary: fostering on the part of the mother, feebleness on the part of the child. As the child grows up, the pattern of fostering and feebleness may persist, or the mother may come to take vicarious pride in the accomplishments of the son, or the relationship may evolve towards assertiveness and submissiveness with either person playing either role. In any case, Bateson asserted, the mother/son relationship is almost always complementary, and if carried over into the son's mar-

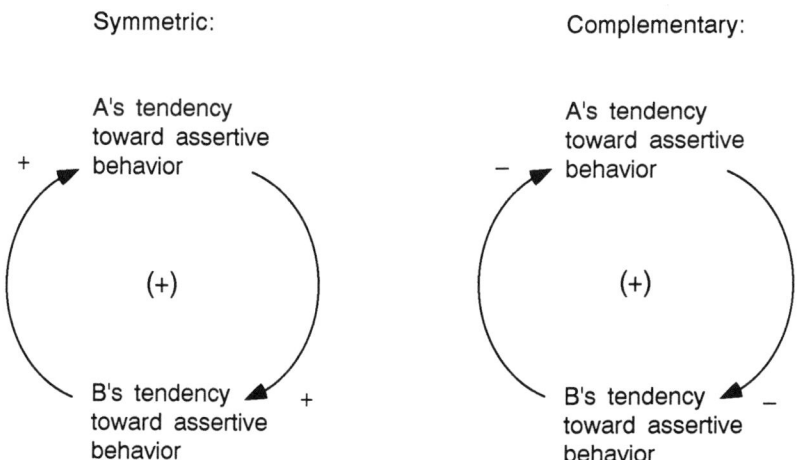

FIGURE 2.26: Symmetric and complementary schismogenesis as positive feedback loops.

riage it can become "the starting point of a schismogenesis which will wreck the marriage" (1936, p. 179). He concluded that viewing such a marital breakdown as cumulative, self-reinforcing, "schismogenic" process would help to explain how the marriage could have been initially satisfactory and yet the breakdown can come to appear inevitable to the participants.

In the area of psychiatry, Bateson hypothesized that the progressive deterioration of a schizophrenic might be due in no small part to the schismogenic character of the person's interactions with others (1936, pp. 179–183). This idea percolated and later emerged, colored by other notions from the field of cybernetics, as the "double bind" theory of schizophrenia (Bateson et al. 1956; see also Reusch and Bateson 1951).

For Bateson politics also has its schismogenic processes (1936, pp. 186–187). He saw symmetric schismogenesis in the intense international rivalries then rife in Europe. He also noted that "class-war" is of the complementary type. Even the megalomaniac dictator and his followers are caught up in a schismogenic process in which the dictator can be pushed closer and closer to an almost psychotic, paranoid state, as his people respond to his behavior in ways that push him toward greater excess. Finally, Bateson suggested that it would be worth investigating the extent to which politicians react to the reactions of their opponents rather than to the more or less objective societal conditions they are supposedly trying to affect.

Controls

Positive feedback loops do not exist without negative loops striving to constrain their tendency toward runaway behavior. Although Bateson did not have a loop concept corresponding to negative feedback, he realized the need for some concept of control or restraint. Cultural differentiation does not usually progress ad infinitum, but reaches a rather steady state. Marriages do not always fail, and even those beset by pathologies of the sort Bateson described sometimes reverse their trend and become healthy. Many schizoid individuals do not experience a complete disintegration of the personality, but maintain a kind of status quo for long periods.

We must therefore think of schismogenesis, not as a process which goes inevitably forward, but rather as a process of change which is in some cases either controlled or continually counteracted by inverse processes (1936, p. 190).

Bateson thus knew that there if schismogenesis were called a "positive" process, there had to be corresponding "negative" processes.

However, it appears that he did not have a loop view of such inverse processes. The closest he came to identifying constraints as circular processes was in his suggestion that symmetrical and complementary schismogenesis probably never occur in purely separate forms. Every symmetric relationship probably has elements of the complementary in it, and vice versa:

For example, the Squire is in a predominantly complementary and not always comfortable relationship with his villagers, but if he participates in village cricket (symmetric rivalry) but once a year, this may have a curiously disproportionate effect in easing the schismogenic strain in the relationship (1936, p. 193).

Thus Bateson, using other terms, suggested that any schismogenic situation is a mix of the two loops shown in Figure 2.26. Putting the two loops together results in some circuits that have negative polarity (one negative link) and may act to control the positive, schismogenic loops. A similar structure emerges from his perception of what he called "diagonal relations" between uncles and nephews in the Iatmul. He saw them as a stabilizing mechanism in an otherwise schismogenetic system, but he did not name these negative loop processes. The positive loop remained his focus.

Bateson suggested that schismogenic processes might be controlled by analogous self-reinforcing processes that simply operate in the opposite direction:

These processes are, like schismogenesis, cumulative results of each individual's reactions to the reactions of members of the other group, but the inverse process differs from schismogenesis in the direction of the change. Instead of leading to an increase in mutual hostility, the inverse process leads rather in the direction of mutual love (1936, p. 197).

Here we have a vicious positive loop counteracted by a virtuous positive loop.

The importance of circular processes

Generalizing from his analyses of schismogenesis in *Naven* (1936), Bateson came to advocate a new definition of the field of social psychology or social anthropology:

When our discipline is defined in terms of *the reactions of an individual to the reactions of other individuals,* it is at once apparent that we must regard the relationship between two individuals as liable to alter from time to time, even without disturbance from outside. We have to consider, not only A's reactions to B's behaviour, but we must go on to consider how these affect B's later behaviour and the effect of this on A (1936, pp. 175–176, italics in original).

These fields, Bateson thus concluded, should be *founded* on the concept of feedback. He came to this conclusion apparently without any knowledge of developments of the feedback concept in engineering and some eight to ten years prior to its emergence in the social sciences. In slight contrast to previous examples, it is a feedback concept with a discrete character. The behavior of individuals is the concern, and it is viewed much like moves in a game, one behavior following in response to another. Yet Bateson goes beyond a theory merely focusing on the "knowledge of results" of a move. The polarity of the resulting loop of information is critical. Schismogenesis is not merely a circular process; it is a self-reinforcing circular process. It is the association of loop polarity with the knowledge of results of behavior that ties Bateson's work to the feedback concept.

Summary

The loop concept underlying feedback and mutual causality has been present in the social sciences for at least the past 200 years. The frequency and significance of its use has been growing. And it appears in the writing of some of our most distinguished scholars. Negative loops are present implicitly in a great range of explanations in the social sciences. Positive loops appear much more explicitly, as circular processes of self-reinforcing phenomena. With the advent of the engineer's concept of the feedback loop in the social sciences in the 1940s, these two patterns of loop thinking could come together.

In Chapter 3 we shall see how these social sciences ideas interact with notions from engineering, mathematics, formal logic, and biology to produce initially two threads of feedback thinking in the social sciences.

Notes

1. Verhulst (1838), cited in Kormondy (1969, p. 69), and Lotka (1925/1956, p. 66); Pearl and Reed (1920, 1921) and Pearl (1922), cited in Lotka (1925/1956, p. 66). For extensive development and applications see Lotka (1925/1956, pp. 64–76).

2. The dominant polarity can be defined rigorously as the sign of dP'/dP, where $P' = dP/dt$. Here $P' = aP - bP^2$, so $dP'/dP = a - 2bP$, which is positive for $P < a/2b$ and negative for $P > a/2b$. Thus the dominance in Verhulst's population equation shifts from the positive loop to the negative loop when P reaches half its maximum (a/b). See Richardson (1984).

3. See, for example, the citations in H. A. Simon (1957).

4. Expanded versions of these early works appeared in Richardson (1947) and (1960). Richardson (1960) is the source of information for this section.

5. The version given here is from Samuelson (1939), with terminology taken from Baumol (1970, p. 170).

6. The direction of the arrow from Y_t to Y_{t-1} is explained by the observation that as time passes the current value of Y becomes the past value, so Y_{t-1} acquires its value from Y_t.

7. See Graham (1977, pp. 221).

8. Tinbergen's purpose and statistical methods make these circularities a real problem. He wanted to estimate parameters statistically. He had computed the two regressions separately, using ordinary least squares. Without the lag he would have had simultaneous equations, which are now known to require more sophisticated statistical machinery. See Pindyck and Rubinfeld (1976), pp. 126–151.

9. For examples see any econometric text, e.g., Pindyck and Rubinfeld (1976, pp. 266–416).

10. The iterative DO-CONTINUE loop from computer programming can be reasonably termed a logic loop because it can be rephrased in terms of IF-THEN statements.

11. Gödel's argument is a rigorous development that is similar to the paradox of "non-self-descriptive adjectives." Imagine placing all the adjectives in the English language in two lists: those that are "self-descriptive" and those that are "non-self-descriptive." The word "short" describes itself, so it goes in the first list. The word "diminutive" does not describe itself, so it goes in the second list. Now in which list does the adjective "non-self-descriptive" belong? If it belongs in the first list, then it describes itself, so it is non-self-descriptive and therefore belongs in the second list. But if it belongs in the second list, then it is non-self-descriptive so it does describe itself and therefore should go in the first list! The paradox is in the form of the contradiction in Cantor's proof, Russell's paradox, and Gödel's proof: "statement p is true if and only if statement p is not true."

12. See Alker (1981) for a strong argument for such a point of view, as well as extensive references to related social science literature.

Chapter 3
Two Feedback Threads

> Man is not the creature of circumstances. Circumstances are the creatures of men.
>
> Benjamin Disraeli, *Vivian Grey*

3.1 Emergence

In the ten years from 1943 to 1953 the engineer's concept of feedback was introduced into the social sciences. The people most responsible for the introduction and spread of the idea in this period were Norbert Wiener, Kurt Lewin, Karl Deutsch, Arnold Tustin, A.W. Phillips, and Herbert Simon. Their seminal publications using the feedback concept helped to define the directions for its future use in the social sciences. In this chapter we shall investigate their work and others' in detail and trace the early spread of feedback thinking in the social sciences.

As we have seen, there were a number of related ideas in the social sciences ready to be connected to the concept of feedback—checks and balances, homeostasis, vicious circles, self-fulfilling prophecies, logic loops, principles of cumulation and progressive differentiation, self-reference, and so on. And there were many other authors who contributed to the mixing of these ideas, as we shall see.

The mixing was not uniform or complete, however. Different individuals emphasized different aspects of the feedback idea and the concepts related to it. These differences become important in the evolution of the feedback perspective in the social sciences. Two lines of feedback thinking emerge from these beginnings. They differ in a number of significant ways that have serious implications for social science. Initially here, however, we shall simply assert that one tends to focus on the role of feedback in communication and control in society, while the other emphasizes the role of feedback loops in dynamic behavior. Wiener, Deutsch, and Lewin are instrumental in the beginnings of the former, while Tustin, Phillips, Simon, and others initiate the latter.

Figure 3.1 sketches the framework of the claim being made. It shows the six intellectual traditions described in Chapter 2, which combine into two main threads to influence the evolution of the feedback concept in the social sciences.

Each of the two feedback threads that emerge in the social sciences is a set of authors and ideas that are interconnected sociologically and/or methodologically. For reasons that will be clearer at the end of this

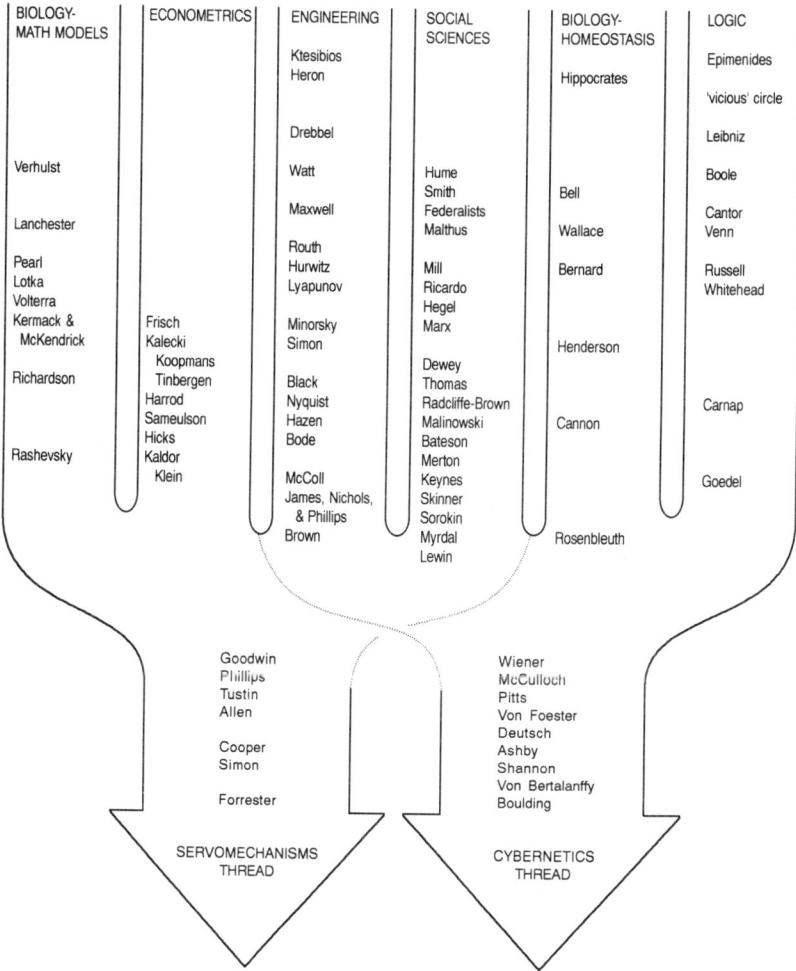

FIGURE 3.1: Representative authors and ideas in the intellectual traditions contributing to the evolution of the feedback concept in the social sciences, showing their relationships to the beginnings of the two feedback threads traced in Chapters 3, 4, and 5.

chapter, I have chosen to label those two lines of feedback thinking in the social sciences the "servomechanisms thread" and the "cybernetics thread." As Figure 3.1 shows, both threads are influenced by ideas associated with the feedback concept in engineering and by loop ideas from the social sciences (the two middle columns in the figure). But the servomechanisms thread has greater similarities, and stronger sociological connections, to mathematical modeling traditions in biology and economics (on the left in Figure 3.1). Similarly, the cybernetics thread can be more associated with people and ideas in formal logic and physiology, including most notably the concepts of self-reference and homeostasis (on the right in the figure).

The differences that emerge and become accentuated in these two feedback threads create different understandings about the role of feedback processes in social systems and the appropriate role of feedback thinking in the social sciences.

3.2 The Cybernetics Thread

The Beginnings

The first published article linking human systems with the engineers' concept of feedback is Rosenblueth, Wiener, and Bigelow's "Behavior, Purpose, and Teleology" (1943). The article contributed a modern slant to the age-old argument about the place of teleological explanations in science. The authors argued that any machine or organism that was controlled by negative feedback deserved to be called purposeful. A scholarly debate ensued, with opponents objecting to what they saw to be the authors' "mechanistic" conception of purposefulness.[1] The debate is of interest here only insofar as it involved the concept of feedback.

Rosenblueth, Wiener, and Bigelow divided all active behavior into two categories: "purposeful" and "non-purposeful." The latter was characterized as "random," while the former implied "behavior directed toward a goal." The authors envisioned two possibilities for goal-directed behavior. Behavior in which signals from the goal modify the action in the course of the behavior they termed "feed-back" or "teleological." Any other purposeful behavior was termed "non-feedback" or "non-teleological."

Their definition of the term "feedback" was the first to reach a wide social science audience:

The expression feed-back is used by engineers in two different senses. In a broad sense it may denote that some of the output energy of an apparatus is returned as input; an example is an electrical amplifier with feed-back. The feed-back is in these cases positive—the fraction of the output which reenters the object has the same sign as the original input signal. Positive feedback adds to the input signals, it does not correct them. The term feed-back is also employed in a more restricted sense to signify that the behavior of an object is controlled by the margin of error at which the object stands at a given time with reference to a relatively specific goal. The feed-back is then negative, that is, the signals from the goal are used to restrict outputs which would otherwise go beyond the goal. It is this second meaning of the term feed-back that is used here.

All purposeful behavior may be considered to require negative feed-back. If a goal is to be attained, some signals from the goal are necessary at some time to direct the behavior (Rosenblueth, Wiener, and Bigelow 1943, p. 222).

For examples of feedback behavior in living organisms, the authors cited an amoeba following an object, a cat pursuing a mouse, a bloodhound following a trail, and a person lifting a glass of water from table to mouth or throwing a stone at a moving target. In each case signals from the goal modify the action in the course of the behavior. The authors acknowledged the possibility of goal-directed behavior that is not modified during the behavior itself, such as a snake striking at a frog or a frog at a fly. The speed with which such actions take place argues that no signals from the goal affect the behavior during the action itself. No information about position, speed, or direction could be fed back and acted upon during the strike, so such behavior, although clearly purposeful, was termed by the authors "non-feed-back" behavior and identified as "non-teleological." It seems quite reasonable that a slight change in the notion of the "behavior" in such situations—expanding it to include the preparation for the strike—would eliminate completely the possibility of purposeful behavior without feedback.

The teleological argument hung on the authors' insistence on a behaviorist point of view. If one focuses on observed behavior and not on hidden (mental or mechanical) operations, then all goal-directed behavior looks teleological. Whatever the merits of their philosophical case, for us the significance of the Rosenblueth, Wiener, and Bigelow paper is that it brought to light in the social sciences the engineers' concept of feedback. The paper contributed directly to the start of an intellectual movement centered on the concept of feedback: cybernetics. "Behavior, Purpose, and Teleology" was first read at a conference in 1942 sponsored by the Josiah Macy, Jr., Foundation devoted to "problems of central inhibition of the nervous system." It led to the organization of the first Macy conference on "Feedback Mechanisms and Circular Causal Systems in Biological and Social Systems," held in 1946. The Macy Foundation went on to sponsor a total of ten such conferences, the last being held in 1953. Over the years the participants included some very respected social scientists, perhaps the most famous of whom were Kurt Lewin, Gregory Bateson, Margaret Mead, and Karl Deutsch.

I want to draw a clear distinction between the field of cybernetics and the line of feedback thinking in the social sciences I am calling the cybernetics thread. We are interested in the field of cybernetics only for its use of the feedback concept. Writings in the field of cybernetics reached a large number of social scientists. The result is that many, if not most, social scientists who make use of the feedback concept are in the cybernetics thread. That conceptual thread of feedback thinking acquires some of its definition from these initial cybernetics meetings and the work traceable to them. It is for that reason that we are investigating these cybernetics writings.

Four characteristics of the Rosenblueth, Wiener, and Bigelow paper tend to reappear in the cybernetics thread in the social sciences. First, the paper is philosophical, addressing a theoretical issue at the foundation of scientific understanding. The feedback concept is employed to make acceptable again the concept of purpose in scientific explanation. Second, the discussion is nonmathematical. There is no evidence of the differential equations or Laplace transforms in which engineers couch their use of the feedback concept. Third, the focus is on the control of behavior. Feedback is seen as the mechanism through which a goal exercises its control over behavior. Fourth, and related to the third, the feedback used in the development of this of the paper is exclusively negative. The existence of positive feedback is noted but is dismissed as having no significance for the issue of purposive behavior. Thus, as the cybernetics thread unfolds, we shall find tendencies toward

- philosophical and theoretical discussion rather than direct application to societal phenomena;
- verbal analyses rather than mathematical models of feedback systems;
- an emphasis on control of behavior, not on dynamic patterns in behavior;
- focus on negative feedback, to the almost complete exclusion of the concept of positive feedback.

A second paper was instrumental in the beginnings of the field of cybernetics. McCulloch and Pitts's "A Logical Calculus of the Ideas Immanent in Nervous Activity" was presented at the same 1942 Macy conference as the Rosenblueth, Wiener, and Bigelow paper. The purpose of McCulloch's work was to find in the biophysics and biochemistry of the brain the bases for human thought. Pitts was a mathematician who brought to the task the machinery of symbolic logic. The "Logical

Calculus" paper was a deep investigation in theoretical neurophysiology that postulated abstract characteristics of neurons and the networks they form together and derived the neural nets necessary and sufficient for propositional logic.

Although the concept of feedback does not appear explicitly in the paper, the authors did make use of the notion of a loop, in two ways. First, they considered the possibility of closed "regenerative loops" of neurons. They used such loops in the course of deriving the full logical potential of neural nets (McCulloch and Pitts 1943, pp. 30–35). In addition, they suggested that a signal reverberating around such a regenerative loop into the indefinite future could account for memory (pp. 29,30). Their second use of the loop notion draws directly on the use of feedback in the Rosenblueth, Wiener, and Bigelow paper, without using the term feedback. They observed that neural nets of sufficient complexity could show purposeful behavior. All that is required, they argued, is a net that responds to incoming signals so as to reduce the discrepancy between those signals and conditions within the net (p. 38). McCulloch and Pitts concluded that "both the formal and final aspects of that activity which we are wont to call mental are rigorously deducible from present neurophysiology." Their logical calculus assured the "formal aspects." For the "final" aspects, meaning observed behavior, they relied on the role of feedback in purposeful activity.

The significance of the McCulloch and Pitts paper in the evolution of the feedback concept in the social sciences is not particularly its own use of a loop notion, but rather the logical calculus itself. The neural net model, composed of binary "off/on" logical devices, reappears in the work of Deutsch, Ashby, Beer, and Klir applying the feedback concept to societal and managerial problems. As will be seen shortly, Deutsch (1948) urges a scholarly exploitation of the analogy between neural nets and social communication networks. Ashby (1956), Beer (1959), and Klir and Valach (1967) advocate the use of such binary models to study what Beer calls "exceedingly complex" systems in preparation for the design and use of a feedback control device for a system. (See section 4.2.)

The Macy Conferences

No records of the early Macy cybernetics conferences were published. Only the sixth through the tenth conferences had published proceedings. The two best general sources of information about the conferences are "A Note by the Editors" in the proceedings of the eighth conference (von Foerster 1951) and McCulloch's "Summary of the Points of Agreement Reached in the Previous Nine Conferences on

Cybernetics" distributed in advance of the tenth conference and appearing as an appendix in the proceedings (von Foerster 1953). From McCulloch's summary prior to the tenth conference we learn that the meetings began chiefly because

> Norbert Wiener and his friends in mathematics, communication engineering, and physiology, had shown the applicability of the notions of inverse feedback to all problems of regulation, homeostasis, and goal-directed activity from steam engines to human societies (McCulloch, in von Foerster 1953, p. 70).

("Inverse" feedback is negative feedback.) We learn that feedback came to defined in these meetings as "the alteration of input by output" (McCulloch, in von Foerster 1953, p. 71). From von Foerster, Mead, and Teuber in their introduction to the 1951 proceedings we get a sense of their enthusiasm for the feedback concept:

> A state reproducing itself, like an organism, or a social system in equilibrium, or a physiochemical aggregate in steady-state, defied analysis until the simple notion of one-dimensional cause-and-effect chains was replaced by the bi-dimensional notion of a circular process (von Foerster 1951, p. xiv).

McCulloch gives an idea of the range of feedback phenomena considered, including regulatory circuits in homeostatic mechanisms, "appetitive behavior" in which part of the negative feedback loop lay within the organisms and part in the environment, cardiac flutter and fibrillation, the control functions of the cerebellum, stretch reflexes in muscles, various pathologies of the nervous system, the polling of public opinion, fluctuations of markets, the banter leading to fights among roosters and boys, arms races initiating wars, and so on. The most complex situations arose in the work of the cultural anthropologists who saw negative feedback structures as the stabilization mechanisms of societies. They reported interwoven, multi-loop systems involving elaborate forms of distinctions and rules with respect to kinship, forms of address, hazing, bullying, praise, blame, and even rituals with respect to eating. There were also examples from ecology reaching as far as the behavior of ant colonies (von Foerster 1953, pp. 71–75). Unfortunately, there are no records of how any of these were represented or discussed.

We learn that the notion of "circular causal processes" was merely the second of the two conceptual models central to the discussions: first and foremost was the theory of information (Shannon 1948, Shannon and Weaver 1949, Wiener 1948). McCulloch sketches the reason for the important shift in focus to information theory and electronic computers:

By [the sixth meeting, 1949], we had already discovered that what was crucial in all problems of negative feedback in any servo system was not the energy returned but the information about the outcome of the action to date. . . . It became clear that every signal had two aspects: one physical, the other mental, formal, or logical. This turned our attention to computing machinery, to the storage of information as negative entropy. Here belong questions of coding, of languages and their structures, of how they are learned and how they are understood (McCulloch, in von Foerster 1953, p. 75).

The titles of papers in the sixth through the tenth conferences show this resulting emphasis on communication, information, and computing. Only three of the thirty paper titles suggest that the concept of feedback is involved.[2] McCulloch, in fact, notes in his summary that after a while "we had become so weary of far-flung uses of the notion of feedback that we agreed to try to drop the subject for the rest of the conference." For our feedback focus it is sad that there were no records kept of the first six conferences in which the concept was apparently flung so far.

Nonetheless, it is from the writings of people associated with the cybernetics movement that most social scientists learn of the concept of feedback. The proceedings of the Macy conferences themselves, however, are not the source. We must look to publications that grow out of the stimulation provided by the Macy meetings.

Early Uses of the Feedback Concept in Social Science Writings

In 1947 feedback publications in the social sciences began appearing, among them works by Kurt Lewin (1947), Karl Deutsch (1948), Norbert Wiener (1948, 1950), W. Ross Ashby (1952), and the general system theorists, notably Ludwig von Bertalanffy (1950a, 1950b, 1951). Together with Rosenblueth, Wiener, and Bigelow (1943) and McCulloch and Pitts (1943), these articles began to define a conceptual feedback thread in the social sciences.

Lewin (1947) is part of a publication on "Frontiers in Group Dynamics." The article is concerned with the problem of "social self-steering." Lewin observes:

Many channels of social life have not simply a beginning and an end but are circular in character. . . . Organized social life is full of such circular channels. Some of these circular processes correspond to what the physical engineer calls feedback systems, that is, systems which show some kind of self-regulation (Lewin 1947, in Buckley 1968, p. 441).

He proceeded to discuss social planning as an example. Ideas, he said, become plans by clarifying an objective, identifying a path to the goal and the available means of getting there, and developing a strategy. The three components reminded Lewin of the gun targeting mechanisms used during the war which linked three entities: (i) the position of the target; (ii) a target-sensing organ; and (iii) an action organ for moving the gun. In aiming at a moving target, the action is continuously steered toward the goal with the help of a sense organ that "seeks" to eliminate divergences between action and goal. Lewin sought the equivalent mechanisms in social life that underlie the steering of social action. What are the social sense organs? What is the steering process?

Lewin's concerns for social planning ranged from small groups to nations. His goal was the improvement of "social practice." It is an applied goal, not theoretical, although his means of reaching it were research and theory-building. He characterized the research needed for social practice as "action-research" in social management or social engineering. If we react in the present day to various undesirable connotations of the notion of social engineering, we should note that Lewin intended the phrase to call to mind the engineering notions of feedback systems and engineering design. If anything, Lewin was reacting against the specters of totalitarianism fresh in the minds of everyone in 1947. He was concerned for the survival of democracy. He believed that it depended less on the development of democratic ideas, which he saw to be widespread and healthy, and more on "the development of efficient forms of democratic social management and upon the spreading of the skill in such management to the common man" (p. 444). Everyone a feedback thinker.

Deutsch (1948) is less action-oriented, more interested in the advances in theoretical understanding that he perceived would come from exploiting the cybernetic analogy among electric networks, nerve systems, and societies. He advocated a new model of organization in machines, minds, and societies to replace the old mechanistic and organismic views. He wanted to view systems as "self-modifying networks." He was motivated by the striking parallels he perceived between behavior in the fields of psychology, neurophysiology, and cultural anthropology and the behavior of engineering feedback devices:

Manmade machines actually operating or designable today have devices which function as "sense organs," furnish "interpretations" of stimuli, perform acts of recognition, have "memory," "learn" from experience, carry out motor actions, are subject to conflicts and jamming, make decisions between conflicting alternatives, and follow operating rules of preference or "value" in distributing their "attention," giving preferred treatment to some messages over others, and making other decisions, or even conceivably overriding previous operating rules in the light of newly "learned" and "remembered" information (Deutsch 1948, p. 390).

To Deutsch the basic pattern that minds, societies, and self-modifying communications networks have in common is feedback. In the paper he reiterated the definition of feedback given in Rosenblueth, Wiener, and Bigelow (1943), including its restriction to negative feedback, and then translates:

> In other words, by feedback is meant a communications network which produces action in response to an input of information and *includes the results of its own action in the new information by which it modifies its subsequent behavior* (pp. 390–391, emphasis in original).

He proceeded to use the concept of feedback to discuss learning, purpose, values, consciousness, will, autonomy, integrity, and freedom (pp. 391–399):

> Learning and purpose—Simple learning is goal seeking feedback, as in a homing torpedo. . . . A more complex type of learning is the self-modifying or goal changing feedback.
> Values—The movements of messages through complex feedback networks may involve the problem of "value" or the "switchboard problem," that is, the problem of choice between different possibilities of routing different incoming messages. . . . The efficient functioning of any complex switchboard requires, therefore, some relatively stable operating rules . . . [T]here seems to be no reason why these operating rules themselves should not be made subject to some feedback process.
> Consciousness—. . . defined, for the purposes of this discussion, as a collection of internal feedbacks of secondary messages. "Secondary messages" are messages about changes in the state of parts of the system. . . . "Primary messages" are those which move throughout the system in consequence of its interaction with the outside world.
> Will—"Will" in all these cases may be tentatively defined in any sufficiently complex net, nervous system, or social group as the set of internal labels attached to various stages of certain channels within the net. . . . Isolating the pattern of "will" in feedback machines may help us to recognize it in men and communities.
> Autonomy—. . . [The will] represents the stored outcome of the net's past now being fed back into the making of present decisions. Without effective feedback of its past, the net's behavior would be determined largely by outside pressures. It would not steer, but drift.
> Integrity—. . . that is, a structure of internal feedbacks, controls, and connections, undisrupted by excessive rates of input.
> Freedom—[The net's] internal rearrangements in response to each new challenge are made by the interplay between its present and its past. In this interplay we might see one kind of "inner freedom." . . . It's outward behavior will be the result of the interplay between the orders transmitted to its effectors, and the feedback data about their results among the pressures of the outside world. In this type of interplay we may see a kind of external freedom for the net to continue its efforts to reach its goal.

This is theorizing on a grand scale, stimulated by the perceived richness of a set of analogies between mechanical, electronic, and human communication networks. Deutsch emphasized that these analogies between electrical networks, nerve systems, and societies, as promising as they appeared to be, were only analogies, implying similarities in some but certainly not all aspects (p. 390). His remarks on the subject echo the reservations of the participants in the Macy cybernetics conferences. In the editors' comments on the influence of the work on "giant electronic brains" we learn that they also saw the feedback concept in social and biological systems as a great analogy:

We all know that we ought to study the organism, and not the computers, if we wish to understand the organism. Differences in levels of organization may be more than quantitative, but the computing robot provides us with analogues that are helpful as far as they seem to hold, and no less helpful whenever they break down. To find out in what ways a nervous system (or a social group) differs from our man-made analogues requires experiment. These experiments would not have been considered if the analogue had not been proposed, and new observations on biological and social systems result from an empirical demonstration of the shortcomings of our models. It is characteristic that we tend to think of the intricacies of living systems in terms of non-living models which are obviously less intricate. Still, the reader will admit that, in some respects, these models are rather convincing facsimiles of organismic or social processes—not of the organism or social group as a whole, but of significant parts.
How this way of thinking emerged in the group is difficult to reconstruct." (von Foerster 1951, pp. xviii–xix)

This way of thinking is what now would be called the modeling approach. All of these authors were familiar with models in science, so their use of the weaker word "analogue" is probably significant. The cybernetic analogues of social systems and communication networks—principally electronic computers and the models that could be formulated with them—were apparently seen to contain less realism than mental models of the same systems.

Although feedback is repeatedly referred to throughout, the quotations from Deutsch's essay indicate that even more central to his thinking are the neural nets of McCulloch and Pitts, Shannon's theory of information, and notions from digital computers. Complex feedback networks, Deutsch says, deal with messages and symbols that can be stored, manipulated according to previously specified rules, and combined to produce results that may or may not have been seen before. "Messages" and "symbols" are defined in terms of changes in a description of the state of the network. A "state description" is defined, somewhat circularly, as "a specification of which of its possible states each element of the network is in." As noted previously, these ideas will continue to be associated with the feedback concept in the cybernetics thread, particularly as it connects with Ashby, Beer, and Klir. Deutsch himself reiterated and extended these views in *The Nerves of Government* (1963).

Norbert Wiener's *Cybernetics*

Cybernetics, subtitled *Control and Communication in the Animal and the Machine*, appeared in 1948 and has been referred to, cited, and acclaimed ever since. One can hardly do better than Stafford Beer's capsule description:

Difficult, quixotic, immensely stimulating (then and now), *Cybernetics* split the scientific world (for those who read it) down the middle. Think of it like this: the great man (he really was) holds forth to his friends after dinner, ruins the tablecloth by scribbling mathematics all over it, sings a little song in German, and changes your life. It is tough going: you have to stay the night (Beer 1972, p. 409).

One reviewer termed it "a beautifully written book, lucid, direct, and despite its complexity, as readable by the layman as the trained scientist."[3] Apparently, he was not a layman.

The concept of feedback is interspersed throughout the book and is the subject of an entire chapter. "Feedback and Oscillation," one of the two mathematical chapters in the book, in its details is surely unreadable by all but a tiny minority of those who encountered the book. The mathematical effort in that chapter is directed to setting out the conditions under which negative feedback in linear systems can result in exploding oscillations. To do it, the chapter manipulates mathematical operators, convolution integrals, complex variables, holomorphic and meromorphic functions, points at infinity, and cardioids in polar coordinates. At one point (in the chapter on "Time Series, Information, and Communication"), Wiener even mentions in passing the one-to-one correspondence between the points on a line segment and the points in the interior of a square—a discovery of Cantor's that Cantor himself found almost impossible to believe.

It is clear that what social scientists learned of the feedback concept from *Cybernetics* they must have gleaned from the comments about it in the nonmathematical parts of the book. It is also clear, however, that from this book they would have gotten an overwhelming impression of the importance of the ideas of cybernetics, including the idea of feedback.

In the Introduction Wiener recounted how he and his colleagues came to realize that the engineer's concept of feedback applied to human activity. Because of the German superiority in aviation at the beginning of World War II, the attention of some scientists was drawn to the problem of improving anti-aircraft accuracy. Wiener saw a way of treating the evasive maneuvering of a plane as a time series that could be manipulated mathematically to improve the gunner's prediction of the plane's future position. To implement the theory, characteristics of the person aiming the gun had to be taken into account mathematically. The effort to do this convinced Wiener and his collaborator Julian Bigelow that feedback is an "extremely important factor in voluntary activity":

When we desire a motion to follow a given pattern, the difference between that pattern and the actually performed motion is used as a new input to cause the part regulated to move in such a way as to bring the motion closer to that given by the pattern (Weiner 1948, p. 6).

Wiener then describes how they thought about feedback in voluntary human activity, discussing the example of picking up a pencil. They were aware of a pathological state known as ataxia, in which the nervous system of a person fails to report, or respond adequately to, the discrepancy between the position of the hand and the pencil, making it impossible to pick it up. In ataxia Wiener saw the symptoms of a weakened or destroyed feedback link in a neural network. Knowing another pathology of engineering feedback systems—that too much negative feedback can cause instability and oscillatory response—Wiener and Bigelow were led to ask the neurophysiologist Arturo Rosenblueth a "very specific question":

Is there any pathological condition in which the patient, in trying to perform some voluntary act like picking up a pencil, overshoots the mark, and goes into an uncontrollable oscillation? Dr. Rosenblueth immediately answered us that there is such a well-known condition, that it is called purpose tremor, and that it is often associated with injury to the cerebellum (p. 8).

The question and the answer were clearly very important to Wiener. There seemed to be no doubt then that some aspects of human activity involve feedback. The three men assembled their new point of view into the paper "Behavior, Purpose, and Teleology" previously discussed, and feedback became a concept common to men and machines.

Wiener's examples in *Cybernetics* of feedback in living systems include the oscillatory phenomenon of clonus in a muscle under tension (pp. 19–21), steering a car on an icy road by frequently testing its tendency to skid (p. 113), various homeostatic mechanisms in the human body (pp. 114–115), and the phenomenon of proprioception in neuromuscular systems (pp. 7, 26, 43, 96). The examples are generally less dramatic and more tightly circumscribed than those encountered in the articles previously discussed. Indeed, our focus on the feedback concept forces us to slight quite considerably what are considered to be the enormously scholarly and philosophical contributions of *Cybernetics*, particularly to our understandings of the nature of information and its characterization as negative entropy.

However, the book added two things to the growing understandings of feedback in the social sciences. First, for those social scientists who did not go back to the original servomechanisms literature, it gave the first pictures of feedback loops. Representative of Wiener's diagrams is Figure 3.2. In the drawings we see clearly the Macy conference definition of feedback as "the alteration of input by output."

Second, Wiener's book rather surprisingly raised the question of the applicability of cybernetics to societal problems. In his Introduction Wiener outlined some of the history of the Macy meetings and observed their growing focus on the importance of communication and feedback in society.

It is certainly true that the social system is an organization like the individual, that it is bound together by a system of communication, and that it has a dynamics in which circular processes of a feedback nature play an important part (1948, p. 24).

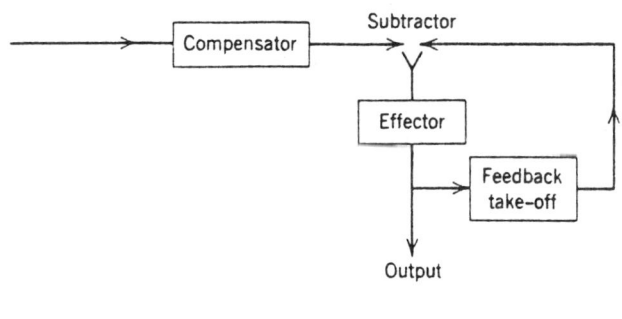

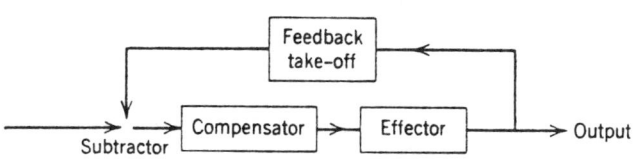

FIGURE 3.2: Generic feedback structures from Wiener's *Cybernetics*. Source: Wiener (1948, p. 112).

He reported that Margaret Mead and Gregory Bateson had urged him to direct his attention toward the application of cybernetics to pressing sociological and economic problems. But he declined, for the following rather cryptic reason:

> To begin with, the main quantities affecting society are not only statistical, but the runs of statistics on which they are based are exceedingly short. . . . For a good statistic of society, we need long runs under essentially constant conditions. . . . Thus the human sciences are very poor testing-grounds for a new mathematical techique (pp. 24–25).

It was an important point to him. He devoted a full two pages to discussing it. What did he have in mind? The differential equations of servomechanisms engineers had been around for hundreds of years, and the modern techniques in the frequency domain involving the Laplace transform dated from the 1930s and were in common use by 1948. What was the new mathematical technique? Why was new mathematics necessary to apply feedback ideas to social systems?

The answers are partly to be found in McCulloch's summary prior to the tenth Macy conference in 1953, previously discussed. McCulloch noted that participants in the meetings found it difficult to detect the causal relations in circular causal systems. Wiener handled this, he asserted, "by pointing out that it was possible to detect causality in the statistical sense by auto- and inter-correlations with lag" (McCulloch in von Foerster 1953, p. 75). Some statistical time-series manipulation would suffice. This is the new mathematical technique Wiener refers to. We can be sure because McCulloch followed his assertion with a direct paraphrase of Wiener: "He doubted the applicability of the method to social problems because of the shortness of our runs of the time series of information concerning human behavior" (p. 74). Now we know that Wiener was worried not about the application of the feedback concept itself to societal problems, but to the use of a statistical technique for determining causal relations in social feedback systems.

This bit of cross-reference detective work is of interest because it is precisely the sort of thing that would never be done by a reader of *Cybernetics*. The social scientist, or anyone else, reading the book would come away with the general impression of the enormous significance of the work and a vague unease that there was something about cybernetics that couldn't be reliably applied to social systems. Biology yes, brains perhaps, but society apparently not. Our runs of statistics are too short; we don't know enough. Is it the feedback concept that cannot be applied? The theory of information? The reader's most likely interpretation of Wiener's remark is that the mathematical and statistical aspects of cybernetic ideas are not applicable. The best that can be done is the sort of verbal analysis and theory-building we have seen in the papers preceding *Cybernetics*. That would be a comforting conclusion to a social science reader, who, like most readers, would simply pass over Wiener's mathematics to hunt for the verbal summaries.

Wiener's less technical book, *The Human Use of Human Beings*, appeared in 1950. Its subtitle is *Cybernetics and Society*. If there was any doubt before, there is now no question that he thinks cybernetics has profound implications for society.

It is the thesis of this book that society can only be understood through a study of the messages and the communication facilities which belong to it; and that in the future development of these messages and communication facilities, messages between man and machines, between machines and man, and between machine and machine, are destined to play an ever-increasing part (Wiener 1950, p. 16).

His repeated characterizations of the feedback concept in this book emphasize the difference between intended action and actual performance.

Control of a machine on the basis of its actual performance rather than its expected performance is known as feedback (p. 24).

In both [the animal and the feedback machine], their performed action on the outer world, and not merely their intended action, is reported to the central regulatory apparatus. This complex of behavior is ignored by the average man and in particular does not play the role that it should in our habitual analysis of society (p. 27).

. . . feedback, the property of being able to adjust future conduct by past performance (p. 33).

Feedback is a method of controlling a system by reinserting into it the results of its past performance (p. 61).

His examples sound familiar: targeting a gun, picking up a cigar, driving a car. An elevator is pictured as a feedback device. It is important, for example, that the doors to the elevator shaft open only when the elevator is actually positioned at the opening. Good elevator intentions would not be enough; something might detain it, and people could tumble into the open shaft.

The things that might interfere with intentions and prevent actual performance from being realized are viewed as "entropic tendencies"— grease in the bearings of a gun turret that gets thick as it gets cold, sand in the bearings, fatigue in muscles, differences in steering effort between a compact car and a heavy truck. The parallel Wiener saw between living individuals and servomechanisms is "their analogous attempts to control entropy through feedback" (p. 26).

Feedback in society was linked to patterns of communication and social structure among the Eskimos, in stratified caste societies and feudal communities, among businessmen and their subordinates, and in ant colonies (pp. 50ff). The focus is on the role of feedback in the control of behavior.

Like Deutsch, Wiener clearly distinguished between lower and higher levels of sophistication in feedback systems, particularly in living systems (pp. 58ff). At the lowest level are the simple feedbacks that tell us about gross successes or failures of performance—did we manage to grasp the object we intended to pick up? At higher levels of sophistication are feedbacks that affect policies or strategies—should I, as a rat running a maze, change my strategy for finding food or avoiding electric shocks? Such learning was seen by Wiener as feedback on a higher level, a feedback of policies rather than simple actions.

At these higher levels, Wiener, Deutsch, and others in the cybernetics thread saw an analogy to Russell's theory of "logical types." The analogy has a danger, however. Russell and Whitehead proposed the theory of types to avoid dramatic contradictions in set theory at the foundation of mathematics. The observation of levels of feedback, however, does not lead to such contradictions. But the observation of the analogy tends to suggest that there are potential problems with the feedback concept in complex systems. The impression is not merely that we can distinguish levels of sophistication among feedback processes, but that we must. Neither Wiener nor Deutsch provided a rationale for the necessity, but once set in motion in the cybernetics thread the distinction among "types" of feedback persisted (see, e.g., Bateson 1958, pp. 292–303).

The Work of W. Ross Ashby

W. Ross Ashby was a principal contributor to the development of the field of cybernetics. His early works (1945, 1946a, 1946b, 1947a, 1947b) were concerned with stability and various aspects of what he referred to as the "dynamics of the cerebral cortex." The Macy meetings initially focused on similar issues, but it appears that Ashby's early work was done independently of Wiener, McCulloch, and their colleagues and did not make use of the concept of feedback. Rather quickly, however, Ashby became connected with the budding field of cybernetics. He appears as the author of two papers in the ninth Macy meeting (1952), and his *Design for a Brain* (1952) contains numerous references to the cybernetics literature existing at that time. In fact, that book assembles the work of his 1946–1947 articles and restates some of it in feedback terms. Ashby's identification with cybernetics became complete with the publication of his *Introduction to Cybernetics* (1956), the first text in the field.

Ashby's use of the feedback concept centered on the question of stability in dynamic systems. In his first note on the subject (Ashby 1945), he observed that constraining a variable in a stable system could easily produce instability. He proved the point using systems of linear differential equations in the matrix form $\dot{x} = A x$. He noted that if the matrix A is

$$\begin{bmatrix} 6 & 5 & -10 \\ -4 & -3 & -1 \\ 4 & 2 & -6 \end{bmatrix}$$

and the third variable (x_3) is held constant, the system composed of the first two variables has the matrix

$$\begin{bmatrix} 6 & 5 \\ -4 & -3 \end{bmatrix}.$$

The latter system has eigenvalues of $+1$ and $+2$ and is consequently unstable, having solutions of the form $x_i = c_1 e^t + c_2 e^{2t}$. Ashby was led to make his observation by considering the effects of controls placed on Britain's economy during World War II. Paradoxically, he warned, price controls could lead to instability rather than stability. Commenting later on the observation, Bateson (1958) gave the example of a tightrope walker who fails to keep his balance if he cannot adjust the position of his balancing pole.

Ashby's first explicit use of the feedback concept appeared in *Design for a Brain* (1952). The book is very interesting for our feedback purpose, because it represents a kind of watershed in Ashby's thinking. The stated purpose of the book was the same as that of his previous articles: he wanted to determine how the brain, viewed as a physiochemical machine, can produce behavior that psychologists would call "purposeful," "intelligent," or "adaptive." Ashby's answer was his notion of "ultrastability." It represents a property of a certain class of "absolute" systems, characterized by the presence of endogenous "step functions." All of this Ashby had argued before in separate articles. In

Design for a Brain he linked his ideas to the feedback concept and shaped the way he would later make use of feedback in his *Introduction to Cybernetics* (1956).

Ashby's path to the concept of ultrastability begins with his assertion that

> The free-living organism and its environment, taken together, form an absolute system (1952, p. 35).

He defined an "absolute" system as one whose behavior over time can be entirely captured, conceptually at least, in a set of equations of the form

$$\frac{dx_1}{dt} = f_1(x_1, \ldots, x_n),$$

$$\frac{dx_2}{dt} = f_2(x_1, \ldots, x_n),$$

$$\ldots\ldots$$

$$\frac{dx_n}{dt} = f_n(x_1, \ldots, x_n).$$

We would recognize such a system of equations as a state-determined feedback system, and in *Design for a Brain* Ashby recognized it that way also:

> Given an organism, its environment is defined as those variables whose changes affect the organism, and those variables which are changed by the organism's behavior. . . . The organism affects the environment, and the environment affects the organism: such a system is said to have "feedback" (pp. 35–36).

An organism is adaptive, Ashby then argued, if and only if the absolute system it forms with its environment is stable, by which he meant that the variables in the system stay within limited ranges (pp. 57–71). To adapt and survive, an organism must respond to the environment (and vice versa) in such a way that "essential variables" in the organism are kept within relatively narrow physiological limits. From his knowledge of the cybernetics literature, Ashby linked the stability of the organism and its environment to the existence of goal-seeking feedback (pp. 49–56).

Ashby's statements about feedback in *Design for a Brain* are particularly significant for the evolution of the concept in the social sciences. He drew what are probably the first published "causal loop" diagrams and gave a very careful description of what such loops mean. His first causal-loop diagram, reproduced as Figure 3.3, captures the information feedback loop implicit in Watt's centrifugal governor. Ashby defined an arrow from A to B in the diagram to represent the idea that "A has a direct effect on B" or "A directly disturbs B."

Ashby was clearly worried that the causal connections in a diagram like Figure 3.3 could seem to be based on "some metaphysical knowledge of causes and effects," so he described very carefully how its causal links could be empirically determined.

The experimenter would fix the variable "velocity of flow of steam." Then he would try various speeds of the engine, and would observe how these changes affected the behaviour of the "distance between the weights." He would find that changes in the speed of the engine were regularly followed by changes in the distance between the weights. He need know nothing of the nature of the ultimate physical linkages, but he would observe the fact. Then, still keeping "velocity of flow of steam" constant, he would try various distances between the weights, and would observe the effect of such changes on the speed of the engine; he would find them to be without effect (p. 49).

The experimenter would have thus established that there is an arrow from "speed of the engine" to "distance between weights" in the causal diagram, but not an arrow in the opposite direction.

Ashby devoted an entire chapter to the determination of causal dependence and independence in this fashion. He noted that, in general, applying the method he outlined for Watt's governor one would have to watch out for "ultimate effects" of a variable change, not just its immediate effects. That is to say, variable A might influence B directly,

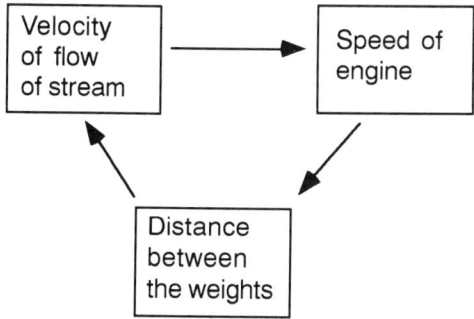

FIGURE 3.3: Ashby's causal-loop diagram for Watt's steam engine and governor. Source: Ashby (1952, p. 49).

and B might have an immediate influence on C, so a change in A has an "ultimate effect" on C, although they would not be directly connected with an arrow in a causal-loop diagram of immediate effects. Ultimate effects could add considerable complexity to the determination of immediate effects using Ashby's empirical *ceteris paribus* scheme. However, the alternative—using a priori reasoning or "thought experiments"—was too close to metaphysical handwaving to meet Ashby's standards for scientific rigor. Any determination of causal structure had to be empirical to satisfy the strong behavioral emphasis of the times. We have already seen a product of that emphasis in Rosenblueth, Wiener, and Bigelow (1943).

In the simple description of determining the causal influences in Watt's governor, we have the key to one of the major differences between the servomechanisms thread and the cybernetics thread in the evolution of the feedback concept. Just four years after Brain, in his *Introduction to Cybernetics*, Ashby was forced to conclude that the feedback concept is "artificial and of little use" in complex systems (1956, p. 54). The statement was one of the puzzles that started me on this investigation of the feedback concept. But viewed in light of Ashby's behavioral prescription for determining causal loop structure, it is easily understandable. It would be hopelessly impossible to hold things constant and systematically vary pairs of elements to determine the causal loop structure of immediate effects in a brain, or an economy, or even a moderately large firm. The feedback structure of a complex system cannot be determined with Ashby's empirical technique. As we shall see in section 3.3, however, some in the servomechanisms thread did not limit themselves to a behavioral approach. They were not so repelled by a priori reasoning and thought experiments. Thus they were not forced to Ashby's dismal conclusion about the value of the feedback concept in understanding complex systems.

A note on Ashby's mathematics

Ashby's *Design for a Brain* shares some of the characteristics that we shall come to associate more with the servomechanisms thread, rather than the cybernetics thread. First, he makes explicit use of differential equations to describe dynamic feedback systems. As we have observed, most of the authors in the cybernetics thread thus far have carried out their analyses verbally, not mathematically, and even the quintessential mathematician Norbert Wiener chose not to use differential equations to model any feedback system in his *Cybernetics*. Second, in addition to his references to Wiener and the Macy conference cyberneticians,

Ashby refers directly to some of the literature of engineering servomechanisms, notably the work of Hurwitz (1895) and Nyquist (1932) on stability. As we shall see, the early authors in the servomechanisms thread all link directly to the servomechanisms literature, while those in the cybernetics thread tend to trace their feedback understandings back to Wiener. In what sense, then, should Ashby be placed in what I am calling the cybernetics thread in the evolution of the feedback concept?

One of the principal differences in the way people make use of the feedback concept in the social sciences involves the notions of stability and dynamics. In the servomechanisms thread, as we shall see, one of the primary concerns is the pattern of dynamic behavior exhibited by a system. In the cybernetics thread, patterns of behavior are almost never discussed. Instead, the question is, "Is the system stable?," and the answer expected is a simple yes or no (perhaps not so simply determined). Ashby shares the yes-or-no focus on stability. Patterns of dynamic behavior are not his concern. In his treatment of dynamics and stability, then, Ashby clearly places himself in what I am calling the cybernetics thread. More than that—he helps to define the focus on stability that characterizes the cybernetics thread. His second book, *Introduction to Cybernetics*, established him as one of the early fathers of the cybernetics movement. Not only was it influential on the field of cybernetics itself; it also helped to spread a particular view of the use of the feedback concept in the social sciences. And the differential equations used in *Design for a Brain* were no longer anywhere in evidence.

Ashby's Introduction to Cybernetics

The subject of Ashby's second book is regulation and control. The feedback concept of the control engineer plays a relatively minor role, however. The book's major concepts are the "state" of a system, "transformations" of the system state, the "black box," and Ashby's notion of "variety." The control problem is how to keep the state of a system at or near some desired state. Ashby's answer is the design of a regulator that counters the possible variety in the inputs to the system. The main requirement for such a regulator is expressed in Ashby's rather famous "Law of Requisite Variety," which holds that "control can be obtained only if the variety of the controller . . . is at least as great as the variety of the situation to be controlled" (Beer 1972, p. 41). Or in Ashby's own more picturesque and cryptic phrasing, "Only variety can destroy variety" (Ashby 1956, p. 207).

Feedback does, in fact, take something of a drubbing in the book. Ashby defined the concept in terms of one of his diagrams of "immediate effects," showing a closed loop of influences:

> When this circularity of action exists between the parts of a dynamic system, feedback may be said to be present (p. 53).

He noted that there is some controversy about the definition of the word. Some want to use it in the general sense of circular causality (this is what Ashby prefers), while others would rather see the term limited to the deliberate "feeding back" of some effect from B to A. He solved the problem, as we noted above, by declaring it irrelevant:

> In fact, there need be no dispute, for the exact definition of 'feedback' is nowhere important. The fact is that the concept of 'feedback,' so simple and natural in certain elementary cases, becomes artificial and of little use when the interconnexions between the parts become more complex. When there are only two parts joined so that each affects the other, the properties of the feedback give important and useful information about the properties of the whole. But when the parts rise to even as few as four, if every one affects the other three, then twenty circuits can be traced through them; and knowing the properties of all the twenty circuits does not give complete information about the system. Such complex systems cannot be treated as an interlaced set of more or less independent feedback circuits, but only as a whole. For understanding the general principles of dynamic systems, therefore, the concept of feedback is inadequate in itself (1956, p. 54).

Even in very simple systems the concept of feedback bothers him. At one point he gave an exercise to determine whether the system given by the transformation

$$x' = \frac{y}{2},$$
$$y' = \frac{x}{2}$$

is stable around (0,0). (x' and y' in Ashby's notation denote the new values of x and y obtained from a transformation). He concluded that the system is stable around the origin, since a point like (10,12) becomes (5,6), which in turn becomes (2.5,3), and so on, heading inexorably to the goal of (0,0) and striving to remain there. Yet, he observed, the system is a positive feedback loop, because an increase in y leads to an increase in x and an increase in x leads to an increase in y. He concluded that trying to deduce stability from simply knowing the polarity of the feedback is not possible, and furthermore,

Feedback can be positive and yet leave the system stable; yet another example of how unsuitable is the concept of feedback outside its particular range of applicability (p. 81).

The conclusion fits Ashby's point of view about feedback, but it is not really justified. The usual way out of the puzzle of goal-seeking positive loops is to introduce the concept of the "gain" of the loop. In Ashby's example, in one trip around the loop a point (x_0,y_0) becomes $(x_0/4, y_0/4)$, or $(1/4)(x_0,y_0)$, for a "loop gain" of $1/4$. One then notes that positive loops with gain less than one are naturally goal-seeking rather than goal-divergent (see, e.g., Milsum 1968, pp. 25–26).

A less usual though more instructive analysis introduces explicit variables Δx and Δy for the changes in x and y, defining them to mean $x' = x + \Delta x$ and $y' = y + \Delta y$. For Ashby's example, $\Delta x = y/2 - x$ and $\Delta y = x/2 - y$. If we sketch a loop diagram of the system with these change variables explicit, as in Figure 3.4, we discover that the feedback structure of the system is not the simple single positive loop that Ashby's transformation notation implies. There are two hidden negative loops. We do not have a "goal-seeking positive loop," but rather a multiple-loop structure containing one positive loop and two negative loops. The dominant polarity happens to be negative because of the chosen parameter values. The result is a goal-seeking feedback *structure*. Incidentally, although this analysis involving Δs may look cumber-

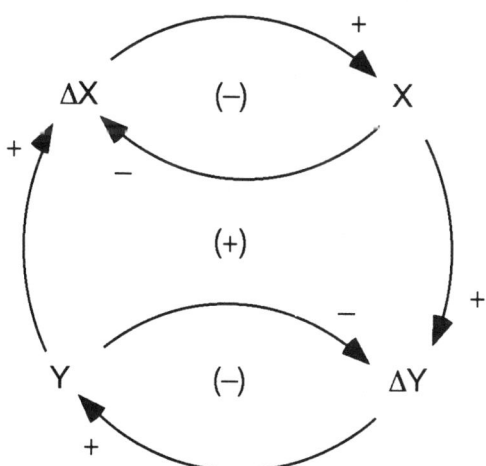

FIGURE 3.4: Implicit multi-loop feedback structure of the system $x' = y/2$, $y' = x/2$.

some, it has the advantage of transferring directly from the discrete to the continuous case by merely replacing Δx and Δy with dx/dt and dy/dt. The conclusion holds in continuous systems as well: there are apparent "goal-seeking positive loops," but in every case there is a multi-loop structure containing at least one negative loop, and the system is goal-seeking because the negative loop dominates.

Ashby took the point of view that complex systems cannot be understood. They must be treated as "black boxes" of unknowable internal structure (1956, pp. 86–117). We can observe the behavior of a black box in response to changes in the conditions of its environment, but we can never observe its internal make-up. Therefore, we cannot determine internal causes of the behavior of a complex system—we cannot determine how, or even whether, a system's internal feedback structure contributes to its patterns of behavior. We are left with observing inputs to the system (its environment), observing its resulting outputs (its behavior), and trying to control the latter through the use of cybernetic principles. The main tool of cybernetic control was, for Ashby, his Law of Requisite Variety. In complex systems, it is not merely largeness that is the problem. "What is usually the main cause of difficulty is the variety in the disturbances that must be regulated against" (1956, p. 244).

Ashby's scheme shows a set of disturbances D impinging on a system T producing a set of outcomes observed in the "essential variables" E. The regulatory problem is twofold: where in the sequence $D \to T \to E$ should the regulator R be placed, and what should its characteristics be, to hold E within some range of desired states? Figure 3.5 shows two possibilities. In Figure 3.5a the regulator gets wind of the disturbances before (or at the same time as) they hit the system and can try to respond in advance to reduce their dysfunctional impacts. Figure 3.5b shows the classic "error-controlled regulator" that has access only after the fact to information about essential output variables.

Ashby argued that case B, the error-controlled regulator, "cannot be perfect" and that regulating disturbances before they hit the system (case A) is preferable. In the error-controlled case,

R gets its information through T and E. Suppose R is somehow regulating successfully; then this would imply that the variety at E is reduced below that of D—perhaps even reduced to zero. This very reduction makes the channel

$$D \to T \to E \to$$

to have a lessened capacity; if E should be held quite constant then the channel is quite blocked. So the more successful R is in keeping E constant, the more does R block the channel by which it is receiving its necessary information. Clearly, any success by R can at best be partial (p. 224).

The implication here is that anticipating disturbances is definitely preferable to responding to difficulties they create in a system. In Ashby's view, ecologists, economists, or psychiatrists who want to regulate the exceedingly complex systems they deal with should work toward anticipatory regulators rather than error-controlled mechanisms.

The significance of Ashby's use of the feedback concept

Readers of Ashby's *Introduction to Cybernetics* would come away with the feeling that the concept of feedback has limited potential for social systems. Societies and corporations and brains are too complex to be understood; it is hopeless to hunt for their internal feedback structures. But they may be able to be regulated or controlled. Feedback

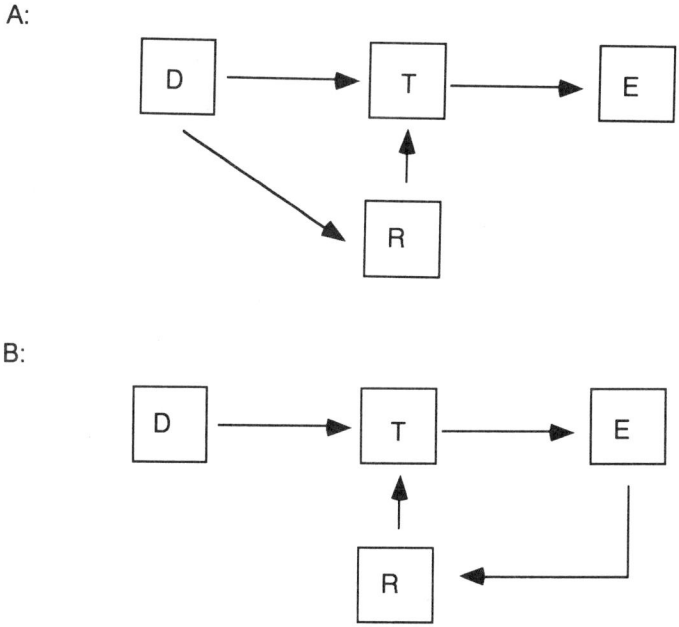

FIGURE 3.5: Different placements of a regulator R for a system T subject to disturbances D and producing output information about essential variables E. Source: Ashby (1956, pp. 222–223).

may be somewhat helpful, but the real key to regulation is to be able to match the variety of exogenous disturbances to the system. To keep a system stable, design a regulator that can insulate the system from the harmful effects of disturbances. The regulator is not the system itself, nor a part of it, but rather an external structure or device added on to the system to suit the regulator's purposes.

These themes—stability, variety and Ashby's law, the black box, the system regulator—become central to the cybernetics movement. Although they are somewhat tangential to our story of the evolution of the feedback concept in the social sciences, we shall hear them echoed and amplified repeatedly, particularly in the work of Stafford Beer on management (section 4.2). The messages they communicate explicitly and implicitly about the feedback concept contribute greatly to the different uses of the idea in the cybernetics and servomechanisms threads. One message in particular stands out: in Ashby's work feedback loops are add-on regulatory devices, not an intrinsic and understandable aspect of the causal structure of social systems.

Feedback and the General Systems Movement

At the time of the Macy cybernetics meetings, another major movement in the social and natural sciences began. Its earliest proponent was the biologist Ludwig von Bertalanffy (1945, 1950a, 1950b, 1951), who gave the movement the name General System Theory. But there were others at the same time who also sought to advance science by proposing testable principles potentially applicable in general to the whole range of systems, from atoms to cells to ecosystems, societies, and solar systems. In 1949, for example, a group of scientists at the University of Chicago (later at the University of Michigan) began to meet regularly to try to assemble an empirically testable general theory of behavior (Miller 1955). The group included, among others, Ralph Gerard (biology), James G. Miller (psychology), John Platt (physics), Anatol Rapoport (mathematics and statistics), Robert Crane (history), David Easton (political science), and Kenneth Boulding (economics). One wonders to what extent this group was actually independent of von Bertalanffy's initial efforts: von Bertalanffy had first expressed his ideas for a General System Theory in a philosophy seminar at the University of Chicago in 1937. Although a letter from Boulding to von Bertalanffy (von Bertalanffy 1968, p. 14) implies the Chicago developments were independent, perhaps the idea just percolated for twelve years.

The feedback concept was implicitly present at the creation of the

General System Theory, and it quickly became recognized explicitly as one of its fundamental building blocks. We shall sketch here the appearance of the concept in early articles by von Bertalanffy and Boulding.

To illustrate a mathematical definition of a "system," von Bertalanffy (1950b) chose a set of simultaneous differential equations:

$$\frac{dQ_1}{dt} = f_1(Q_1, Q_2, \ldots, Q_n),$$

$$\frac{dQ_2}{dt} = f_2(Q_1, Q_2, \ldots, Q_n),$$

$$\cdots\cdots\cdots$$

$$\frac{dQ_n}{dt} = f_n(Q_1, Q_2, \ldots, Q_n).$$

This is Ashby's absolute system. We would now recognize such a thing as a dynamic feedback system, with information from the various states Q_i feeding back to influence their own rates of change, creating the closed loops characteristic of mutual causal systems. Von Bertalanffy did not observe the closed-loop property of such a system in general. However, he did make use of the feedback idea at one point in the article, in his discussion of "finality."

A system that approaches a steady state can be written in terms of its distance from that final state. Von Bertalanffy gave the example of an equation for the length of an animal over time,

$$x = x^* - (x^* - x_0)e^{-kt}$$

where x represents the length of the animal at time t, x^* is its final steady-state length, x_0 is its initial length, and k is a growth constant.[4] The problem is that the system looks teleological—the "purpose" of the growth of the animal is to attain x^*, although the animal cannot "know" the goal. Von Bertalanffy observed that many formulations in the sciences have an "apparently finalistic character." He cited in particular Le Châtelier's principle in physical chemistry and Lenz's rule of electricity. "All these principles," he said, "express that in case of disturbance, the system develops forces which counteract the disturbance and restore a state of equilibrium" (1950b, p. 75).

Sounding very much like Rosenblueth, Wiener, and Bigelow, von Bertalanffy argued that "true finality or purposiveness," in which behavior is determined by foresight of the goal, is only one of a number of kinds of finality. Another kind operates in the case of the growing animal. The differential equation that produces the equation for x above is

$$\frac{dx}{dt} = E - kx,$$

where E represents metabolic activity contributing to growth (anabolism) and kx represents metabolic activity opposing growth (catabolism). The steady-state final value, $x^* = E/k$, is not a goal perceived, but an end implied by the ongoing processes of anabolism and catabolism. Von Bertalanffy asserted that every system that reaches a time-independent condition behaves this way (1950b, p. 78).

Feedback was mentioned in the context of a second kind of finality, which von Bertalanffy called "directiveness based upon structure." Here he linked to feedback the concept of homeostasis and the regulation of energy and materials in an organism:

These regulations are governed, in a wide extent, by feedback mechanisms. Feedback means that from the output of a machine a certain amount is monitored back, as "information," to the input so as to regulate the latter and thus to stabilize or direct the action of the machine. . . . Feedback mechanisms appear to be responsible for a large part of organic regulations and phenomena of homeostasis, as recently emphasized by cybernetics (1950, p. 78).

At this point he referenced Wiener (1948) and the proceedings of the Conference on Teleological Mechanisms (Frank et al., 1948), so we can surmise that some of his thoughts on teleological mechanisms were indeed influenced by the line of thinking initiated by Rosenblueth, Wiener, and Bigelow (1943).

However, von Bertalanffy argued for yet a third basis for organic regulation and dynamic teleology—his concept of "equifinality":

. . . the fact that the same final state can be reached from different initial conditions and in different ways. This is found to be the case in open systems, insofar as they attain a steady state. It appears that equifinality is responsible for the primary regulability of organic systems—i.e., for all those regulations which cannot be based upon predetermined structures or mechanisms and were regarded therefore as arguments for vitalism (p. 79).

Equifinality and the theory of the "open system" were two of the founding concepts of von Bertalanffy's General System Theory (1950a). In his view they went beyond the concept of feedback. They promised to explain organic regulation in situations where feedback structures were apparently not yet in place, as for example, early embryonic development.

In a paper given at a symposium on General System Theory at a meeting of the American Philosophical Association in 1950, von Bertalanffy explored at length the distinctions between feedback and his notions of equifinality and the open system. Both cybernetics and the theory of open systems aim at the same goal, he said, which is "scientific explanation and laws of organic 'wholeness' and teleology." In both cases, a system in a steady state responds to disturbances with forces that try to reestablish the steady state. But, according to von Bertalanffy, the causal basis is different:

> According to the [open] system conception, "teleological" behavior results from dynamic interaction within a unitary system that attends certain conditions of equilibrium or steady state. . . . According to the feedback scheme, "teleological" behavior . . . is due to pre-established structural arrangements (von Bertalanffy 1951, p. 353).

The argument is subtle and tends to hide in von Bertalanffy's complex phrasing, but the idea can be illustrated using the example of animal growth cited above. There are two ways the differential equation could be written, and to von Bertalanffy they mean dramatically different things:

(a) $\quad \dfrac{dx}{dt} = E - kx,$

(b) $\quad \dfrac{dx}{dt} = k(x^* - x).$

As noted above, in (a) the equilibrium state is $x^* = E/k$, which one can find by setting dx/dt to zero and solving for x. Thus (a) and (b) represent exactly the same dynamic phenomenon: either could be rewritten to become the other. But (b) says the goal x^* is, in some sense, "known" to the system. In (a) the goal is only implicit, a final value that, in von Bertalanffy's words, merely "attends" the final equilibrium state. It is the result of "dynamically interacting elements," in this case the anabolic and catabolic processes represented by E and kx. In (b) those processes are not explicitly represented.

The issue is important to von Bertalanffy, but before stating why we should observe that he offered three more reasons for preferring to view biological phenomena not as feedback systems but as open systems of dynamically interacting elements. First, "the model scheme in feedback is essentially a machine theory" (p. 353); second, it is restricted to closed systems (p. 354); and third (a startling objection), "feedback mechanisms are essentially one-way causal chains" (p. 354).

In support of the first objection von Bertalanffy cited numerous references to machine analogies and automata in the cybernetics writings. His justification of the second objection is a cryptic reference to a statement of McCulloch's about "activity in closed circuits" (p. 354). The passage suggests that von Bertalanffy may have confused the concept of a closed loop of circular causality with his own notion of a "closed system." The latter is a system that exchanges no material or energy with its environment, an entirely distinct and independent idea from the notion of a closed sequence of causes and effects. Alternatively, von Bertalanffy may have equated information with "material" and "energy," and thus found information loops equivalent to materially closed systems. His third objection appears at first to be a claim that feedback systems do not involve two-way circuits of mutual or circular causality. But there were too many closed-loop diagrams of homeostatic mechanisms and feedback loops around at the time to believe that von Bertalanffy was unfamiliar with the loop concept. Instead, by the phrase "one-way causal chains" he apparently meant to argue that feedback models do not allow the interaction of many elements to move the system toward its final equilibrium state. He seems to be claiming that only one goal in a feedback system, perhaps comprising many states, determines the system's response to a disturbance. For biological systems the interactions of many elements compose the system's response (p. 354).

In sum, von Bertalanffy saw the cybernetic view to be based upon "structural arrangements" while his open system view was based upon "dynamic interactions" (p. 353). All of the distinctions he described had serious implications:

> Where we have homeostasis, we must unveil the mechanisms involved. But where we have purely dynamical regulations, the consideration of phenomena as 'homeostasis' leads us astray since we look for mechanisms when there are none. This is a positive danger which has muddied biological experimentation and theory in many cases (p. 358).

Where we do not have homeostasis, where instead we have "systems of dynamically interacting elements," similar confusions arise. Von Bertalanffy criticized in particular the use of the feedback concept that appeared in a paper on "circular causal systems in ecology" (Hutchinson 1948),

> . . . exposing the regulations in population systems on the basis of modern mathematical theory, as developed by Volterra and others. This has nothing to

do with feedback in the proper sense. For Volterra's theory is essentially kinetic. . . . Ecological equilibrium is not feedback, as this notion is used with respect to technical devices, such as Watt's governor, thermostats, servomechanisms, etc., or to proprioceptive control, homeostasis, and so forth (p. 358).

Von Bertalanffy recognized that his concerns were fundamentally semantic in nature, but he strongly believed the semantic distinctions should be preserved. In terms of the differential equations (a) and (b) above, a blending or confusion of the two results in a misunderstanding of the causal structure implied. In (b), the case of homeostasis, we can look for the goal state x^*, we can try to determine how it is known within the system, and we should try to find out how a discrepancy between x and x^* generates pressures for change. In (a), however, we should look for the dynamic processes, E and kx; we would look in vain for an explicit goal state x^*. Which representation to use, which causal interpretation to make, depends upon our view of reality. To von Bertalanffy's concern I would add that the representation may even feed back subtly to influence our view of reality.

We now have the likely source of Ashby's observation of a disagreement about the definition of the word "feedback." We also have a seriously argued point of view that would take offense at the broad use of the word "feedback" in this investigation. Our interest, as I have said, is the loop concept underlying feedback and mutual causality. I have chosen to use the words "feedback" and "feedback loop" to stand for this longer but undoubtedly more accurate phrase. In my use of the term "feedback," I am not signaling von Bertalanffy's distinction. For us, von Bertalanffy's "open systems of dynamically interacting elements" are, like Ashby's absolute systems, feedback systems, and they are so by definition. On the other hand, in applications of the loop idea I wholeheartedly subscribe to von Bertalanffy's drive for causal clarity. There is a real difference in meaning between equations (a) and (b) above, even if their dynamic behavior is identical and each can be readily transformed into the other. Such distinctions in meaning should be preserved, even as I continue to use the word "feedback" in the broad sense to stand for all instances of closed-loop, circular causal structure.

Boulding's "general empirical theory"[5]

The economist Kenneth Boulding was one of the early social scientists attracted to General Systems Theory and, along the way, to the feedback concept. His first discussions of feedback appear in a paper laying out a conceptual framework for social science (Boulding 1951). The framework he described was briefly stated and very general, focusing on two broad categories—"theories of the individual" and "theories of interaction." He identified the first building block of a theory of the individual in the social sciences with the concept of behavior itself. For the second most general notion, he picked homeostasis, viewing it as shown in Figure 3.6 as a closed-loop, feedback process. He repeated the figure and some of the discussions of homeostasis in Boulding (1953), where he explicitly connected the concept to feedback. The latter reference undoubtedly reached a wider audience.

Boulding listed six essential parts of any homeostatic mechanism. The "datum" in his diagram is the object of stabilization—he gave the usual example of the temperature of a house to be controlled by a thermostat. Something must receive the datum and transmit information about it to what he called an interpreter, which decides what action to take. The interpreter then directs some other part of the system to adjust, and the effect of that action is transmitted back to alter the datum. Boulding thus viewed the homeostatic process as a closed loop of information and action, but he did not use the word feedback.

He gave two examples of homeostatic mechanisms from the social sciences, which are remarkable not so much for their accuracy or insight as for their extreme aggregation.

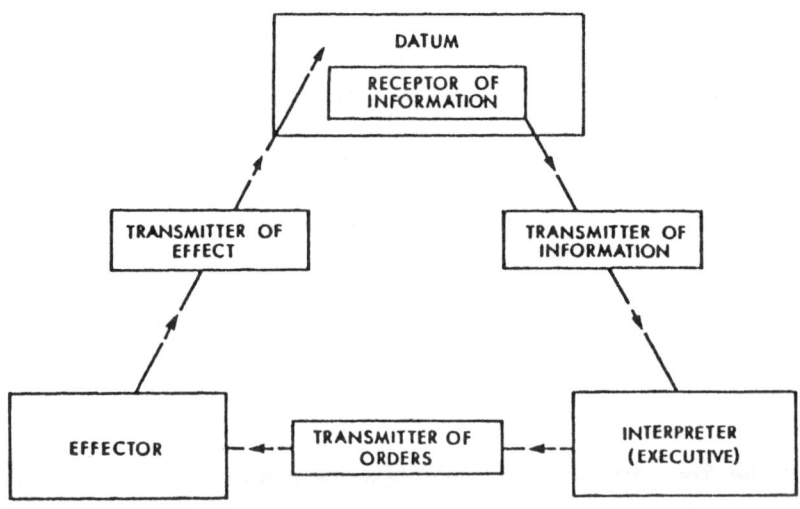

FIGURE 3.6: Boulding's conception of the general homeostatic mechanism. Source: Boulding (1951, p. 60; 1953, p. xxviii).

The firm is an organism consisting of an interpreter (the executive), who receives information through the accounting system, the budget system, the market-research system, Kiplinger letters, gossip on the golf course, and so on. As a result of his interpretations of this information certain decisions are made which are transmitted to the production and sales organizations and result in certain actions. The actions produce effects on the datum (sales, prices, and profits), which in turn are picked up by the receptors and transmitted to the executive as information, which again produce decisions. The simplest possible theory of the firm is the theory of "homeostasis of the balance sheet," in which we suppose that there is some composition of the balance sheet which the firm wishes to maintain. . . .

At the political level we can suppose also that there is some 'state' of an organism, such as a nation, at which discontent is a minimum. A divergence from this happy condition results in discontent, which is eventually transmitted through the political and constitutional machinery into a need for action on the part of government (the effector)—the action presumably being directed toward reducing the discontent and restoring the body politic to contentment (1951, p. 61).

He claimed that any homeostatic mechanism produces cycles, and suggested the reasons could be the lag between the receptor and the effector, too little sensitivity to changes in information, or even too much sensitivity. As examples he cited the business cycle, which oscillates, he said, because of "inadequate homeostatic mechanism in the monetary and fiscal system"; the "war cycle," which results from "the absence of homeostatic mechanisms in international relations"; and political swings from liberalism to conservatism (p. 62). He gave no justification for these claims and, in fact, considered the concept of homeostasis applied this broadly to be only the very first approximation to a theory of the individual.

Boulding felt that considerable progress could be made in many of the social sciences, particularly his own field of economics, by investigating what states are being maintained by individual homeostatic mechanisms. The concept would become linked to dynamics: he cited in particular growth and decay, learning and accretion, and survival and evolutionary development.

The social sciences have given far too little attention to [questions about homeostatic mechanisms]; especially this is true of economics, where the theory of the firm remains at a level of static equilibrium (maximization), is only just beginning to advance toward a theory of a homeostatic mechanism, and has scarcely begun to include theories of growth and decay, life cycles, and learning. It is hardly an exaggeration to say that in the world of the economists man never learns anything; he just knows it already. Thus there is practically no discussion in economic literature with which I am familiar of how business men become aware of the nature of the demand which faces them (p. 62).

Boulding thus advocated a dynamic view of social systems based upon the homeostatic mechanism as a closed-loop, feedback process. In his concern with dynamics he resembles authors we shall encounter in the servomechanisms thread. But his emphasis on homeostasis (and implicitly, negative feedback), his biological/information theoretic terminology (effectors and receptors), the lack of mathematical representations of dynamic structures, and the general philosophical, theory-building tone of his essay place him squarely in the cybernetics thread in the evolution of the feedback concept. His brief linking of circular homeostatic mechanisms to learning, survival, and evolution echo the high-level theorizing of Deutsch and Wiener. Moreover, there is no evidence that he draws his understandings of closed-loop mechanisms from any engineering literature. To the contrary, his most direct intellectual forebears in this essay appear to be the physiologist Walter Cannon and those social scientists who drew inspiration from him.

The drive to identify generally applicable theoretical constructs that is evident in Boulding's essay integrated him naturally into the General Systems movement. He spent a year with von Bertalanffy and Anatol Rapoport at the Center for Advanced Study for the Behavioral Sciences (von Bertalanffy 1968, p. 14) and shortly thereafter wrote an article in *Management Science* on General Systems Theory as the "skeleton of science" (Boulding 1956). Here his connection to the cybernetics thread became explicit. He sketched a "system of systems" identifying nine levels of theoretical discourse, outlined in Table 3.1. Feedback appears in the third level in his hierarchy and is presumably present to some degree in all the levels above the third.

TABLE 3.1 Boulding's hierarchy of systems (abstracted from Boulding 1956, pp. 89–94)

Level	Characteristic unit	Summary description
(1)	framework	static systems
(2)	clockwork	simple dynamic systems
(3)	thermostat	control mechanisms and cybernetic systems
(4)	cell	open systems, or self-maintaining structures
(5)	plant	genetic/societal systems
(6)	animal	mobile, teleological systems with self-awareness
(7)	human individual	animal systems with self-consciousness
(8)	human society	social systems with self-consciousness
(9)	transcendental idea	ultimates, absolutes, and inescapable knowables

Without delving any more deeply at this point into the history of General Systems Theory, we should observe that Boulding interpreted the concept of feedback in essentially the same circumscribed sense as von Bertalanffy.

> [The level of the thermostat] differs from the simple stable equilibrium system mainly in the fact that the transmission and interpretation is an essential part of the system. As a result of this the equilibrium position is not merely determined by the equations of the system, but the system will move to the maintenance of any given equilibrium, within limits (1956, pp. 90–91).

Thus if $dx/dt = k(x^* - x)$ and $k > 0$, the equilibrium position is $x = x^*$, whatever value of x^* is picked. The implication is that the system goal, x^*, is set outside the system of equations describing a feedback system. Like von Bertalanffy, Boulding prefers to limit the word "feedback" to homeostatic mechanisms with explicit goals. It is not clear at what level in his hierarchy Boulding would place a system in which x^* is set endogenously by other equations in the system. The emphasis in systems above the third level is on their characteristics as living systems. Closed loops of mutual or circular causality are never mentioned.

Miller's Living Systems

Undoubtedly the most complete development of general systems theory is James Grier Miller's *Living Systems*. Published in 1978, this encyclopedic 1100-page work is ahead of our story at this point, but it contains valuable references to the significance of feedback in the behavioral sciences and is built upon work first appearing in the 1950s (e.g., Miller 1953, 1955).

Bringing together the work of a generation of general system theorists, Miller (1978) placed feedback prominently among his list of basic concepts for the study of living systems. It is defined in a section titled "Steady state," and is presented, along with concepts such as power, conflict, and purpose, as an essential component of adjustment processes. Although the 1978 work contains coverage of feedback thinkers who define the idea somewhat differently, Miller defined feedback as the Macy meetings had, in terms of input and output and channels of information:

> The term *feedback* means that two channels exist, carrying information, such that channel B loops back from the output to the input of channel A and transmits some portion of the signals emitted by channel A.
> . . . When the signals are fed back over the feedback channel in such a manner that they increase the deviation of the output from a steady state,

positive feedback exists. When the signals are reversed, so that they decrease the deviation of the output from a steady state, it is *negative feedback* (1978, 36).

His citations here are to Rosenblueth, Wiener, and Bigelow (1943), Ashby (1952), and Vickers (1959a).

Summary of the Beginnings of the Cybernetics Thread

The publications discussed in this section served to introduce the engineer's concept of feedback to the social sciences. How the concept was handled in them, the problems addressed, the associations, even the language used, all added to the communication of the nuances of feedback to its new audience. Some of the tendencies initiated here persist for a considerable time, as we shall see. Those tendencies can be summarized as follows:

- Feedback is defined in terms of input and output. It is seen as the influence of output back on input.
- The use of the feedback concept is limited to loops of negative polarity.
- Feedback is viewed as the mechanism of homeostasis and control.
- The negative feedback concept is not associated with the general concept of mutual or circular causal processes.
- The concept of feedback is used to address philosophical and theoretical questions relating to control.
- The stability of a feedback system, and the conditions producing instability, are a central concern.
- Feedback analyses are usually verbal rather than mathematical or pictorial.
- Feedback is viewed as information transmitted in messages. The concept is associated with communication networks and information theory.
- Feedback is associated with the creation of intelligent machines and the automation of human functions.

In the preceding discussions, one can find exceptions to some of these tendencies within the cybernetics thread. They are tendencies, or perhaps ideal types, not absolutes. Others pick up these tendencies, however, and a discernible thread of thinking in the social sciences emerges as the feedback concept evolves over time. But not everyone who makes use of the feedback concept in the social sciences shares or emphasizes these same tendencies. In section 3.3 we shall investigate the beginnings of a more or less independent line of thinking in the evolution of the feedback idea.

3.3 The Servomechanisms Thread

Engineering Ideas in Economic Dynamics

Around 1950 a different feedback thread began appearing in the social sciences. The first people associated with this thread were economists who were personally familiar with the engineering literature on servomechanisms. They were concerned with understanding economic fluctuations. They focused on dynamics. They drew upon the engineering servomechanisms literature to aid their analyses of model behavior and to formulate and test theories for the control of economic cycles.

Physical models

This line of feedback thinking in economics had its beginnings in the use of electrical, mechanical, and fluid devices to simulate economic dynamics. In the 1930s, Frisch (1933), Kalecki (1935), Kaldor (1940), and others had formulated dynamic models to explain business cycles. The difficulty of the mathematics required to analyze the behavior of such models put severe constraints on the theories that could be considered. The system of equations had to be low order, nonlinearities usually had to be eliminated, computations were laborious, and analyses had to be redone from scratch to test a parameter change.

Phillips (1950) reported on the construction and use of a fluid model as a simulator of economic cycles. Levels of water in containers in the model represented economic quantities. The fluid "economy" oscillated as the water flowed in and out of the containers in the simulator, conveyed by tubing containing valves controlled partly by the system's own water levels. While Phillips made no mention of feedback in this paper, the fluid device he discussed was clearly a closed-loop model of economic dynamics. The fluid model could be adjusted somewhat to represent and test parameter changes in the analogous economic system, but structural changes were obviously much more difficult. A significantly different theory would require the construction of a wholly new fluid device.

A natural alternative was the analog computer, capable of showing the dynamics of electrical signals on an oscilloscope. A series of economic simulation studies along this line were reported (see Morehouse, Strotz, and Horwitz 1950; Strotz, Calvert, and Morehouse 1951; and Enke 1951). Perhaps the most interesting use of the technique was by Strotz, McAnulty, and Naines (1953), who investigated the dynamics of the nonlinear model of economic cycles proposed by Goodwin (1951b).

Goodwin's model was a version of the multiplier/accelerator. The usual accelerator assumed for induced investment, $K'(t) = k\,Y'(t - \theta)$, was passed through a function ϕ which saturated for high and low values of $k\,Y'(t - \theta)$. Through some approximations and some rather laborious mathematics Goodwin had found that the system produced a limit cycle. Using analog simulation, Strotz, McAnulty, and Naines found many different limit cycles from the structure. They naturally concluded that use of such simulation techniques would contribute considerably to economic analyses. It is interesting to note that Allen (1956, p. 302) later points out that additional high frequency cycles are characteristic of the solution of a model with one or more discrete lags such as the $t - \theta$ in Goodwin's model. Allen concludes that the high frequency solutions are an essentially spurious result of the "unrealistic assumption of a fixed-time delay."

The feedback thinking of R.M. Goodwin

The first serious in-depth attempt to make explicit use of the feedback concept in economics is in an article by Richard Goodwin (1951a) that appeared in Hansen's *Business Cycles and National Income*. As noted above, Goodwin had formulated a model of the business cycle, also published in 1951. But it was in the Hansen book that Goodwin exposed the role of the feedback concept in his thinking.

Goodwin began by drawing a clear distinction between exogenous and endogenous theories of economic dynamics:

At once the oldest and the simplest are the exogenous theories in which the cycle is maintained by perpetually alternating "outside" disturbances. . . . Whether or not the economy has dynamic elements, it will oscillate with the same period (duration of cycle) as the exogenous disturbance. . . . Schumpeter's [innovation] theory of the cycle partly fits into this category. . . . By contrast there are the endogenous or self-generating theories that create, by virtue of their own structure, the alternations of expansion and contraction (1951a, pp. 419–420).

He believed the endogenous point of view was the most fruitful approach. And at the core of endogenous theories of economic oscillations he identified the concept of the feedback loop.

He took his definition of feedback from the servomechanisms text by James, Nichols, and Phillips (1947):

A mechanical or electrical system with feedback is one in which the output of some part of the system is used as an input to the system at a point where this can affect its own value. A servo system is a feedback system in which the actual output is compared with the input, which is the desired output, and the driving element is activated by the difference of these quantities (James, Nichols, and Phillips 1947, cited in Goodwin 1951a, p. 437).

He noted that human beings often served as feedback devices, as when they steered a vehicle or adjusted a furnace to maintain the temperature of a house. Feedback or servo systems are important for understanding business cycles, Goodwin argued, because of their potential to overshoot and oscillate.

If the person is unskilled or the feedback mechanism badly designed, the system may "hunt" its desired output or target but never find it (p. 438).

Goodwin endeavored to describe the economy as a feedback or servo system that has this hunting property. Conceptually his approach contrasts significantly with the view expressed by Frisch (1933) and others that the economy would settle down and stop oscillating if all exogenous disturbances were somehow eliminated.

In this article Goodwin described his feedback theory of the business cycle in words rather than equations. He identified two principal mechanisms responsible for the system's tendency to overshoot its equilibrium condition. First was the common notion of the lag in the construction of capital:

Ships, for example, continue to be started so long as freight rates are such as to give a present value of ships greater than their cost of construction (p. 440).

He noted that this mechanism had been central to all business cycle theories, tracing back to Aftalion (1927). However,

both Aftalion, excusably, and Keynes, inexcusably, left out of account a much more violent, positive, and hence, unstable, feedback through the multiplier (p. 445).

Net investment, he observed, contributes to national income. To reach a new level of desired capital, net investment must increase, but that very increase pushes up national income even further by the investment multiplier. That leads to still more desired capital.

Intuitively, it is fairly clear that this is a distinctly unstable factor. It is inherently unstable because the feedback is positive: a deficiency of capital leads to investment, but this leads to a greater deficiency instead of to its elimination. (Goodwin 1951a, p. 441).

Goodwin observed that his reasoning had an apparent circularity: income controls investment; investment controls income. But for those unfamiliar with feedback systems, he pointed out that that was hardly a logical defect, but was basic in the nature of the loop idea:

It is, in fact, a central feature of feedback systems that the system controls itself, and thus its circularity lies in its automatic quality, which constitutes its peculiar character (p. 442).

His quantitative version of this model (Goodwin 1951b) would have led to exploding oscillations without the intervention of a nonlinearity that placed a "floor" and "ceiling" on investment decisions. He interpreted the nonlinearity as a change in the structure of those decisions:

Actually, when the oscillations get too large . . . the previous laws of behavior cease to hold (i.e., we have a non-linearity) and something ceases to function as before (p. 439).

The thinking here is remarkable for a number of reasons. First, unlike the previous uses of the engineer's concept of feedback in the social sciences, Goodwin's use is not limited to negative loops. He sees a naturally occurring positive loop (more than one, in fact) and makes it the central feature of his theory. Second, unlike Tustin and Phillips and some others that follow him in the servomechanisms thread, he makes use of the feedback concept to *conceptualize,* that is, to develop economic theory. He does not limit himself to using the engineer's concepts and techniques for the analysis of model behavior. Third, he has completely translated his feedback thinking from model terms into real terms. For example, when net investment falls to zero and national income drops, forcing down desired investment, Goodwin suggests "Our error-sensitive devices, the entrepreneurs, report an error, and command a decrease of capital" (p. 441). And finally, he sees feedback structure as the endogenous determinant of the observed behavior of the economy. Such an explicit statement of the role of feedback in an endogenous point of view about dynamic behavior does not appear again until the work of Forrester (1958, 1961).

In spite of the richness of the feedback view expressed by Goodwin, credit for the introduction of the feedback concept and servomechanism theory into economics is usually given to Tustin (1953) and Phillips (1954, 1957), to which we now turn. (See Athans and Kendrick 1974 and Cochrane and Graham 1976.)

Feedback as the Mechanism of Economic Systems

Arnold Tustin shifted the focus of this engineering work in economic dynamics to the mathematics of engineering control theory. In *The*

Mechanism of Economic Systems (1953) he exposed the implicit feedback nature of some of the thinking in Keynes's *General Theory* (1936) and the formal models of Hicks, Kalecki, Goodwin, Clark, Klein, and others.

To establish his point of view and purpose, Tustin summarized Keynes's basic equilibrium model of the economy in a loop diagram, shown in Figure 3.7. The advantage of such a diagram, Tustin claimed, was that it exposed "closed sequences of dependence"—chains of causal relationships that close on themselves, forming loops. He phrased his definition of feedback in terms of these closed sequences:

> The variation of any one quantity in a closed sequence causes variation of all the other quantities in the sequence, and because the sequence is closed the quantity that first varied suffers further variation, the repercussion round the closed sequence. This repercussion is referred to as "feed-back." (1953, p. 5).

Tustin held that the existence of such a closed loop was equivalent to interdependence between quantities. The resulting feedback loop idea was important, he said, because

> The behaviour of systems that have closed sequences is characteristically different on that account, because only such systems are capable of self-excitation (1953, p. 5).

Tustin's use of the feedback idea contrasts significantly with the writings of Wiener, Deutsch, Lewin, and the others described earlier. His purpose was not theory building but theory testing. In fact, he applied his techniques to existing economic theories generated by others. Tustin's purpose was to study the causes of dynamic behavior.

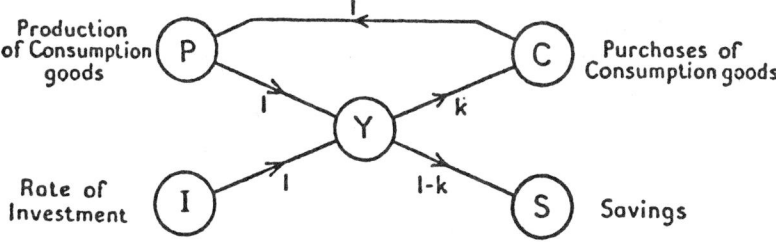

FIGURE 3.7: Loop diagram of Keynes's basic equilibrium model of the economy. Source: Tustin (1953, p. 6).

Economic systems are certainly systems with interdependence. The problem to be investigated is whether the closed sequences of dependence are such as to account for business cycles (1953, p. 5).

Could the feedback structure of an economic system generate the oscillatory behavior observed in business cycles? To answer such a question, Tustin applied the mathematics of servomechanism engineering to economic models that embodied closed causal sequences. His work thus differs from the writings in the early cybernetics thread in its use of explicit differential equations to represent dynamic social systems. To Tustin mathematics was indispensable in this sort of analysis:

The general notion that "interdependence" is a key to the understanding of the behaviour of economic systems is of course a very old one. It was, for example, precisely what Engels was attempting to express by his "dialectic relationship," but the consequences of interaction in a complex system escape purely verbal analysis. They call for the use of mathematical notation (1953, p. 6).

We shall see this theme repeatedly invoked in the servomechanisms thread.

Tustin recognized the difficulties and limitations of using the methods of servomechanism engineers to analyze nonlinear systems, and he believed strongly that economic systems are nonlinear. The "employable population" is limited by the ceiling of full employment, for example, and the rate of net investment "cannot become significantly negative" (p. 123). He saw the possibility that the turnarounds and amplitudes of business cycles could be related to nonlinearities, particularly lower limits on the net rate of investment.

For this reason he was sceptical of the linear econometric modeling approach being developed by Lawrence Klein (1950). Attempts to extrapolate predictions from such linear models had been disappointing. Tustin suggested that the failure traced in part to the way the linear model structure affected the statistical estimates of parameters. Parameters selected to best fit a linear model to data from a truly nonlinear system would be "fictitious" values. They would make the solution of the resulting system of linear equations a "just-selfmaintaining oscillation," capturing the "fundamental" of the oscillation but missing its nonlinear character:

Any linear system can have only exponentially damped or cumulative or constant oscillatory free motions. Systems of the kind considered, if "fitted" over one cycle, must have a principal mode that is approximately sinusoidal, of the appropriate period, higher modes being rather highly damped. Thus the extrapolated motion, in the assumed absence of further disturbance, would, after a fraction of a cycle, approximate to a pure sinusoid, and could not therefore represent the peculiarities of the actual wave-shape in so far as these were due to non-linear phenomena (pp. 124–125).

Prediction by extrapolation from a linear model could not therefore be expected to match reality very well. Nonlinear models were necessary to match nonlinear reality.

Besides simply not fitting reality well, linear models lacked a structural feature that Tustin implied was important:

> The basic assumption of a linear model may however be expressed by saying that the dependencies correspond with (hypothetical) differential equations that constantly apply at all times (p. 125).

The implication is that a given equation in a system of nonlinear differential equations does not constantly apply at all times. In fact, nonlinear models have the property that they change their active structure over time.[6] This was the property of nonlinearity that Goodwin had expressed.

Tustin was pointing out that a linear system does not have the property of being able to change its structure endogenously. The influence of any equation in a linear system or any term in a linear equation on the behavior of the system remains constant for all time t. To Tustin it was therefore not surprising that linear macroeconomic models such as Klein's (1950) had trouble when extrapolated beyond the interval of fit. They were not capable of mimicking the changes in the strengths of influences in the real system over the course of the business cycle.

Tustin went so far as to suggest that the amplitude of the business cycle might change not by random accident but by some such endogenous structural feedback effects:

> It might therefore seem, at least at first sight, that the particular amplitudes of business cycles that have been experienced were fixed by the "accident" that the rate of opportunities to invest has fallen short of the rate of investment required for full employment of the population at various times by just so much and no more. The interesting view has however also been expressed that a feed-back or self-regulating sequence controls the amplitude of the business cycle, because the acceptable rate of interest and the willingness to take risks both in the long run reflect the degree of instability actually experienced (p. 124).

Such endogenous structural variation surrounding the decision to invest would presumably have to be captured in nonlinearities.

Unfortunately, nonlinear systems are not only hard to estimate statistically, but are also usually impossible to solve in closed form. Not even the powerful mathematics of engineering servomechanisms that Tustin promulgated was much help. Tustin therefore supported the analysis of nonlinear systems using simulation methods. Physical simulation models and analog computers looked the most promising at the time (p. 127). What remained was for someone to build a good nonlinear model of the macroeconomy.

A.W. Phillips and Stabilization Policy

A.W. Phillips, the father of the curve bearing his name, was also an early advocate of viewing economic systems from the feedback point of view of engineering control theory. In two articles, Phillips (1954, 1957) used mathematical methods of control theory to analyze policies for stabilizing the business cycle. Because of this work, he is seen, with Tustin, as the founder of a control theory approach to macroeconomics (Cochrane and Graham 1976).

Phillips saw the need to address policy questions about the stabilization of production and employment from a dynamic perspective. Static equilibrium analyses could not expose possibly undesirable transient consequences of a proposed policy. His engineering background led him to study the use of proportional, integral, and derivative (PID) control mechanisms to stabilize models of economic cycles. As the basic oscillating mechanism in the economy, Phillips (1954) postulated the multiplier/accelerator theory, which he represented in an engineer's block diagram shown in Figure 3.8.

The multiplier appears as the positive feedback loop at the bottom of the figure. The smaller positive loop on top of the multiplier loop is the accelerator. The top portion of the figure represents the PID control structure. The symbols in the diagram are defined as follows:

> P aggregate production (units per year), measured from an initial equilibrium value
>
> E aggregate demand (units per year), also measured from an initial equilibrium value
>
> D differential operator
>
> k acceleration coefficient
>
> l "marginal leakage" ($1 - l$ = the marginal propensity to spend)
>
> L_a exponentially distributed lag in the effect of accelera-

tion (dP/dt) on aggregate demand. The coefficient of the lag is a; its time constant is 1/a.

L_p exponentially distributed lag representing the delay in adjusting aggregate production (P) to meet aggregate demand (E).

u an exogenous test input (a unit step down) disturbing the model from equilibrium and stimulating its potentially oscillatory behavior modes

ε error signal, the difference between actual and desired production

f_i, f_p, f_d adjustment factors for the PID control structure

The block diagram represents a system of differential equations. Samuelson's original combination of the multiplier and accelerator theories had been in terms of difference equations (Samuelson 1939). Phillips argued in favor of the more continuous representation:

The response of production to changes in demand is assumed to be gradual and continuous. For aggregative models this is more realistic than the usual

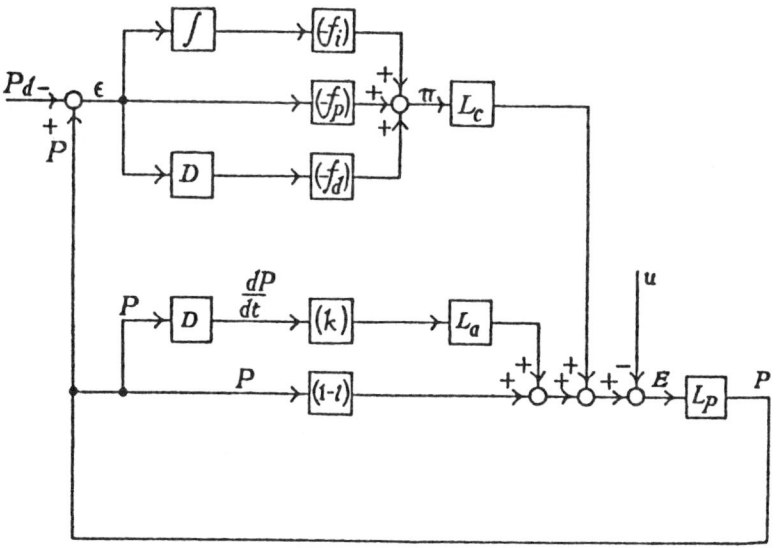

FIGURE 3.8: Block diagram of multiplier/accelerator model. See text for explanation of symbols. Source: Phillips (1954).

assumption that production changes in sudden jumps. Even if each producer were to have a rigid production plan which he altered only at intervals of several months, the planning periods of the thousands of individual producers would overlap, and the response of aggregative production to a sudden change in aggregate demand would consequently be more nearly approximated by a continuously changing variable than by one changing only at discrete intervals of time (Phillips 1954, p. 291).

To discuss stabilization policies, Phillips included a constant desired level of production, P_d (see Figure 3.8). The discrepancy or "error" $P - P_d$ was then used to activate the control mechanisms designed to bring production P in the model economy smoothly up to the desired level after a disturbance from equilibrium.[7]

Phillips interpreted the three classical mechanisms of control theory in this economic setting. Proportional control, adding $-f_p \cdot (P - P_d)$ into aggregate demand, represents the idea that the demand for corrective action is simply proportional to the error $P_d - P$. Integral control, in which the pressure for corrective action is proportional to the accumulated error, represents mounting pressure for corrective action as a given error persists over time. Phillips suggested the example of monetary policy:

> There may, however, be a tendency for monetary authorities, when attempting to correct an "error" in production, continuously to strengthen their correcting action the longer the error persists, in which case they would be applying an integral correction policy (pp. 297–298).

Derivative control adds corrective action that is opposite to the rate of change of production. The faster production is rising (or falling), the more a derivative control policy attempts to slow its rise (or fall).[8]

Phillips demonstrated in this economic context what the control engineers already knew about PID controllers. Proportional control by itself fails to bring production up to desired production. It leaves a steady state error, which gets smaller if the magnitude of the proportional control factor is increased but never goes away completely. Integral control solves the steady state error problem, but it adds dramatically to the instability of the system. Overshoot and oscillations can easily result from integral control action even if the original system was stable to begin with. (That is why Phillips expressed some concern about monetary policy that might have the character of integral control action.) Finally, by opposing the direction of change in the system, derivative control adds considerable stability, adding damping to an oscillatory system and usually slowing its response to changes. Phillips noted that the acceleration principle, acting proportionally to the rate of change of production, works just like derivative control but in the opposite direction. Hence, it ought to the significantly destabilizing (p. 302).

In his final and most complex model, Phillips added to the multiplier/accelerator a variable price level and a variable interest rate, both responding endogenously to changes in aggregate production. He discussed in detail the stabilizing and destabilizing effects of various quantities and loops in the model. The effects of the price structure on the dynamic behavior of the model under various assumptions suggested to him that monetary policy had a chance of contributing significantly to economic stabilization (p. 315).

Phillips (1957) repeated some of this analysis with a multiplier model including inventory adjustments to absorb differences between aggregate demand and aggregate production. His concern in this later article was with the time forms of the lags and delays in the system. Working with analog simulations of his models, he found that different lag distributions affected his conclusions about the proper mix of proportional, integral, and derivative control mechanisms to achieve stabilization.

A number of Phillips's conclusions in these two articles are important for our story of the evolution of the feedback concept in the social sciences.

It is quite possible that certain types of policy may give rise to undesired fluctuations, or even cause a previously stable system to become unstable (1954, p. 290).

If there is a long delay before corrective action is taken or before it begins to have an appreciable effect, it is better that the effect, when it does come, should be gradual rather than sudden (1957, p. 276).

It is unlikely that the period needed to restore any desired equilibrium conditions after an economy has experienced a severe disturbance could be much less than two years, even assuming that the regulating authorities use the policy which is most appropriate to the real system of relationships existing in the economy. As these relationships are not known quantitatively, it is unlikely that the policy applied will be the most appropriate one; it may well cause cyclical fluctuations rather than eliminate them. (1957, p. 276)

Intuitions about dynamic processes may be dangerously misleading and need to be carefully tested (1957, p. 276).

It is necessary to study the properties of more realistic models in which nonlinear relationships, growth trends, multiple objectives and multiple disturbances are incorporated (1957, p. 277).

These conclusions correspond closely to the points of view of Tustin and Goodwin, and differ considerably from the use of the feedback

concept apparent in the cybernetics thread, whose beginnings were sketched in section 3.2. They emphasize the following characteristics:
- The focus is on the dynamics of feedback systems.
- Formal dynamic models are employed.
- Feedback loops are seen as an intrinsic part of the real system, not merely as possible mechanisms of system control.
- Positive loops are present in the analyses, along with negative loops.
- The dynamic behavior of a feedback system is considered to be difficult to discern without the aid of formal mathematical models.
- Well-intentioned policies are seen to have the potential to create or exacerbate the problem behavior they were intended to cure.
- The patterns of behavior of a dynamic system are traced to its feedback stucture.
- The importance of nonlinearities is emphasized, but linear models are employed because of the lack of tools appropriate for analyzing nonlinear models.
- The work is directed toward policy analysis.

Early Developments of the Servomechanisms Thread in Economics

The work of Tustin and Phillips was picked up and developed by R.G.D. Allen (1955). Allen was the first to include the engineering approach to economic modeling in a textbook (Allen 1956).

For his economics audience Allen characterized feedback in terms of the familiar multiplier effect. He presented the structurally identical block diagrams reproduced in figure 3.9 and defined the meaning of the top diagram by the equations

$$Q_2 = k_{12}Q_1 + k_{32}Q_3,$$
$$Q_3 = k_{23}Q_2.$$

Substituting the second into the first, he derived

$$Q_2 = \frac{k_{12}}{1 - k_{23}} Q_1.$$

He noted that in the absence of the loop from Q_3 back to Q_2, Q_2 would merely be $k_{12}Q_1$. He then pointed out that the same arithmetic applied to the other looping diagram in Figure 3.9 yields

$$Y = \frac{1}{1 - c} A,$$

which an economist would recognize as the multiplier effect of consumption on autonomous expenditures A producing economic output Y. Allen noted that to an engineer the multiplier effect is thus a consequence of "a feed-back" (Allen 1956, p. 281).

Allen saw two major advantages to looking at economic models through the spectacles of a feedback engineer. The metaphor is quite apt, for the first advantage he saw was indeed visual. Loop diagrams figured prominently in Allen's text. He used them as shown in Figure 3.10 to show the structural similarities of four dynamic macroeconomic models. The block diagrams here are Allen's representations of then current economic models of business cycles. Allen observed that all but the Kalecki model obviously consist of essentially the same two loops. There are differences in the way lags are treated, and one (the Hicks model) uses difference operators rather than differential operators, but the loop structure of all the models is essentially the same. The two loops in common to all the models are positive feedback loops, although Allen did not point out any polarities. The Kalecki model differs in the structure of the investment decision B and the investment I that results after a lag θ. The additional loop in his model embodies the assumption that the investment decision B includes consideration of the existing capital stock K. In another departure from the other models, Kalecki assumed that desired investment depends upon the existing capital stock and past aggregate production Y, not on the rate of change DY (or ΔY) as in the acclerator principle. To Allen, a major contribution of these loop diagrams was the ease with which structural similarities and differences could be perceived.

A second advantage of the engineer's approach to economic models

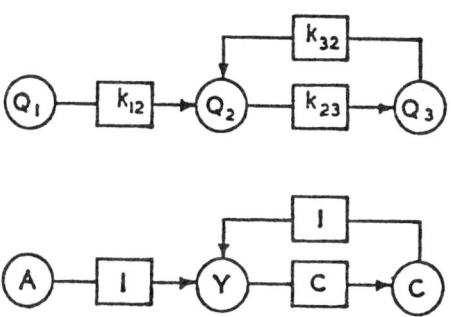

FIGURE 3.9: Loop diagram used to define feedback in terms of the multiplier effect. Source: Allen (1956, p. 281).

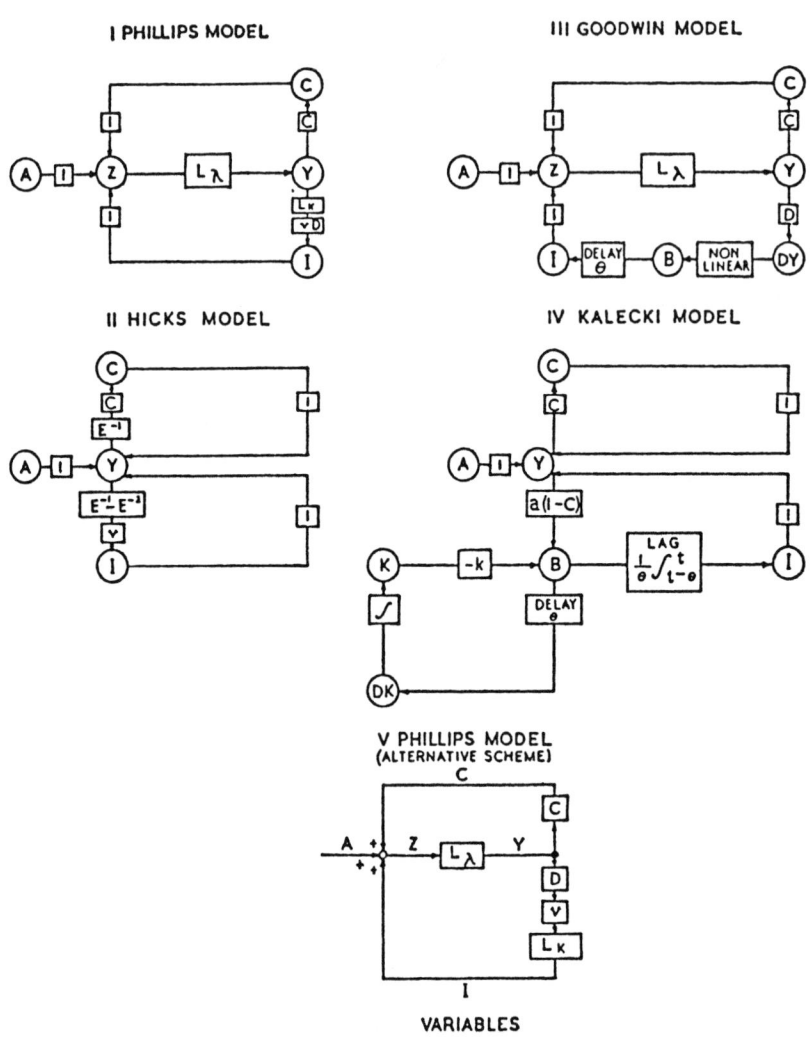

FIGURE 3.10: Loop structure of several models of business cycles. Source: Allen (1956, p. 285).

for Allen came from the mathematical techniques. The engineer's use of operator notation and Laplace transforms replaces all kinds of time variations (derivatives, integrals, lags) with algebraic expressions. Lags of many different types can be handled, from discrete delays to high-order exponentially distributed lags. The problem of finding the frequency and damping characteristics of an assumed macroeconomic structure reduces to the algebraic solution of a polynomial equation. One then sees clearly that the frequency and damping characteristics in a linear model are determined by the assumed structure of the system (p. 289). On the negative side, the polynomial equation yielding the inherent oscillatory characteristics of a system usually can be solved only by approximation. Furthermore, the exact time paths of the different variables in the system, including relative amplitudes and phasings, require additional calculations or simulation. And again, we see the criticism of the assumption of linearity. Allen believed that nonlinear elements were characteristic of real economic systems and had to be incorporated in economic models (p. 305). In sum, he concluded that economists interested in models having practical applications had much to learn from feedback engineers (p. 312).

It is significant that at this stage in the evolution of the servomechanisms thread there was little attempt to use the feedback concept as a tool for model conceptualization. Goodwin's work aims in this direction, but for Tustin, Phillips, and Allen economic models were, in a sense, taken as given. Engineering servomechanisms ideas were invoked to analyze behavior and suggest stabilization policies. These authors tended not use the concept of the feedback loop to alter the way they viewed the world. Significant progress in that direction would come later. A hinting of a beginning in that direction, however, can be found in the work of Herbert Simon.

Feedback in the Work of Herbert Simon

The mathematics of servomechanisms entered the social sciences also through some early work of Herbert Simon. In an article entitled "On the Application of Servomechanism Theory in the Study of Production Control" (1952), Simon explored the potential for the use of differential equation models, Laplace transforms, characteristic polynomials, and stability criteria in the study of a particular type of human system. Unlike Tustin's book, Simon's article made no attempt to teach the mathematics of servomechanisms, but rather used engineering tools to investigate the design of a human control system. Simon's emphasis was on formulating the problem in the language of servomechanism theory, determining the criteria for evaluating the merit of a control system, and surveying the general basis in servomechanism theory for approaching such problems (Simon 1952, p. 104).

The problem he chose for his investigation concerned the planning of production to try to meet incoming orders and to keep inventory near some desired or optimal value. He noted:

This system obviously possesses the characteristics of a servomechanism.... It has a feedback loop: error → planned production → actual production → inventory → error. The error initiates a change in planned production in such a direction as to reduce the error (p. 99).

Simon represented this closed-loop system in block diagram form and in systems of equations. Production minus shipments, for example, was the rate of change of inventory:

$$\frac{dI}{dt} = PR - OR.$$

The order rate OR was determined outside the system and given as a function of time. In the second and more complex of the two systems Simon analyzed, the production rate PR was taken to be a lagged version of desired production DPR:

$$PR = K_1 DPR.$$

The latter was then set to be some function of the order rate and the discrepancy between desired and actual inventories:

$$DPR = K_2(DI - I) + K_3 OR.$$

Here K_1, K_2, and K_3 represent mathematical operators, each of which may be a constant or a linear mathematical operation such as differentiation, integration, or a combination of such things. K_1, for example, was taken to be an operator representing a production delay. Simon interpreted K_2 and K_3 as the human "decision rules" in the system. Together those operators determined the production policy. The task was to use the mathematics of feedback engineering to help pick an optimal decision rule for desired production.

After deriving decision rules that minimized costs under various assumptions about order patterns and production costs, Simon concluded:

> The basic approach and fundamental techniques of servomechanism theory can indeed be applied fruitfully to the analysis and design of decisional procedures for controlling the rate of manufacturing activity (p. 114).

Adaptive behavior

The focus of Simon's use of the feedback concept in production control was on human decision making. The mathematics of servomechanism theory makes the production control work look different from his earlier work on *Administrative Behavior* (Simon 1947), but both can be viewed as parts of a single, long-term research interest in human problem solving. In many different contexts he repeatedly asked how do people solve problems and make decisions? How can understanding of the processes involved enable people to do better? The feedback concept entered that line of study even more fundamentally in an article Simon wrote in 1954. In it he identified feedback as the basic mechanism of adaptive behavior.

The background to that use of the feedback concept lay partly in Simon's own work on decision making in organizations and in related work done by his colleague W.W. Cooper. Simon (1947) had argued that there are limits on the ability of administrators to act and to make correct decisions. Skills, habits, and reflexes that have reached the point of being used unconsciously tend to limit one's range of options. Values and visions of purpose also influence decisions and limit possible actions. Knowledge, or rather the lack of perfect knowledge, is a further limiting factor. The number of alternatives to investigate and the information needed to evaluate them are too vast for one individual to handle in a completely rational manner.

> Individual choice takes place in an environment of "givens"—premises that are accepted by the subject as bases for his choice; and behavior is adaptive only within the limits set by these "givens" (1947, p. 79).

The "rational economic actor" and the "efficient administrator" are not economic or administrative principles, but goals. Real actors fall short. Their rationality is "bounded," their behavior adaptive.

Cooper (1951) made a direct connection between this line of thinking and the theory of servomechanisms. The theory of the firm developed by Marshall and others to address macroeconomic issues treated the firm as a unit having certain desirable properties. It kept marginal price equal to marginal cost to maximize profits, and so on, but how it did so was not specified. Cooper, on the other hand, was interested in the mechanisms within the firm. In his view, in real firms with many factors of production and several layers of management, the classical profit maximization principles are hard to put into practice. The effort to control operations, for example to maximize profits, comes to focus on standards, such as standard costs, and on the extent of variations from the stated norms. Comparing actual performance relative to a standard and taking action to reduce the difference creates a negative feedback loop. In addition, Cooper also placed the setting of the standards within a self-adjusting loop.

Control acts in both directions. Operations control the reporting system, and the reporting system controls operations. . . . Standards may be set by staff agents . . . as a means of controlling actual costs; but actual costs are also used to control standards (Cooper 1951, pp. 92, 95).

The total process was pictured in the diagram shown in Figure 3.11.

Cooper was explicit about the presence and function of feedback in such a system. On a cost accounting and reporting system based on standard costs, he noted:

The entire operation may be viewed as analogous to a servomechanism in which the accountant serves as a filter (with certain criteria built into it) in the feedback (p. 94).

Several times in the same article Cooper represented the behavior of

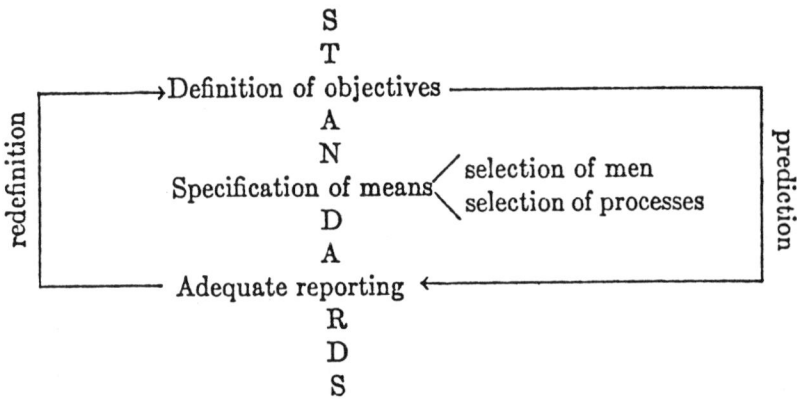

FIGURE 3.11: Loop structure summarizing the processes of control in a firm. Source: Cooper (1951, p. 92).

agents in the firm in simple mathematical models that he identified as negative feedback control systems.

Simon (1954) expanded on this connection between feedback, mathematical models, bounded rationality, and adaptive behavior. To sketch the idea of adaptive human behavior and to contrast it to the model of the rational economic actor, he constructed a simple hypothetical setting in which a manufacturer uses a single raw material from which to fabricate a single product. From x units of the raw material, the manufacturer can make y units of the product [$y = y(x)$]. The costs depend upon on how much is bought [cost = $C(x)$], and the price charged depends on how much of the product is made [price = $P(x)$]. How much raw material does the manufacturer buy? The classical profit maximization criterion gives a unique optimal answer, the value of x that makes the derivative of profit equal to zero:

$$D_x[y\ P(y) - x\ C(x)] = 0.$$

Simon suggested that a more realistic picture of the situation would capture the dynamics of the manufacturer's adjustments in the raw material x in seeking to reduce the discrepancy between actual profit and this optimal condition. He proposed the following model of adaptive behavior:

$$\frac{dx}{dt} = b\{D_x[y\ P(y) - x\ C(x)]\} \quad (b > 0).$$

If the derivative of profit is equal to zero, there will be no change in x. If the profit derivative is positive, marginal costs are below marginal returns, so profit can be increased by increasing production and dx/dt should be positive. And similarly, if the derivative of profit is negative, production should be decreased to reach maximum profit.

Simon then observed that this model is a goal-seeking, negative feedback process familiar to servomechanisms engineers:

$$\frac{dx}{dt} = b(\theta_s - \theta_0),$$

where

$\theta_s = 0$ is the desired condition (the "set point"),
$\theta_0 = -\{D_x[y\ P(y) - x\ C(x)]\}$ is the actual condition,
$(\theta_s - \theta_0)$ is the "error" between desired and actual.

The loop is negative because a deviation from the desired optimal profit condition is counteracted over time by the adaptive process that always adjusts x to move in the direction of higher profits.

Simon's purpose in the article from which this feedback development is taken (Simon 1954) was to show the gains that might come from translating social science theories into mathematical models. His observation of the feedback nature of the adaptive profit maximization model was almost an aside in the discussion. In the remainder of the paper two other dynamic models are formulated and discussed, one dealing with motivation and learning and the other with a theory of group interaction developed by George Homans (1950). Although both of these are models of closed-loop feedback systems, Simon chose not to make anything of that fact in the discussion. Nonetheless, the importance of the paper for us here is the association it makes of the feedback concept with the dynamics of adaptive human behavior. As Simon himself noted two years earlier,

It might be pointed out that the notion of a servomechanism incorporating human links is by no means novel. . . . The idea of social, as distinguished from purely physiological, links is relatively new (Simon 1952, p. 95).

He credited Goodwin (1951a) with the insight.

Simon apparently turned away from the line of feedback thinking expressed in these two articles. The effort to understand the behavior of people in organizations and economies was just one aspect of his interest in understanding how individuals solve problems. He turned to the digital computer as a simulation tool, and became a central figure in the field of artificial intelligence (Newell, Shaw and Simon 1958, 1959; Newell and Simon 1956, 1963, 1972). The feedback concept, as a circular causal process with positive or negative polarity, is hard to discern in this later work. It is conceivable, however, that it is the basic idea behind Newell and Simon's "means-end" analysis (see the argument in Weizenbaum 1976, p. 249). In that notion, the problem-solver sees a goal (an end or "desired object") and a current situation ("present objects") that differs from the goal and searches for some means of reducing the discrepancy between them. This search strategy, coupled with some action intended to bring present objects closer to desired, looks something like a negative feedback loop. However, analyses of such problem-solving situations do not employ the concept of signed circular causal loops. The specific details of a given problem-solving situation are much more significant in explanations of problem-solving strategies and heuristics. Consequently, we will not pursue the link between feedback and means-end analysis or artificial intelligence. (See section 5.2 for some later feedback-related work of Herbert Simon.)

Industrial Dynamics

In 1958 in an article appearing in the *Harvard Business Review*, Jay W. Forrester announced a new approach to the study of industrial management problems. In the spirit of the work of Goodwin, Tustin, Phillips, and Simon, but arrived at from an independent direction, Forrester's "industrial dynamics" is a modeling approach based on the feedback concept that uses digital computer simulation to trace out model behavior. The use of the computer altered dramatically the character, scope, and size of the models that could be investigated. The nonlinearities that those earlier authors had wanted to incorporate were handled easily by digital computer simulation. The number of equations that could be handled was much greater. Consequently, there was a significantly greater potential for realism in the model structure assumed.

Forrester founded his new approach on four recent developments: advances in computing technology, growing experience with computer simulation, improved understanding of strategic decision making, and developments in the understanding of the role of feedback in complex systems (Forrester 1958). The components of Forrester's approach reflected his own experience as a servomechanisms engineer, digital computer pioneer, and manager of an extensive R&D effort. He had spent the years from 1940 through 1946 associated with MIT's Servomechanisms Laboratory. The activities of the lab and particularly its founder and director, Gordon Brown, were strong influences on his thinking. (Forrester dedicated one of his later works, *World Dynamics* to Brown, and Brown wrote the introduction to Forrester's *Collected Papers*.) The result of this experience was that feedback emerged as a basic characteristic of Forrester's view of the world. His definition of the concept was correspondingly broad:

Systems of information feedback control are fundamental to all life and human endeavor, from the slow pace of biological evolution to the launching of the latest satellite. A feedback control system exists whenever the environment causes a decision which in turn affects the original environment (Forrester 1958, p. 4).

His examples in the 1958 article included the mandatory reference to a thermostat as a feedback device, as well as several unusual references to corporate feedback loops:

In business, orders and inventory levels lead to manufacturing decisions which fill orders and correct inventories.... A profitable industry attracts competitors until, to use the economist's terms, the profit margin is reduced to equilibrium with other economic forces.... The competitive need for a new product leads to research and development expenditure that produces technological change (pp. 4–5).

The linking of the feedback concept here with human decisions echoes Simon's interpretation of mathematical operators as human "decision rules" in a system. To my knowledge, however, Forrester and Simon developed their ideas independently, based on their own experiences and knowledge of the servomechanisms literature. Here we have a case of a methodological congruence in the absence of a direct connection between feedback authors.

Thus one advantage Forrester saw in the use of the feedback concept was realism. Feedback loops were to him simply a part of reality. Ignore them and you ignore a vital part of reality:

Feedback theory explains how decisions, delays, and predictions can produce either good control or dramatically unstable operation. It relates sales promotion to production swings, purchasing and pricing policies to inventory fluctuations, and typical life cycles of products to the need for research (p. 6).

In Forrester's use of the feedback concept, we see a strong resemblance to Goodwin's endogenous point of view:

Typical manufacturing and distribution practices can generate the types of business disturbances which are often blamed for conditions outside the company. Random, meaningless sales fluctuations can be converted into annual, seasonal production cycles. Advertising and price discount policies of an industry can create two- and three-year sales cycles (p. 6).

In the same phrases we hear more explicitly the claim voiced above by Phillips that policies intended to solve dynamic problems could, instead, actually create or at least exacerbate them.

The approach was "systemic" and like the work of Goodwin, Tustin, Phillips, Allen, and Simon, it was focused on dynamics:

The company will come to be recognized not as a collection of separate functions but as a system in which the flows of information, materials, manpower, capital equipment, and money set up forces that determine the basic tendencies toward growth, fluctuation, and decline. . . . It is not just the simple three-dimensional relationships of functions that counts, but the constant ebb and flow of change in these functions—their relationships as dynamic activities (p. 26).

One can read into these phrases Gordon Brown's emphasis on the importance of "synthesis" in the design and understanding of a feedback system (see section 2.1).

Forrester's emphasis on particular physical flows in a company, and by implication the stocks or accumulations of those flows, was something of a departure from the other authors in this feedback thread. (Recall that in the various macroeconomic models of Goodwin, Tustin, and Phillips inventories conserving the flows of production and consumption were usually not present, and the stock of capital, even when it was present in a model, was not used to produce the flow of goods.)

To illustrate his approach, Forrester showed the results of modeling and simulating the behavior of a production-distribution chain. A factory supplies a product to a set of distributors who, in turn, ship to a set of retailers of the product. Orders for the product flow back upstream from the retailers to the distributors and finally to the factory. Each link in the chain is trying to do the best it can to maintain inventories a ship with minimal lead times, but the system as a whole shows a marked tendency to amplify minor ordering disturbances at the retail level and oscillate. An initial 10 percent increase in customer orders at the retail level eventually caused production at the factory at its peak to be a full 40 percent above its initial value.

The goal of Forrester's analysis was understanding how the structure of system gives rise to its observed behavior. He clearly saw the system itself potentially at fault:

The system, by virtue of its policies, organization, and delays, tends to amplify those retail sales changes to which the system is sensitive (p. 12).

Forrester showed that the system produced persistent oscillations of identifiable period even if the customer orders varied in a completely random (patternless) way. Frisch (1933) and Wicksell had long ago pointed out that possibility in dynamic systems, but Forrester observed that many of us could be fooled by such behavior:

examination of the factory production curve could easily lead the casual observer who lacked the retail sales information to conclude that a seasonal sales pattern was present (p. 14).

Simulation allowed experimentation with policies designed to improve the problematic behavior. Speeding up clerical delays, such as faster order handling, was found to yield only very slight improvement. Changing the ordering procedures, however, by transmitting retail orders directly to the factory, reduced the oscillatory tendencies considerably. This later change, it should be noted, alters the information feedback structure of the system. Advertising, another structural change, was found to have the potential to aggravate the production cycle. In Forrester's production-distribution chain model it considerably lengthened the period of the swings. Yet Forrester concluded that advertising and pricing policies could be used to stabilize the swings if management chose to do so.

The book, Industrial Dynamics

Where his predecessors had seen some potentially useful techniques to apply to dynamic problems, Forrester saw a new field emerging. He expressed his view in remarkably complete form in *Industrial Dynamics* (1961). The book lays out in its entirety an approach to dynamic problems that is founded on the feedback perspective and digital computer simulation. Forrester (1958) sketched the outlines of the approach. *Industrial Dynamics* filled in the details. It spends 170 pages describing and analyzing three dynamic simulation models. It contains ten chapters on principles of model formulation. And it concludes by sketching the potential future of the approach and its impact on management and management education.

In spirit, *Industrial Dynamics* is closely aligned with the work of Goodwin, Tustin, Phillips, and Simon. Forrester, in fact, cites both Tustin and Phillips in his bibliography, although in both cases it is for the purpose of disagreeing with minor points in their work. (Tustin had argued analog simulation models; Forrester saw the richer promise of digital computer simulation. Phillips had claimed he found his models to be sensitive to the form of delay he used; Forrester's work convinced him that such a result would only be likely in very simple systems.) Minor disagreements aside, all of these authors were concerned with using ideas from engineering feedback control theory to investigate dynamic socioeconomic systems. All believed that explicit mathematical models of feedback systems were necessary to deduce reliably the dynamic implications of an assumed feedback structure. All saw feedback structure as a source of dynamics. All were concerned with economic and corporate problems. All made use of continuous models, either in the form of differential equations or, in the case of Forrester, in digital computer simulations of continuous systems.

But in its details Forrester's work was a significant departure. Phillips, for example, had argued in favor of representing the multiplier/accelerator in continuous form, in differential rather than difference equations. Aggregation of the decisions of thousands of economic actors would lead to apparently continuous aggregate behavior. Forrester made the same aggregation argument, but went further. He argued for a continuous *point of view.*

Discreteness of events is entirely compatible with the concept of information-feedback systems, but we must be on guard against unnecessarily cluttering our formulation with the detail of discrete events that only obscure the momentum and continuity exhibited by our industrial systems (1961, p. 64).

Forrester advocated a focus on the persistent feedback system structure that underlies discrete events.

Even major executive decisions represent a rather continuous process; they are reached after a period of consideration; advance actions may be taken in anticipation of the probable outcome; action is not taken immediately after the decision; and the decision is interpreted and "smoothed" and produces gradual change as it overcomes resistance and the inertia of other persons in the organization (p. 65).

Forrester also argued that the dynamics of a continuous flow model are easier to understand, so initial modeling efforts, at least, should be directed toward a continuous representation. Discreteness, in the form of noise or discontinuities, could be added to the continuous structure later if necessary. But this pragmatic point seems like an afterthought in Forrester's argument. His primary focus was on the continuities in socioeconomic systems.

A continuous-flow model helps to concentrate attention on the central framework of the system (p. 65).

Thus the representation of the system in a continuous model was more of a consequence than a cause of Forrester's continuous point of view.

The endogenous point of view expressed by Goodwin was amplified considerably by Forrester in a chapter devoted to a discussion of exogenous variables. Forrester took exception to the use of the endogenous time series in econometric modeling, citing Klein (1950) as an example. Variables that are truly exogenous in the real system can, of course, be modeled as time series. But variables that are actually coupled to the real system, influencing and being influenced by the rest of the system, must be modeled as endogenous variables embedded in information-feedback loops. Without that coupling, the model would fail to exhibit "modes of operation" that stem from the feedback effects the coupling generates (p. 112). Some dynamic patterns would be missed or misrepresented. Worse, an exogenous time series could "drive" the model, making its behavior more a consequence of the time series than of the information-feedback structure of the system itself. The dynamic character of the actual system would be obscured in the model, and that would vitiate for Forrester the purpose of building the model.

Simulation in Industrial Dynamics

Forrester's use of digital computer simulation led to several departures from the work of Goodwin, Allen, Tustin, Phillips, and Simon.

Most obviously, differential equations were not employed. It would be closer to say that integral equations were used, but they appeared as discrete approximations in the DYNAMO simulation language. An integration or accumulation in DYNAMO appears in the following form (1961, p. 76):

$$IAR.K = IAR.J + (DT)(SRR.JK - SSR.JK).$$

where

IAR = inventory at retail (a stock accumulation),
SRR = shipments received at retail (the inflow rate to inventory),
SSR = shipments sent from retail (the outflow rate of inventory),
DT = computation interval for the simulation.

The subscripts J, K, and JK denote the immediate past (J), the present (K), and the time interval between (JK). Thus the equation states that the current value of the inventory (IAR.K) is computed from its most recent past value (IAR.J) plus changes from inflows and outflows that occurred in the intervening time (SRR.JK − SSR.JK). The simulation parameter DT is a very small increment of time, the length of simulated time between the previous computation (J) and the present (K). The connection to differential equations is easy to see by rearranging:

$$\frac{IAR.K - IAR.J}{DT} = SRR.JK - SSR.JK,$$

which, for small values of DT, is a discrete approximation of the derivative

$$\frac{d}{dt}(IAR) = SRR - SSR.$$

In place of the more formal calculus concepts of derivatives and integrals, Forrester introduced the terminology of "rates" and "levels." Levels are accumulations or integrations of rates. Rates are rates of change of levels. A given rate, such as SRR, is one component of the derivative of the level to which it is connected. The actual derivative of a level is the sum of all of its inflow rates minus its outflow rates.

The use of digital computer simulation to trace through time the behavior of a dynamic system makes it easy to incorporate nonlinearities. Indeed, computational ease in spite of model complexity and nonlinearity was the point of simulation. In contrast to the work of Tustin, Phillips, Allen, and Simon, Forrester's models were insistently nonlinear. To illustrate the concept of nonlinearity, Forrester noted that the availability of a stock of goods in an inventory affects the shipment rate, but the extent of the effect varies. When inventory is near the desired level, there will be essentially no impact of availability on shipments. The firm will ship according to its incoming order rate. When inventory is very low, however, availability will constrain shipments, and orders will have to accumulate in an order backlog, awaiting units to ship. Thus the effect of the availability of inventory on shipments cannot be captured in as a simple linear function of inventory (pp. 105–107, 147). Another example of nonlinearity cited by Forrester involved a saturation in the additional production obtained by adding to the firm's work force. If production equipment is limited, a point will be reached where additional workers do not actually produce more and may even produce less if the additional numbers interfere with efficient operations (p. 106).

Rates as decisions

Forrester's theory of decision making was based explicitly on the feedback concept. Figures 3.12, 7, and 8 show the way he pictured the role of feedback. In figure 3.12 Forrester sketched a decision stream in the simplest possible feedback structure, a single loop. He argued that real systems are much more complex multi-loop structures, such as that shown schematically in Figure 3.13. Finally, in figure 3.14 he sketched the elements of a theory of decision making stated in terms of feedback and his concepts of rates and levels.

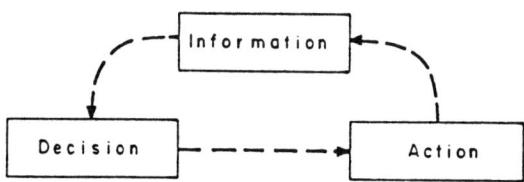

FIGURE 3.12: Decisions and information feedback.
Source: Forrester (1961, p. 94).

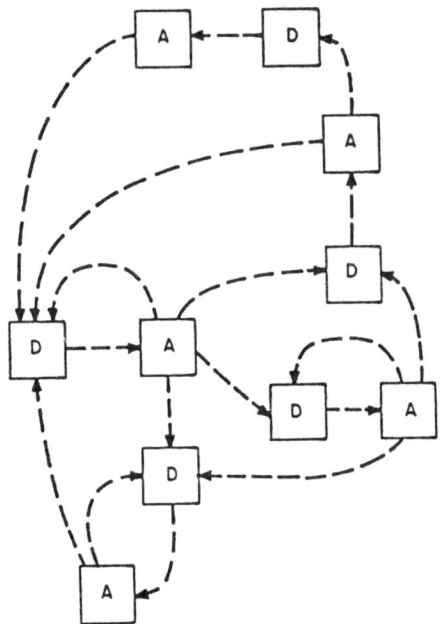

FIGURE 3.13: Multi-loop decision-making system. Source: Forrester (1961, p. 94).

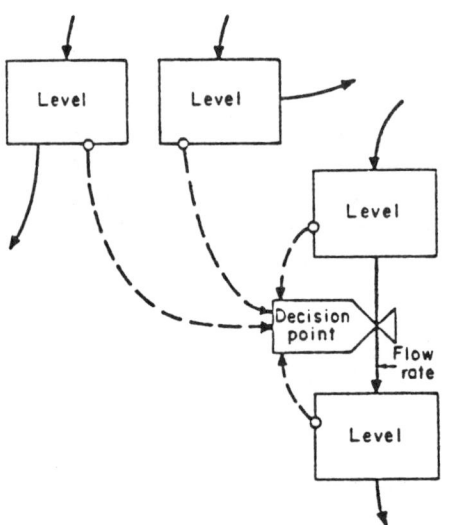

FIGURE 3.14: Schematic view of decision making in terms of rates, levels, and information feedback. Source: Forrester (1961, p. 95).

Echoing Simon, Forrester saw rates of change as the decision points in a system. Information about the system, stored in the system levels, brings pressure to bear at a decision-point (a rate), resulting in action (a flow) that changes a system level and results in a new state of the system. The loop structure of this theory of decision making is obvious: information originates at system levels, travels to decision points, and returns in the form of action that changes system levels. (See Figure 3.14.)

In more detail, Forrester argued that decisions involve three components: apparent actual conditions, desired conditions, and corrective action. He saw natural limitations on such a process. First, appearances can be deceiving. The conditions that actors in the system perceive can be different from actual conditions. Second, not all information in the system is available. For example, information about the apparent actual conditions in the system can come only from levels in the system. Rates are not instantaneously knowable within the system. Even the world's cleverest economist cannot know today's GNP; the average annual production rate as of last month's figures is probably the best one can do. Information about rates of change must accumulate over time in various kinds of averaging processes before it can be perceived and used as a basis for future corrective action. Third, constraints in any system limit the possible range of corrective action. Desired conditions may not be achievable. Inventory might be so large as to cause desired production to be negative, but actual production can go no lower than zero. The limitations Forrester saw on decision making are very similar to those perceived by Simon.

Consistent with his choice of essentially continuous differential (or integral) equations, Forrester chose to view decision making as a continuous process (1961, p. 96). Again, however, the continuous point of view seems to precede the selection of continuous versus discrete equation formulations.

[Decision making] is a conversion mechanism for changing continuously varying flows of information into control signals that determine rates of flow in the system. The decision point is continually yielding to the pressures of the environment (p. 96).

He noted that this continuous point of view was the result of looking at the decision process "from a very particular distance"—closer than the stockholder who is ignorant of the internal pressures of the firm, but not so close that we are concerned with the precise mechanisms of human thought or a single action of an individual. The continuous viewpoint is more the perspective of the upper level manager:

The superior is close enough to know how desired goals are established. He is in a position to observe and probably provide the information sources to be used by the subordinate to determine his concept of actual conditions. He knows in general the guiding policies and the manner in which the subordinate decision maker would respond to various kinds of circumstances (p. 96).

The concept of viewing decision making from a certain distance also appeared in Forrester's focus on "policy." A policy is the rule structure within which individual decisions are made. In Forrester's wording,

Policy is a formal statement giving the relationship between information sources and resulting decision flows. It is what has often been referred to in the literature as a decision rule. In physical systems, particularly in the field of servomechanisms, the corresponding term is "transfer function" (p. 97).

Forrester's focus was thus not on individual decisions, but on the policy framework producing continuous "flows" of decisions. He linked the feedback concept of servomechanisms to this level of decision making. The attempt was not to model any single isolated decision but rather to capture the flow of decisions resulting from the policy structure of the firm. There is a "distancing" here as well—not so close that we see each individual decision, but not so far away that we fail to capture the guiding policy framework within which day-to-day decisions are made.

Forrester saw both an overt and an implicit policy structure in a firm. "Overt decisions" are the conscious decisions people make to manage the firm. "Implicit decisions," on the other hand, are unavoidable results of the physics of the system. Being out of stock results in the decision not to ship, even if orders are pouring in. The decision in this case is forced by the state of the system. Forrester distinguished overt or "free" decisions from implicit ones in order to emphasize that there frequently may be a difference between what is desired and what can actually be achieved. An overt decision to hire additional people may be countered by a lack of available people. The system thus makes an "implicit decision" of its own that holds down the hiring rate in spite of the overt managerial decision to increase it (p. 102).

The net result of Forrester's view of the decision process is the connection of the feedback concept with decision making at a certain level of aggregation and continuity. Forrester did not attempt to model individual decisions or to claim that feedback theory could predict how an individual actor would behave in a given setting. He advocated moving back from the system far enough for individual actions to blur into a continuous flow of activity. From that slightly more distant vantage point, individual decisions would merge into patterns understandable through the overt and implicit policy structure of the firm.

For Forrester, that policy structure was characterized by the existence of information-feedback.

The use of feedback for conceptualization

It should be noted in the foregoing discussion that Forrester used a feedback point of view not merely for the analysis of a model, but for model conceptualization and formulation. Feedback is present at the creation of a Forrester model, and its presence makes a difference in the way reality is perceived. In this sense, Forrester's use of the feedback concept is allied with the work of Goodwin and Simon and differs somewhat from the efforts of Tustin, Phillips, and Allen. The latter authors made use of the methods of servomechanisms engineers to deduce the behavior of economic models. (To be sure, Phillips conceptualized the structure for the PID controller he added to various models of the business cycle, but such a control structure was a standard piece of engineering method at the time, not an original conceptual advance.) Goodwin and Simon, however, altered their views of the world because of feedback. These authors, and Forrester, took more than the ideas or the methods of feedback engineers. They took a feedback point of view—a different sort of lens through which to see the world.

Summary of the Beginnings of the Servomechanisms Thread

The uses of the feedback concept in the work of Goodwin, Tustin, Phillips, Allen, Simon, and Forrester are similar. These authors share a family resemblance that sets them somewhat apart from the authors discussed in section 3.2. In one sense, they are linked rather mechanically, by citations. Allen built upon the work of Goodwin, Tustin, and Phillips, and Simon and Forrester each cited at least two of these earlier authors. It is perhaps also significant that these authors tended not to cite any of the authors in what I am calling the cybernetics thread. Simon alone referenced Wiener. We have the beginnings of a thread of thinking in the social sciences defined by the citation evidence.

In a larger sense, however, these authors are linked and distinguished significantly in the way they made use of the feedback concept. They focused on dynamics. More specifically, *they focused on the role of feedback in creating the patterns of movement in dynamic systems.* Tied to that emphasis on dynamics are the following characteristics common to their work:
- The patterns of behavior of a dynamic system are traced to its feedback structure.

- Formal dynamic models are employed.
- The dynamic behavior of a feedback system is considered to be difficult to discern without the aid of formal mathematical models.
- Feedback loops are seen as an intrinsic part of the real system, not merely as possible mechanisms of external system control.
- Positive loops are present in the analyses, along with negative loops.
- Well-intentioned policies are seen to have the potential to create or exacerbate the problem behavior they were intended to cure.
- Nonlinearities are perceived to be a persistent characteristic of real socioeconomic feedback systems. Consequently, they are considered to be a necessary characteristic of reliable formal models of such systems.
- The work is directed toward policy analysis.

From these authors emerges in the social sciences a way of using and thinking about feedback that differs from the cybernetics thread. I have chosen to call this separate line of thinking the "servomechanisms thread." It is a chain of people and ideas in the social sciences that are connected over time and that share some or all of the feedback points of view listed above. The name "servomechanisms" is intended to call to mind (a) the closeness of each of its early authors to practical control theory engineering; and (b) the tendency within this thread to try to derive the dynamic behavior of a socioeconomic system from its feedback structure, in much the same way that the behavior of a servo system is seen by an engineer to depend upon its feedback control structure. I do not mean to suggest that authors in the servomechanisms thread are doing engineering. Far from it: the servomechanisms thread, like the cybernetics thread, names a chain of ideas and people in the social sciences. And I also do not mean that authors in the servomechanisms thread take the view that socioeconomic systems are simple automatic mechanisms. Some of the early models could perhaps be justly accused of being simple and mechanical, but their simplicity was forced by the mathematical tools available at the time. The simulation models in *Industrial Dynamics,* each with more than 50 equations and at least 10 state variables (levels), could hardly be called simple. And their numerous nonlinearities, capable of endogenously shifting active structure as conditions change, give these models a decidedly nonmechanical, lifelike character.

Two additional, budding characteristics of the servomechanisms thread deserve notice at this time. They appear frequently in the work of Goodwin, Simon, and Forrester:

- Feedback is seen as the mechanism of systems that adapt over time.

- The concept of feedback is employed to conceptualize system structure.

These conceptual tendencies overlap somewhat with thoughts in the cybernetics thread. But in the servomechanisms thread they are associated with formal models and dynamics.

3.4 Origins and Prospects

Observing the formation of two distinct threads of feedback thinking in the social sciences, it is natural to ask why they coalesced differently and in what directions they point. We have already suggested answers to both of these questions. Before summarizing them, however, I should point out that answers to the first, the question of origins, are speculative at best, and one set of answers to the second, the question of future directions, are provided in the following chapters. Here we summarize to bring together the prehistory outlined in Chapter 2 with the developments to be traced in Chapters 4 and 5.

The sources of the differences between the servomechanisms and cybernetics threads appear to lie in the backgrounds of the scholars instrumental in first bringing the feedback concept to the attention of social scientists. Most obviously, there are direct connections from early writers in the servomechanisms thread in the social sciences to the actual engineering servomechanisms literature. Phillips did his undergraduate work in engineering. Tustin was an engineer. Forrester was for fifteen years a practicing servomechanisms engineer, a student and colleague of Gordon Brown, the founder of MIT's Servomechanisms Laboratory. Even Herbert Simon had a direct connection: his father was an engineer. As Simon recounts:

He was involved in designing control systems and built some of the gun turret controls for World War I battleships, and light controls for theaters, and things of this sort. It later became servomechanism engineering. I knew perfectly well what he was doing. I don't mean I ever went into it, but I knew what he was doing. And towards the end of World War II, when the word servomechanism became popular, the idea of feedback was one of many cybernetic ideas in the air which were obviously relevant to what I was after. And I remember asking him for a reference to a good book to get me started in the servomechanism literature (Herbert Simon, quoted in McCorduck 1979, p. 130).

In contrast, only a very few of the early writers in the cybernetics thread were engineers, and few had direct knowledge of the engineering literature. Wiener was a mathematician and statistician. It is relevant to point out that shortly after publishing *Cybernetics,* he wrote

Extrapolation, Interpolation, and Smoothing of Stationary Time Series (Wiener 1949). The work deals with analysis and prediction of stochastically varying time series, and shows us one of the components of the field of cybernetics as Wiener intended it. Of additional significance, I think, is the fact that Wiener had studied mathematical logic under Bertrand Russell. Rosenblueth was a physiologist, a student and colleague of Cannon, the father or midwife of homeostasis. Among these early cybernetics writers, Bigelow and von Foerster were, I believe, the only engineers. McCulloch was a neurophysiologist. Pitts was a mathematician and logician, a student of Rudolph Carnap. Lewin and Deutsch were very influential social scientists (Deutsch is still active). I do not know what Ashby's background was, but it obviously contained considerable mathematics and some connection to the engineering literature. Von Bertalanffy was a biologist somewhat in the tradition of Lotka. Boulding is an economist. Claude Shannon, whose work in communication theory so stimulated the early cybernetics writers, wrote a graduate thesis on the application of Boolean algebra (a formal mathematical logic) to switching circuits. The mathematician John von Neumann and the economist Oscar Morgenstern participated in the early Macy cybernetics meetings and shortly thereafter collaborated on their pathbreaking book on the theory of games. In addition, von Neumann at that time was heavily involved in bringing an early digital computer, ENIAC, into being.

The differences in backgrounds here are suggestive. It requires only a small jump to the conclusion that an engineering background, or direct ties to that servomechanisms literature, created a mindset about the use of the feedback concept that differs from the mindset established by mathematics, formal logic, neurophysiology, and even social science. (That conclusion will be unexpectedly reinforced in section 4.7, where we will encounter a psychologist in the cybernetics thread who turns away from many of the established traditions in that thread, and we will find that he did his undergraduate work in physics.)

These backgrounds suggest that it was natural for the early writers in the cybernetics thread to emphasize feedback as the mechanism of homeostasis, to focus on feedback in communications networks (such as neural networks and computer circuitry), to stress the role of messages and symbols in feedback processes, and to take a largely discrete, often linguistic view of feedback processes. Ashby's emphasis on the ability of a controlled system to respond to and counteract a variety of incoming signals seems like a reasonable development of the emphasis on stochastic processes in Wiener's work.

An additional characteristic of the early writers in the servomechanisms thread deserves mention, however. All were concerned with matters relating to industry and economics. It is reasonable to suggest that the emphasis on dynamics that helps to characterize the early work in the servomechanisms thread stems not from dynamics in engineering servomechanisms but from economic dynamics. Put another way, the servomechanisms thread may have started with an association of the feedback concept with dynamic behavior because of the obvious dynamic nature of the economic and industrial problems these writers were investigating. On the other hand, these early writers in the servomechanisms thread may have selected such problems because they perceived them to have the dynamic character appropriate for a feedback perspective.

Thus, the backgrounds of the principals might have started these two feedback threads on their separate ways in the social sciences, or perhaps the nature of the problems the scholars were most interested in were more influential. There is also a likely third possibility that stems from differences already existing within engineering by the 1940s, to which I referred obliquely in section 2.1.

Two Kinds of Feedback Engineers

There was not a single discipline of feedback engineering, but actually two different engineering problem areas that made use of feedback.[9] They come to us today as control theory and communication theory. In looking back on these in 1960, one of the principals, Heinrich Bode, suggested that even within engineering the implications of their differences have not been clearly perceived.

The modern feedback and control field is the fusion of what were originally two quite different technological areas. One parent area is represented by the typical mechanical regulator or elementary control circuit, such as the household thermostat. This is a very vast as well as a very old field. . . . The second parent area is much more recent. It resulted essentially from the efforts of communications engineers, and particularly telephone engineers, to provide extremely high quality amplifiers. . . . The two parent fields were originally quite different in character and emphasis (Bode 1960, pp. 2–3).

In designing a regulator, Bode noted, the control engineer was interested in reaching and holding a given operating point. The details of transient behavior were of interest only so far as objectionable behavior characteristics that came as a consequence of the control mechanism had to be designed away. The communications engineer, however, was primarily concerned with transient behavior of an input signal. The goal was to design an amplifier that could reproduce, with larger amplitude but with extreme faithfulness, the transient contortions of any incoming wave form.

World War II brought these two engineering areas together. The servo control systems on the guns of warships, for example, had to have capabilities that blended the achievements of both control engineers and communications engineers. Aiming at another ship is something like the basic regulator problem, achieving and maintaining a set point, while aiming at an attacking plane is more like trying to track a wildly varying wave form. Bode described the resulting blend as "a sort of 'shotgun marriage' between two incompatible personalities." Looking back on it, he observed:

> The two fields are really quite different. In the orthodox regulator problem, only one or a few modes of behavior are envisaged. Great stress is laid on analyzing exactly what the system may do in these circumstances, or perhaps in devising particular components which will aid it in its operation.
> The heart of communication engineering, on the other hand, resides in the fact that we are dealing with the response of relatively complex systems to very complex ensembles of messages which are individually unpredictable but which can be dealt with in terms of some general defining characteristics. . . . Both the communication engineering and the regulator approaches are, of course, of great importance, but they are obviously quite different in fundamental intellectual texture. . . . The marriage has lasted twenty years; perhaps an amicable divorce is in order (pp. 16–17).

From the point of view of what we have observed in the two feedback threads in the social sciences, a startling and potentially significant fact stands out. The two engineering fields differed in their perceptions of the sources of dynamics in the system. In a servomechanism, it is *the process of regulation* itself that shapes the dynamic behavior of the system in response to some arbitrary and largely uninteresting exogenous disturbance. In an amplifier, it is *the incoming wave form* that shapes the major characteristics of the output of the system.

That distinction closely parallels differences we have identified in the uses of the feedback concept in the social sciences. Social scientists we have identified in the cybernetics thread defined feedback in terms of input and output and tended to see feedback as tool for controlling systems in the face of exogenous disturbances. In contrast, social scientists we have placed in the servomechanisms thread tended to see feedback loops as an internal aspect of the structure of social systems and to see circular causality as an endogenous source of a system's behavior. It seems quite natural that social scientists more acquainted with servomechanisms engineering would evolve the endogenous point of view of dynamic behavior that we have seen in the work of Tustin and Phillips and emphasized particularly in the work of Goodwin and Forrester, while social scientists more acquainted with the use of feedback in communications engineering would evolve more of the input/output point of view that we have seen in the cybernetics thread.[10]

Of less dramatic interest, but still perhaps significant, is the fact that the mathematics used in these two engineering endeavors was different. The control engineer might be able to represent a system with a system of differential equations of fourth order or even less, but often nonlinearities would be important and would make the analysis very difficult. The communications engineer, in contrast, would work with systems perhaps as high as 50th order, but nonlinearities could be ignored. One could often work with circuit equations in the "steady state" form (Bode 1960, p. 3).

Thus we have the distinct possibility that our two feedback threads in the social sciences mirror something of a split in the use of feedback in engineering. The cybernetics thread emphasizes aspects more connected to communications theory, particularly in its concern for randomly varying system inputs. The servomechanisms thread is more associated with control theory as, for example, in its emphases on behavior generated by the regulatory system itself and on the significance of nonlinearities.

But more importantly, something of the difference in worldview that is discernible between our two threads shows up in differences between the communication theory and control theory points of view in engineering as well. The significant behavior of variables in a communication traces most directly to the behavior or shape of the incoming "wave form," the input signal. Social scientists in the cybernetics thread tend to look outside the system for the causes of system behavior and for the regulatory structures in social systems designed to maintain control, homeostasis, and adaptability. In contrast, the significant behaviors in a servomechanism are the transients that are triggered by an input to the system but whose pattern over time traces not to the input but to the internal structure of the control system. Similarly, the social scientist in the servomechanisms thread tends to see dynamic behavior as a consequence of a system's internal feedback structure. The fundamental points of view that help to characterize the feedback threads in the social sciences may well have origins in engineering itself.

These are very suggestive ideas, but they must remain at the level of speculations in this investigation. We shall have to be content with three quite reasonable hypotheses for the origins of the differences in the use of the feedback concept in the social sciences:
- the background of the social scientists first drawn to the feedback concept,
- the nature of the problems they preferred to address,

- a split in the use of the feedback concept within engineering itself.

Whatever the reason, by 1960 a fundamental difference in point of view separates the two feedback threads in the social sciences, a difference that continues to the present. In the servomechanisms thread, there is a strong tendency to focus on dynamic behavior and to regard it as endogenously generated by positive and negative feedback loop structures that naturally exist in complex systems. In the cybernetics thread, there is a strong tendency to focus on homeostatic mechanisms, stability, and the use of feedback for control and regulation in the face of stochastically varying system inputs.

Prospects

We should expect that these tendencies, and others listed in sections 3.2 and 3.3, would continue to be associated with the feedback concept as it moves over time through the social sciences. Social scientists who connect with the feedback concept through writings or scholars in the cybernetics thread ought to reflect tendencies that we have observed in the use of the feedback concept in that thread. Similarly, social scientists who learn about feedback from the servomechanism thread ought to reflect tendencies of that thread. Over time, however, as feedback ideas become more pervasive in the social sciences, we should expect a blending of the two threads. Some natural selection process ought to take place, guided by whatever governs the ecology of ideas.

We should also expect an initial dominance of the perspectives of the cybernetics thread. The social scientists associated with it were more diverse and more famous, and the questions to which they initially applied it had an aura of enormous significance. It is, in fact, true that what most social scientists know of the feedback concept comes directly or indirectly from the cybernetics thread in the social sciences.

In the following chapters we shall explore the implications of these observations. We shall be interested in evidence that supports the existence of two distinct threads of feedback thinking in the social sciences, evidence that contradicts that thesis, and analysis of the implications of differences between the two threads. Throughout Chapters 4 and 5, we shall occasionally observe strengths and weaknesses in uses of the feedback concept that we encounter, and in Chapter 6 we shall try to draw together a number of those issues and implications for the future evolution of the feedback concept in the social sciences.

Notes

1. For a collection of articles in the debate, see Buckley (1968), pp. 219–255.
2. Macy Conference contents:

6th Conference—March 24–25, 1949
John Stroud, Psychological Moment in Perception; Lawrence S. Kubie, Neurotic Potential and Human Abilities; Heinze von Foerster, Quantum Theory of Memory (discussion), Possible Mechanisms of Recall and Recognition; Norbert Wiener, Sensory Protheses

7th Conference—March 23–24, 1950
Ralph W. Gerard, Some of the Problems Concerning Digital Notions in the Central Nervous System; J.C.R. Licklider, The Manner in Which and Extent to Which Speech Can Be Distorted and Remain Intelligible; Claude E. Shannon, The Redundancy of English; Margaret Mead, Experience in Learning Primitive Languages Through the Use of Learning High Level Linguistic Abstractions; Heinze Werner, On the Development of Word Meanings; John Stroud, The Development of Language in Early Childhood; Lawrence S. Kubie, The Relationship of Symbolic Function in Language Formation and in Neurosis

8th Conference—March 15–16, 1951
Alex Bavelas, Communication Patterns in Problem-Solving Groups; Ivor A. Richards, Communication Between Men: Meaning of Language; Lawrence S. Kubie, Communication Between Sane and Insane: Hypnosis; Herbert G. Birch, Communication Between Animals; Claude E. Shannon, Presentation of a Maze-Solving Machine; Donald M. MacKay, In Search of Basic Symbols

9th Conference—March 20–21, 1952
Gregory Bateson, The Position of Humor in Human Communication; Lawrence S. Kubie, The Place of Emotions in the Feedback Concept; W. Ross Ashby, Homeostasis; J.Z. Young, Discrimination and Learning in Octopus; John R. Bowman, Reduction of the Number of Possible Boolean Functions; Ralph W. Gerard, Central Excitation and Inhibition; W. Ross Ashby, Mechanical Chess Player; G. Evelyn Hutchinson, Turbulence as Random Stimulation of Sense Organs; Walter Pitts, Investigations on Synaptic Transmission; Henry Quastler, Feedback Mechanisms in Cellular Biology

10th Conference—April 22–24, 1953
W. Grey-Walter, Studies on Activity of the Brain; Y. Bar-Hillel, Semantic Information and Its Measures; Yuen Ren Chao, Meaning in Language and How It Is Acquired

3. John B. Thurston, *Saturday Review of Literature*, cited on the paperback cover of the second edition of Wiener's *Cybernetics* (Cambridge, Mass.: MIT Press, 1961).

4. Von Bertalanffy used l instead of x; I have used x to avoid confusion between l (el) and 1 (one). Actually, in this reference (von Bertalanffy 1968, p. 76), the equation was written with a misplaced parenthesis:

$$l = l^* - (l^* - l_0 e^{-kt}).$$

I have substituted the correct solution because there is no doubt that the error was typograhpical.

5. The phrase is Boulding's own early version of the general systems idea, taken from a letter from him to von Bertalanffy (von Bertalanffy 1968, p. 14).

6. Individual terms in a nonlinear differential equation can change their degree of influence on the behavior of a system. An example is the famous van der Pol equation

$$x'' + a(x^2 - 1)x' + x = 0$$

in which the damping effect of the middle term varies endogenously as x changes over time.

7. Phillips used $P - P_d$ as his error term. More common practice is to use desired minus actual, $P_d - P$. Phillips followed the latter convention in his later article (Phillips 1957).

8. Geometrically, integral control is corrective action that is proportional to the (signed) *area* between the desired and actual production curves. Proportional control is proportional to the (signed) *distance* between the two curves. Derivative control is proportional to the difference in the *slopes* of the two curves.

9. Brown (1948, pp. 92–94) corroborates this observation, which I first came across in Bode (1960).

10. Had I known about this particular distinction between communications and servomechanisms engineering in advance of discovering different feedback threads in the social sciences, it would not seem so significant to me. But coming upon the engineering difference after seeing the difference in the social sciences adds considerable confidence to the ideas being proposed in this investigation.

Chapter 4
Developments in the Cybernetics Thread

> To be in hell is to drift; to be in heaven is to steer.
> George Bernard Shaw

4.1 Introduction

It is likely that by the mid-1950s almost all serious social scientists had heard of the feedback concept. To catalog even half of the times it has appeared would be well beyond the scope of this book. Instead we shall concern ourselves in this chapter and the next with what might be called "significant" appearances of the idea—uses of the feedback concept in the social sciences that profoundly affect point of view, approach, or conclusions.

The test for significance of the feedback concept in a given work might be the extent to which the work would have been different without the use of the idea. The most significant uses of the feedback concept are those that not only put a particular stamp on the work in which they appear but also continue over time to affect other works in the same fashion. It is these most significant uses of the feedback concept that we shall try to concentrate on. Even in this restricted range, I doubt that it is possible to be sure one has found all significant uses of the concept. The examples in this chapter and the next encompass virtually every discipline in the social sciences, including economics, management, psychology, political science, anthropology, sociology, organization studies, and interdisciplinary social science. Nevertheless, the most we can guarantee for the works discussed in these two chapters is that they are among the most significant attempts to make something of the feedback concept in the social sciences.

In this chapter we shall treat those uses of the loop concept underlying feedback and circular causality that grow out of the early writings in the feedback thread sketched in section 3.2, which I am calling the cybernetics thread. It should be emphasized again that the label refers not to cybernetics itself but to a line of feedback thinking in the social sciences that traces to cybernetics origins.

4.2 Management Science and the Work of Stafford Beer

Stafford Beer is a British operations researcher whose work has been profoundly influenced by the emergence of cybernetics. Almost all of his numerous publications deal with varying aspects of what he calls "management cybernetics." Indeed, so close is his identification with cybernetics that he is one of the few social scientists we shall discuss who could legitimately be called a "cybernetician." The audience he attempted to reach, however, consists of a much more general public: those who study management, those who manage, and to some extent the real general public who are usually the subjects of management. The avowed purpose of his writings is to improve the science of management by the use of cybernetic ideas, including, among others, the idea of feedback.

Management Cybernetics

It was natural for Beer to try to marry cybernetics and management. He saw them potentially paired as theory and applications of the same discipline—later on he neatly characterized cybernetics as the "science of control," and management as the "profession of control" (Beer 1972/1981, p. 17). He saw enormous promise in the concept of cybernetic control in industry and government. The control he envisioned is not repressive authoritarian restrictions backed by management sanctions, but rather the natural self-regulatory character common to biological systems. Beer sought homeostatic mechanisms in industry. The grand goal he envisioned for management cybernetics is the design of industrial systems as what he called cybernetic systems (1959, p. 23).

He first outlined his approach in *Cybernetics and Management* (1959). The book appeared at about the same time as Jay W. Forrester's first publications in industrial dynamics (see section 3.3). Both are concerned with the art and science of management. Both are founded on the feedback concept. They are even linked somewhat through the works they cite in common, notably Tustin's *Mechanisms of Economic Systems* (1953) and Porter's *Introduction to Servomechanisms* (1950). But the gulf between the two is dramatic. Beer's introductory classification of systems (reproduced in Figure 4.1) begins to reveal their differences.

To set out his point of view, Beer loosely divided systems on one dimension into the simple, the complex, and the "exceedingly complex." On another dimension he classified them as either deterministic or probabilistic. Beer asserted that the fifth cell in this scheme, complex probabilistic systems, is the province of his own multidiscipline of operations research. The focus of cybernetics, he said, is the sixth cell: exceedingly complex, probabilistic systems.

Beer placed such structures as brains, national economies, and corporations in the class of exceedingly complex systems. The defining characteristic of such systems is that they "cannot be described in a precise and detailed fashion" (p. 12). They also happen to be, by their nature, probabilistic. The class of exceedingly complex deterministic systems, Beer asserted, is empty. In his view it is pointless to try to approach exceedingly complex systems as if they were either describable or deterministic. The contrast to Forrester's approach to industrial dynamics is vivid. What Beer asserts cannot be known in a corporation, Forrester attempts to describe precisely in deterministic equations.

Beer's approach to the management of exceedingly complex systems involves three components: the use of feedback to create homeostatic mechanisms so that the system can be made self-regulating; the use of statistics and information theory to deal with the inherently probabilistic, uncertain nature of the system; and the use of the concept of the "black box" to deal with the system's extreme complexity (1959, p. 49; see also pp. 28–57). Figure 4.2 reproduces Beer's sketch of the resulting "cybernetic factory." A feedback path for homeostasis appears in the lower left. Probabilistic and statistical considerations are evident, if cryptically, in the "Markovian randomizers" and "conditional probability registers." And there are two black boxes, one for the environment and one for the company itself. The diagram is perhaps hard to

Systems	Simple	Complex	Exceedingly complex
Deterministic	Window catch	Electronic digital computer	Empty
	Billiards	Planetary system	
	Machine-shop lay-out	Automation	
Probabilistic	Penny tossing	Stockholding	The economy
	Jellyfish movements	Conditioned reflexes	The brain
	Statistical quality control	Industrial profitability	The Company

Figure 4.1: Stafford Beer's classification of systems based on degrees of complexity and uncertainty. Source: Beer (1959, p. 18).

grasp, but the influence of Ashby is unmistakable in the general approach and the specific concepts of ultrastability and variety.

Beer's book is clearly a development and extension of Ashby's ideas for a management audience. However, that audience would have trouble applying the ideas and building a cybernetic company. The book is only an outline, as Beer acknowledged, a sketch of what the cybernetic approach ought to be able to accomplish in corporate management.

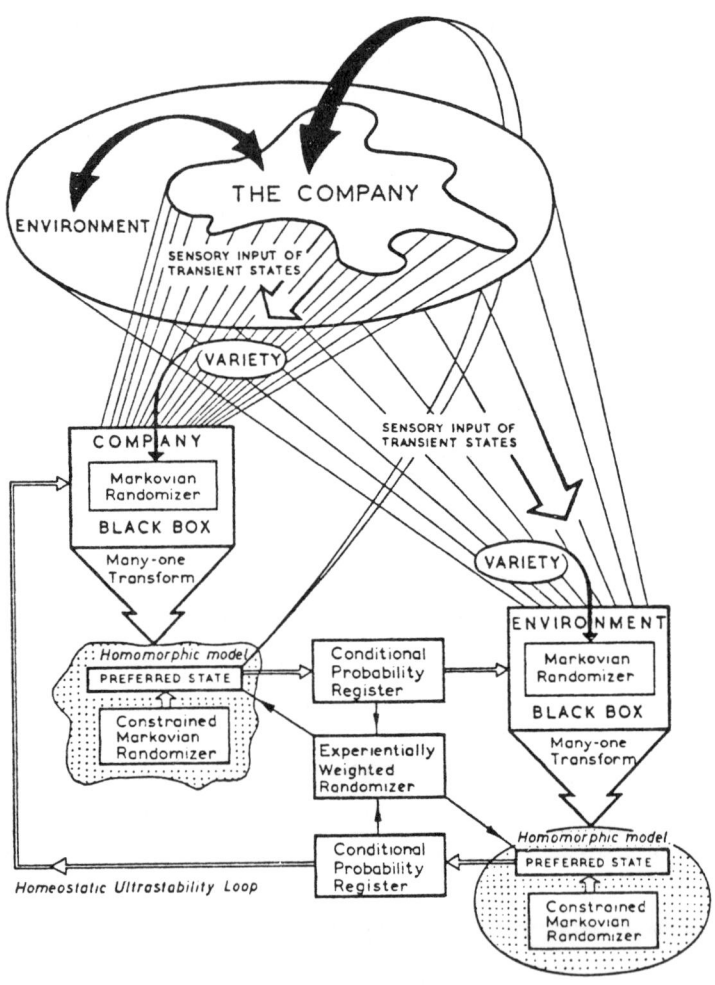

FIGURE 4.2: Stafford Beer's sketch for a "cybernetic factory." Source: Beer (1959, p. 150).

The message of the book is that there is, in the offing, a potentially powerful approach to management that promises to ensure corporate ultrastability—the continued meeting of goals (profitability, growth, survival) in the face of an unpredictable or even hostile environment. The approach involves feedback, as would any capable control system, but feedback is only the foundation. For Beer, the most important idea built on that foundation is Ashby's concept of variety and the Law of Requisite Variety.

Requisite variety appears in the sketch for the cybernetic factory in the "homeostatic ultrastability" feedback loop. The information that circulates in the loops does not have the full mind-boggling variety of the two "part-systems," the real company and its environment. It is information about a selected number of essential variables in each part-system. The device that reduces variety for the cybernetician is the black box:

The set of facts which is the actual setting of each part-system has to be considered in this problem of adaptation. These facts are multitudinous: many of the factors involved cannot even be stated, still less measured. So each part-system is exceedingly complex: indefinable. This variety is therefore handled as a black Box. The input to each part-system black Box is the output of the other. And the output of each is a homomorphism. That is to say, the sub-set of the part-system previously called the "output measure" is a homomorphic model of the part-system itself, expressed in that special set of factors which concerns the other part-system. . . . It is a device for absorbing variety (1959, pp. 143–144).[1]

A black box is an unknowable system or "machine" whose inputs and outputs are all that can be observed. Beer (1959, p. 50) talks about such a system with eight inputs and one output, each of which can take on the values of 0 or 1. There would be 2^8 or 256 patterns of 0s and 1s in the inputs. Since the box is "black," we cannot know the functional relationship between the inputs and the output, so any value (0 or 1) of the output can go with any of these 256 input patterns. This black box can therefore simulate 2^{256} "distinguishable machines." It has enormously high "variety." It can also reduce a lot of complexity (variety) in its inputs to a single judgment, an output of 0 or 1.

It is still not terribly clear what a black box looks like in practice or how to build one into a control scheme (nor is it clear why Beer capitalized "Box" and not "black"). The binary examples given by Beer suggest that the idea harks back to the neural nets of McCulloch and Pitts (1943).[2] However it is formulated, its purpose in Figure 4.2 is to reduce the complexity of the information circulating in the control loop without endangering the mechanism's essential homeostatic qualities. It is also clear that Beer sees limited potential for the feedback concept as a control mechanism without the ideas of requisite variety and the black box:

> The importance of the black Box technique in cybernetics cannot be overrated. An exceedingly complex system has to be controlled: it has high variety, and that variety cannot be ignored. It must be represented, and deployed homeostatically. Often one hears the optimistic demand: "give me a simple control system; one that cannot go wrong." The trouble with such "simple" controls is that they have insufficient variety to cope with the variety in the environment. Thus, so far from not going wrong, they cannot go right. Only variety in the control mechanism can deal successfully with variety in the system controlled. This vital conclusion must be adopted as an axiom in the process of destroying the variety of (that is, controlling) complex systems. The principle has been named by Ashby the law of Requisite Variety (Beer 1959, p. 50).

In addition to reducing variety to manageable proportions in the corporate control loop, the black box in some sense substitutes for the understanding of causes and effects:

> The methods we should use to handle exceedingly complex systems are those of input-manipulation, output-classification; they are not those of "cause-and-effect" analysis (p. 52).

To illustrate Beer gives the examples of a child who shakes and rocks the side of the crib until it falls, and an adult twiddling the dials on a radio (pp. 51–52). Neither, presumably, understands the real cause-and-effect relationships that tie the input manipulations to the output observed, but each may manage to achieve the desired result.

The corporate manager is in a similar situation:

> Consider the currently orthodox methods of industrial control: they rely implicitly on cause-and-effect analysis. Up to a point this is useful: non-standard processes can be examined, unduly high costs highlighted and so forth. But from the point of view of cybernetic control, the whole approach is basically wrong. It deals with a homomorphism of the real situation, one in which cause-effect relationships are assumed to hold. But they do not hold. Consider this example: we are supposed to believe that the cost of a product varies as a discoverable inverse function of the level of output. Any economist will give us the theory; any manager will explain that the "law" does not work. The system is too complex, and its connectivity too "black" to endow the idea of "x is a discoverable function of y" with meaning. When we identify a system as an absolute black Box we are in effect saying that f(x,y) is by definition undiscoverable: an enigma. The arguments of this paragraph are of paramount importance (pp. 52–53).

Thus Beer's claim is that the cybernetic approach can enable management to control a corporation without understanding the exceedingly complex cause-and-effect structure of the company or its environment.

Beer's use of the feedback concept thus follows naturally along the lines of the early writers in the cybernetics thread, summarized at the end of section 3.2. Only negative feedback is considered, and it is seen as the mechanism of homeostasis. Corporate control is the motivation for the use of the concept. Corporate stability, in the form of Ashby's notion of adaptive "ultrastability," is the goal. The treatment is essentially verbal. The mathematics that does appear consists of brief references to probability, combinatorics (for computing variety), information theory, and boolean algebra. Differential equations describing the structure and dynamics of feedback systems are never mentioned. Feedback appears in a single closed control loop connecting a corporation and its environment. It is seen as a property of a management system that must be "designed in" to achieve corporate ultrastability. A system's internal closed-loop structure of information feedback and mutual causality is aggregated into the concept of "complexity" and is not a subject for investigation.

At the heart of the approach Beer outlined in *Cybernetics and Management* is a fundamental philosophical difference from the servomechanisms thread. The difference is not that one takes a probabilistic point of view and the other views things deterministically—Forrester (1961), for example, frequently subjected his models to random influences. The critical difference is that one management thread seeks to use the concept of feedback to control a complex system without understanding, while the other tries to use the concept of feedback to understand complexity. What Beer says people cannot know, Forrester tries to describe precisely in quantitative terms. While both would say they focus on information feedback, one completely ignores the system's internal information feedback structure that is the primary focus of the other. And in the other direction, the concept of variety that one sees as the key to feedback control in complex systems is nowhere evident in the work of the other. The difference is at the level of worldview—what a person considers knowable and essential to know. It is conceivable that a thinker centered in either of these management threads would view the other's attempts to be doomed from the start.

Beer's Later Work

The foregoing summary and conclusions sound right in view of the evidence in *Cybernetics and Management,* but they do not appear to hold up over time. Beer continued to develop and write about his cybernetic approach to management. His most complete statement appeared in his discursive text, *Decision and Control* (1966). His most personal, even passionate statement of it was expressed in *Platform for Change* (1975). And in the second edition of *Brain of the Firm* (1981), we get the most complete picture of what the approach looks like in practice, for in it Beer discussed his great experiment applying management cybernetics to the operation of the government in Chile. In all three books Beer appears to have altered somewhat his view of the use of causal feedback models with explicit internal structure.

Decision and Control communicates a mixed message about causal models. In the extensive sections on "Controlling Operations" and "Controlling Enterprises," Beer reiterates and significantly amplifies the ideas described above, emphasizing black boxes and the inability to understand complexity. But in discussing "The Outcome for Industry" he presents diagrams of explicit corporate policy structure and praises the potential of "modelling the total system."

First, an overview of the chapters on control: Beer investigated at length the problem of controlling production in a plant consisting of many machines and work orders. The problem as he sketched it was to forecast accurate job times and "quote delivery promises that could be met" (p. 338). He critiqued the standard operations research solutions to the problem because of the enormous detail required in his particular setting of the problem, the necessary simplifications that degrade the forecasts, the difficulty of adjusting OR mathematical models to changing conditions, and the inherent weakness of the "ad hoc control" mechanisms that would be invoked to learn from experience and improve the forecasting process over time (pp. 305–311). Orthodox techniques fail to deal adequately with the proliferation of variety in such a system.

He then sketched a "cybernetic model" for the solution of the problem. In several stages, which we will have to shortcut, he worked up to the diagram shown in Figure 4.3. Here the W_i are the "world situation" in the production system at successive points in time. The BB_i are black boxes. The interesting part of the diagram for us is the lower left, where M1a and M1b represent explicit models (white boxes!) of the real system. Beer described them as follows:

M1a is a structural model of the world situation. Just because a full-scale analytical model of so complex a system eludes the OR man, there is no reason to ignore altogether the more evident constraints of the system.... This kind of structural modeling of events and their basic relationships ... is the standard modelling process of operational research. As usual, most of the model M1a can be written down in terms of mathematics or statistics. But other aspects of the structure of the world situation are more difficult to handle, and call for the more sophisticated language which is specially adaptable to the statement of relationships—logic. ...

Next it is necessary to deal with the problem of quantifying the structural model. To do this, numerical data will be required: actual quantities that are descriptive of the world situation. This is attempted in the parametric model M1b (pp. 313–316).

The data he chose to put in the parametric model M1b were optimal performance criteria (p. 317).

The details of Beer's formulations are too complex to go into in detail.[3] The point in this context is that Beer sees some role for white box models in production control. It is a severely circumscribed role, however. Indeed he emphasizes that the models M1a and M1b can be poor, and cybernetic control system will learn to compensate properly over time (pp. 329–330). Further, white box models could never do the job alone:

> We have always recognized the possibility of producing analytical models of gross physiology (as in M1a) and of quantified patterns (as in M1b). But as cyberneticians, we have also found it necessary to say that an entirely analytical account of this kind, however exhaustively conducted, will not give the complete description of the world situation that its controller would like to have. This is not say merely that the task is too great; it is to say that the task is theoretically impossible, and that a means must be found for proceeding rather differently from this point (p. 320).

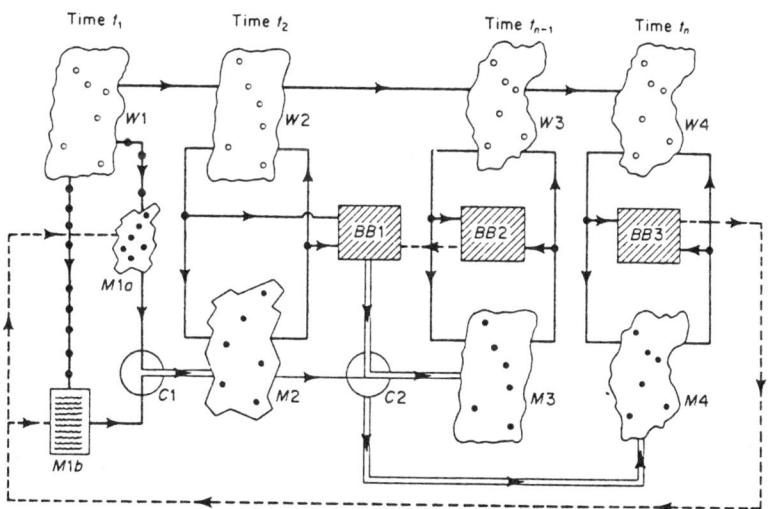

FIGURE 4.3: Beer's schematic diagram of the design and operation of a cybernetic control system. Source: Beer (1966, p. 328).

He cited the inability of optimization models to maximize more than one criterion at the same time. (Thus we learn what sort of models he had in mind.) At a higher level of profundity and difficulty he cited the necessity of what might be termed the ability of the model to be able to talk about itself, to "rewrite" its own structure. Building on Russell's theory of logical types and Gödel's incompleteness theorem, Beer asserted:

> We know that the language of each higher order control system must be a language of higher order than that of the world situation. Otherwise, the language will be "incomplete"; that is to say it is possible for things to happen of which one cannot adequately speak, and about which one certainly cannot decide. The model at M2 is expressed in the language of the world situation, albeit that use has been made of a good deal of formal science. But this language, like the control loop, is incomplete; and the language used at the closure must be one in which it is possible to assert that the descriptive language of M2 simply will not do (p. 321).

Beer's solution to the incompleteness dilemma he perceived is embodied in the "homeostatic connections" introduced in Figure 4.3 between M2 and W2, and especially the black box BB1 that monitors their interaction and supplies information to the system.

> The use of a black box of inherently high variety permits the entrance to the modelling line of unanalysable high-variety components from the behaviour of the world situation line. In doing so, it proposes the closure of the metasystem, and offers a new mode of description in a meta-language (p. 321).

It is not clear whether this theorizing about meta-language is meant to be rigorous or merely associative and analogical. What is clear, however, is that Beer sees at most a minor role for white box models of known structure and parameters in the cybernetics approach to control of operations. Furthermore, the white box models he incorporates are apparently optimization models, not models linking feedback structure and dynamic behavior as in the servomechanisms thread.

Beer's sketch for the cybernetic control of whole enterprises (a firm, a quasi-monopolistic industry, a national health service) follows similarly, but apparently reverts back to no role for white box models. The discussion concludes with the diagram reproduced as Figure 4.4. There are some new words: "coenetic," adopted from Sommerhoff (1950) and meaning "common causal influence," and the "algedonic loop," a term coined by Beer to represent the feedback of reinforcement (pain, pleasure, reward). On the whole, however, the diagram has familiar rings, and we will not discuss Beer's development of it.

The "Outcome for Industry" gives a very different impression, however. Here he advocates building "ecosystem" models for a firm and its environment, and his illustrations contain explicit representations of feedback structure. He describes a sequence of models of a company and its market, beginning with a simple, single feedback loop linking the two sectors as black boxes and moving through several progressively more disaggregate renderings. Figures 4.5 and 4.6 show the first

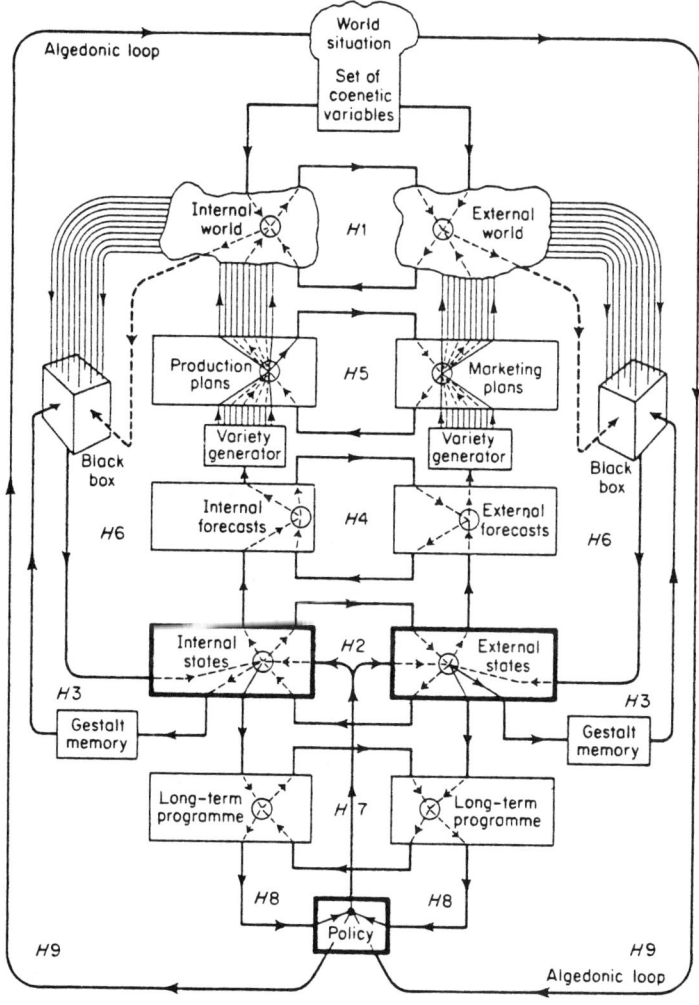

FIGURE 4.4: The cybernetic enterprise. Source: Beer (1966, p. 392).

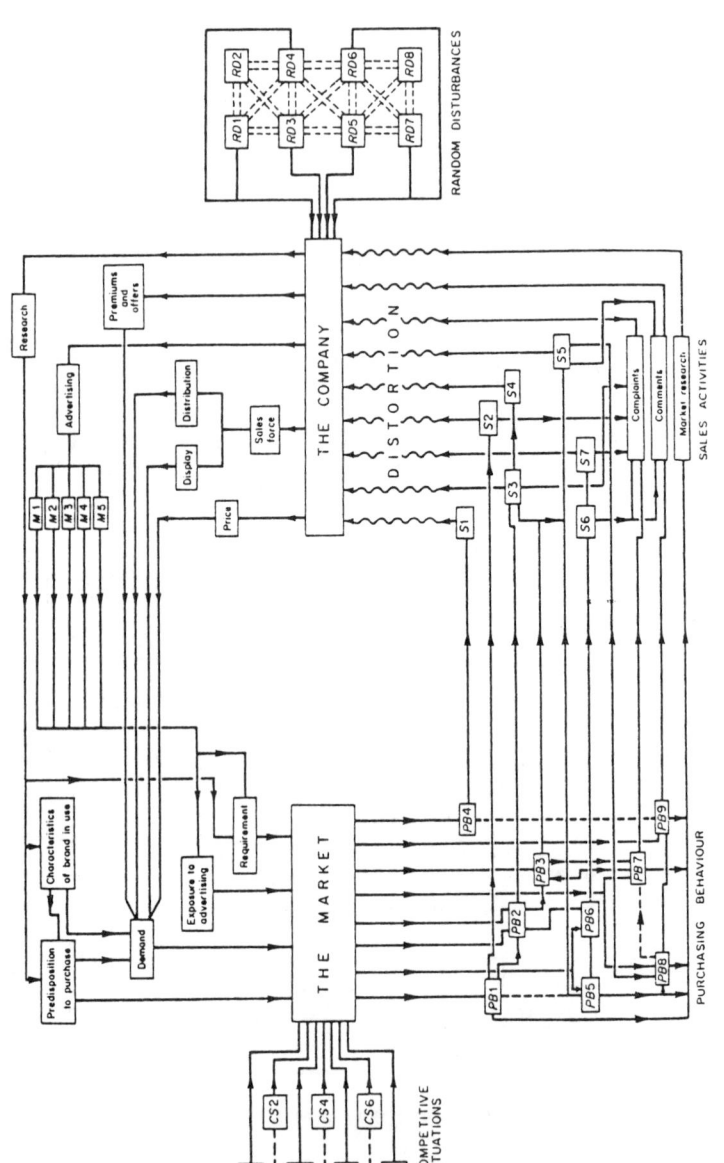

FIGURE 4.5: Disaggregated model of a company and its market. Source: Beer (1966, p. 410).

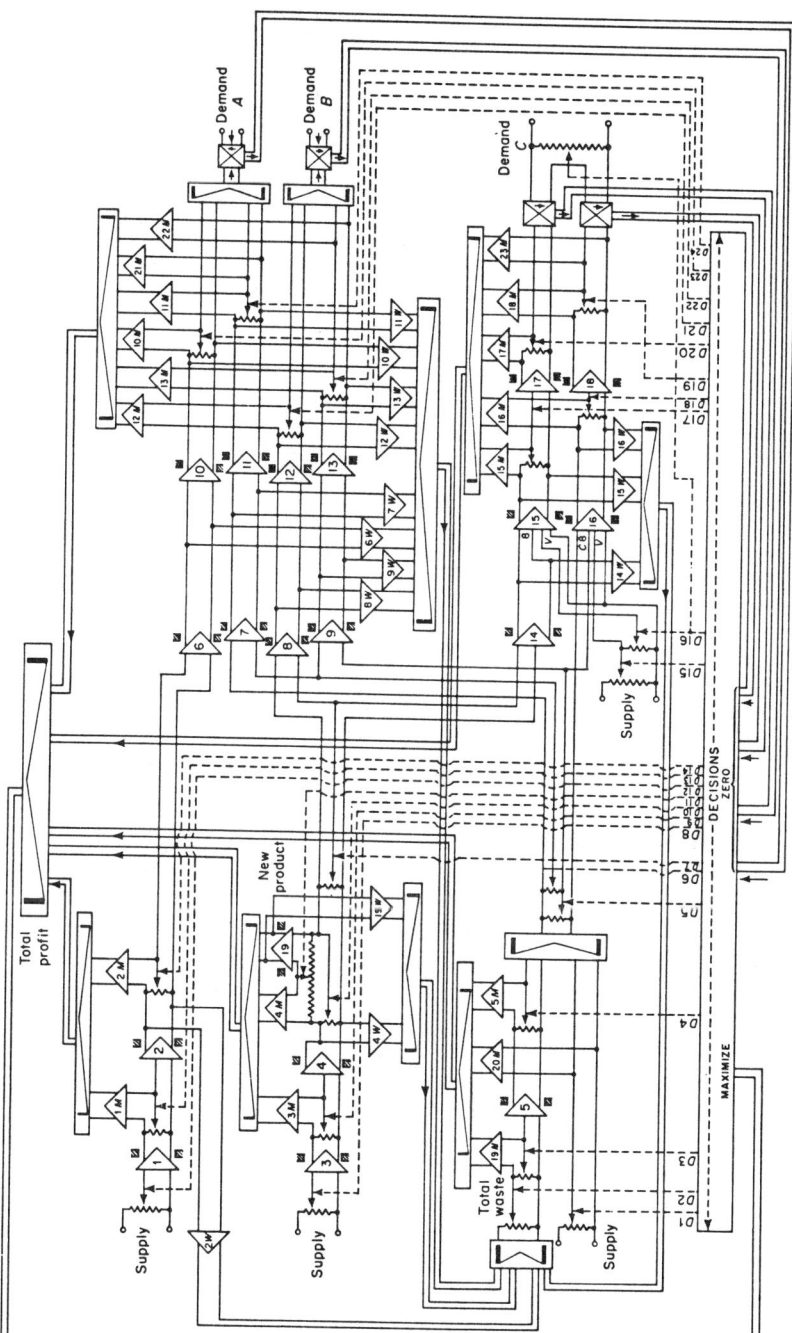

FIGURE 4.6: Further disaggregation of the model shown in Figure 4.5, symbolized for analog simulation. Source: Beer (1966, p. 419).

disaggregation of the single loop model and the final, most detailed structure.

Of Figure 4.5 Beer says:

> The very high variety which the market is capable of generating emerges in the bottom left-hand corner as a mass of sub-systems representing kinds of purchasing behavior.... The torrent of high variety information which eventually reaches the company through its sales returns (bottom right-hand corner) is rather seriously distorted.... But fortunately this model does not call for any causal analysis of the sort normally undertaken.... In the top right-hand corner of the diagram is the organization of the company's affairs over which the management has direct control. It too is pumping high variety onwards: promotion, towards the market.... In the top left-hand corner of the diagram appears an attempt to unravel the mechanism by which the market is influenced....
>
> Figure [4.5], then, is a diagram of a generalized model from cybernetics of the company considered as a sentient ecosystem (p. 411).

He then stressed that the company and the market were still being captured as black boxes. The emphasis is on the interactions, because "it is the interactions which embody homeostatic equilibrium."

And then he seemed to step away from the entire black box tradition he has advocated:

> But of course, in any given study, it may be very necessary to resolve the structure of these two black boxes, either more or less, according to further models as required (p. 411–412).

He proceeded to describe a series of models that "resolve" the structure into progressively greater detail. The next resolution (which I have not shown) broke the system down into a supply environment, a demand environment, and a representation of the company's internal activity, including finance, labor, and the policy focus of the study—production needs and investment in capital plant. The model constructed at this point derived, he said, from general systems theory. From it "quite a lot could already be said about the stability of the system, and about the vulnerability of the organism." The black box idea is still prevalent:

> Note that anything inferred at this point is in fact inferred from the ecological relationships, and not from information about the content of all the little boxes. Although many more of them are shown, so that more structure is resolved, the individual boxes are still black (p. 417).

Perhaps the proper interpretation of this is as a big "white box" containing little black boxes connected together by information feedback paths.

Then a still further resolution became possible, and the model pictured in Figure 4.6 was developed. It was based on "the science of servomechanics."

From this could be computed how one locality of the system would stabilize itself when perturbed by inputs from another locality. In fact, any large-scale mixed system that is a going concern has features, call them intrinsic governors, which tend to make the total system self-regulating. If managerial systems were not basically self-regulating, it would never be possible to manage them. They would generate too much variety for us to cope with (p. 418).

Presumably, there would be little difference in principle between this model and one of Forrester's digital simulation models of corporate policy structure and dynamics.

What happened? Did Beer's thinking evolve over time to a point of view closer to that described for the servomechanisms thread? Or was he always of this frame of mind? An answer to these questions can be approached by recalling that there are two characteristics that define the conceptual threads in this study, the methodological dimension and the sociological dimension (section 1.3). The latter refers to the communication network of scholars, which is at least partially expressed in who and what an author cites as references. We have already explored at considerable length the methodological dimension and found evidence for a shift in thinking over time. An investigation of Beer's citations reveals that he had one foot, or a toe at least, in the servomechanisms thread from the beginning.

The bibliography in Beer (1959) contains four references that would be classified in the servomechanisms thread, among twenty in the cybernetics thread. The authors are shown in Table 4.1. Porter was the author of an introduction to servomechanisms (Porter 1950). Nicholas Rashevsky was a mathematical biologist in the tradition of Lotka who had had a significant influence on Herbert Simon (see Simon 1954). For the others: Beer cited the application of servomechanisms to production control in Simon's *Models of Man* and Tustin's *Mechanism of Economic Systems*, both of which we have discussed in section 3.3. Most of the authors listed in the citations from the cybernetics thread are also familiar; those that we have not mentioned before (Bowman, George, Pask, Quastler, and Walter) wrote about neural networks or were connected directly to the cybernetics conferences. In addition, Beer cited Alan Turing and Stephen Kleene, whose work in logic, computability, and finite automata connects with a central interest in the field of cybernetics itself.

Beer (1959) had indeed discussed the work of Tustin and Simon. He used Tustin's feedback translation of Keynes to introduce the concept

TABLE 4.1 Authors cited in Beer (1959) classified by feedback thread

Cybernetics thread	Servomechanisms thread
Ashby	Porter
von Bertalanffy	Rashevsky
Bowman	Simon
von Foerster	Tustin
George	
McCulloch and Pitts	
von Neumann	
Pask	
Quastler	
Shannon and Weaver	
Walter	
Wiener	
(Kleene)	
(Turing)	

of feedback in the economic context (pp. 33–35). Simon's work on production control appeared in a section entitled "The Analog of Mechanism" toward the end of Beer's book (pp. 170–176). There Beer outlined what Simon had done, praising it as "perhaps the deepest exploration of the possibilities" of applying "the full mathematical apparatus of servo-theory" to the problem of production control (1959, p. 170).

Thus Beer is loosely connected sociologically to the servomechanisms thread in the evolution of the feedback concept in the social sciences. However, the methodological emphasis in his early work, including the first 400 pages of *Decision and Control*, unquestionably places him in the cybernetics thread. At the same time he lauded Simon's work, he concluded that there were dangers "from too facile an adoption of servo-models for the analogue of mechanism in cybernetics" (1959, p. 169). He had two major criticisms: (1) servo-models are "closed systems" while cybernetic systems must be "open systems" (pp. 168–169); and (2) a servo-control system cannot "discuss itself" and adjust to new conditions imposed by its environment if the need arises (pp. 169–170). I find both claims to be incorrect,[4] but the deeper issue is this: Readers of *Cybernetics and Management* would not be moved to build dynamic models of explicit causal feedback structure. They would be oriented toward the use of the concepts of homeostasis, variety, the black box, and ultrastability for achieving control in management situations. For these reasons, Beer is centrally in the cybernetics thread. Because of his wide knowledge of scholarly literature—and his citations show that he is amazingly well read—he appears to have kept somewhat abreast of the other feedback thread in the social sciences. And over time he appears to expand his methodological emphasis to include more elements of the servomechanisms thread.

Platform for Change (Beer 1975) provides some corroboration for that claim. In a speech on "Management in Cybernetic Terms" prepared for UNESCO and published in 1972, Beer makes an explicit link to the servomechanisms thread in the following comment:

Now although we may think of a large organization, such as a firm, as having a single output (labelled profitability for example), and although we may imagine a large feedback loop which would adjust the inputs to the system so as to keep the output at a desired level, this is no more than a conceptualization of how management works. It is inconceivable that the complexities of a company's operations could actually be stuffed into a single mechanism of this kind. But we may start by this method to account for the way the business works. Then we ask questions about the way the system breaks up—into divisions of the company and divisions of the market, for instance, and model those. Thus we shall gradually be able to devise a more complicated model, redolent with feedback loops, which is of practical value. In doing all this we pass from the notion of a straightforward feedback mechanisms to the notion of multiple loop systems. Outstanding in this area of study is the work of J.W. Forrester (1961, 1969) (Beer 1975, p. 107).

Both the content of this paragraph and the reference to Forrester's work signal Beer's awareness and acceptance of the approach of the servomechanisms thread. One might also suggest further evidence of a shift in the direction of the servomechanisms thread in that Platform contains the first graphs of variables over time that I know of in Beer's books.

If we pursue the matter to the second edition of *Brain of the Firm* (Beer 1981), we learn that Beer had blended the two feedback threads in his own work as early as 1971–1973. "Project Cybersyn," his bold attempt to apply management cybernetics to the complete reorganization and governmental management of a national economy, included as one of its parts an explicit, causal feedback model of the Chilean economy, called "Checo." Checo was written in DYNAMO, the simulation language developed and used in Forrester's system dynamics methodology. It was to be one of four components of Project Cybersyn; see Table 4.2.

Cybersyn (an abbreviation of "cybernetics" and "synergy") was intended to carry out the Marxist program of Allende and place the

TABLE 4.2 Components of "Project Cybersyn" for the reorganization and management of the economy of Chile, 1971–73 (summarized from Beer 1981, pp. 245–256, 261–270)

"Cyberstride"	A set of statistical filters assembled in a computer program. Its purpose was to monitor tens of thousands of indices at all levels in the economy, alert firms and sectors to "incipient change" before it happens, and to provide an "arousal filter" for governmental corrective action.
"Cybernet"	An electronic communications network designed to pass information rapidly among major firms and sectors and the government.
"Checo"	A DYNAMO simulation model of the Chilean economy, intended eventually to be a hybrid digital/analog model.
"Operations Room"	The central governmental office which coordinates the whole, gathering and processing the information provided by Cyberstride, Cybernet, and Checo, and deciding on corrective action.

economy in the hands of the workers of Chile. Beer also intended it to be a showpiece of cybernetics management, incorporating real-time monitoring, sophisticated data processing, computer simulation, and rapid anticipatory action to keep the economy stable and on plan. Both Allende's government and Beer aimed to organize the economy so that down at the level of the workers it was as self-correcting as it could be. Beer's image for the structure was the hierarchical information processing and control structure of the human organisms shown in Table 4.3—hence the title, *Brain of the Firm*.

In Beer's view every viable system is organized in such a hierarchy of five systems. Furthermore, complex viable systems are composed of viable subsystems that again have this five-level structure. To be viable, a division of a firm should be organized this way. The entire firm as a set of interrelated divisions should also have the same sort of structure. A viable sector of an economy should have the same structure. Naturally, an entire national economy, to be viable, should be similarly organized. Viewed from the elevated level of the national economy in the hierarchy in Table 4.3, an individual firm or sector operates at the level of systems one, two, and three. The functions of its own high-level systems four and five are seen as part of the autonomic processes of the more aggregate system.

TABLE 4.3 Beer's hierarchy of components of a viable system, modeled after the human nervous system (summarized from Beer 1981)

System level	Physical analog	Function
System One	organ, muscle	action
System Two	spinal cord	signal processing and transmission
System Three	pons/medulla	autonomic control
System Four	diencephalon, ganglia	monitoring the system and its environment; sensing; arousing systems three and five
System Five	cortex	policy formation

Checo, the DYNAMO model of the Chilean economy, was intended to contribute to system four in the management of the entire economic system. In Beer's words:

> We need a simulation model of the industrial sectors and their interaction embedded in an environment that takes account of investment capability in terms of both foreign exchange and domestic savings. The emphasis will be on structure (which can be changed—by inserting new feedback loops for example) and on the dynamic interplay of the factors modelled (which produces "multiplier" effects of crucial importance). This model would be used to mediate between the detail of current performance (arising from the 3-2-1 monitoring system) and the current structural situation (as reflected in the Operational Plan) on the one hand, and the formulation of strategy on the other. System Four, in short, is a mediator between Systems Three and Five. With a simulator of this kind, we can investigate the nature of the trapped states in which the economy is currently enmeshed, and which appear to be functions of a metasystem that extends beyond the national boundaries. . . . Cybernetic considerations certainly suggest that new structure, involving new information pathways and the harnessing of motivational factors, will be needed to achieve Chile's radical political goals. The simulator will be the government's experimental laboratory (p. 266).

Beer wanted to have a running model in six months. The only way to accomplish that was "to make use of the immediately available DYNAMO compiler that has been extensively developed over many years by J.W. Forrester of MIT" (p. 266). Thus we appear to have in Beer's work a linking between the two feedback threads, both methodologically and sociologically (in citations).

The modeling effort in Cybersyn was not very successful, however. The time frame for model development was woefully short, and the team Beer had assembled to do it saw itself as "learning the trade." No one was ready to put much faith in the simulation results (p. 267). But more telling to Beer was the fact that "the data were untrustworthy":

As is usual with national figures, they were out of date; and, also as usual, they were differentially lagged. Too little attention has ever been paid to these, and associated, dangers in the origins and development of DYNAMO simulations that have achieved world-wide attention, and continue to do so. Obviously, my own plan was not to rely on such "national statistics" any more. I wanted to inject information in real time into the Checo programs via Cyberstride. Thus any model of the economy, whether macro or micro, would find its base, and make its basic predictions, in terms of aggregations of low-level data—as has often been done. But Checo would be updated every day by the output from Systems 1-2-3 and would promptly re-run a ten-year simulation; and this has never been done. This was one of my fundamental solutions to the creation of an effective Three-Four homeostat; it remains so, but it remains a dream unfulfilled (pp. 267–268).

It is clear that something of the two feedback threads comes together in Beer's work. Explicit causal feedback models are apparently added to the cybernetics concepts Beer and Ashby had emphasized since the 1950s. To the traditional focus on negative feedback loops as the mechanism of homeostasis and control is added a recognition of the natural reality and significance of positive feedback loops in corporate and economic systems (see, e.g., Beer 1981, pp. 235, 285). There are still some differences, carrying potentially significant messages—tendencies to focus on control, to look outside a system for the sources of its problems (an exogenous point of view enlightened by feedback), to emphasize theoretical and philosophical questions relating to control, and to couch dynamic analyses in words rather than differential equations (Beer still never reveals a formal model).

There remains the possibility that Beer's use of the feedback concept encompassed tendencies in both feedback threads from the beginning. It is hard to reconcile his claims about complexity and causal modeling in *Cybernetics and Management* with such a conclusion, but it might be argued that Beer had simply overemphasized what was to him the exciting new cybernetics insights at the expense of ideas that received more attention in the servomechanisms thread.

In the end, what really matters here is the set of messages communicated to others in the social sciences by these writings. Beer's loudest messages advance the ideas placed in motion by the early writers in the cybernetics thread. His readers are likely to be stimulated to try to advance those ideas still further, and that is the point of pursuing distinctions between feedback threads in the social sciences. It is fitting to leave the last word on the question of where Beer fits to Beer himself:

There is, we have often insisted, no "right" or "correct" model of anything at all; there are only more or less appropriate models for particular purposes. If, however, we are to approach Plato's desideratum[5] of discussing not some one particular aspect of an organized whole, but the whole of it at once, we need a model able to encompass not only the firm but all its interactions. The methodology to be followed has been expounded, and its critical point ought to be reiterated. It is this. Since we cannot account for the entire detail of a system as big as a firm, the model must be a many-one homomorphic reduction. Requisite variety can be obtained in such a model, by including in it data generators. But any conclusions drawn for managerial purposes from this model must be supported by a level of variety commensurate with their own variety. This in general can be done, either by resolving in more detail those parts of the system which relate to the decision, or by generating variety stochastically in a simulation (1966, pp. 406–407).

I interpret this paragraph to say, of course we can try to do what those in the servomechanisms thread do, but requisite variety will always be a thorny managerial problem.

4.3 The TOTE Unit

Three renowned psychologists, George Miller, Eugene Galanter, and Karl Pribram, spent the academic year 1958–1959 at the Center for Advanced Study in the Behavioral Sciences, during which time they conceived and wrote *Plans and the Structure of Behavior* (1960). Their fundamental concern in the book was to discover whether cybernetic ideas have any relevance for psychology (1960, p. 3). Looking back over the whole sweep of psychological theory, they were disturbed by a "theoretical vacuum between cognition and action" (p. 11), between the "images" we have and the "plans" we make and execute. They attempted to fill this theoretical gap with a version of the feedback loop they called a "TOTE unit." Their book is nothing short of an attempt to place the feedback loop at the foundation of psychology.

Background

Loop-like notions had existed in psychological theory for a long time. We have previously observed them briefly in proprioception, kinesthesia, drive-reduction and tension-reduction theories of motivation. Skinnerian stimulus-response-reinforcement, and so on (see section 2.5). Miller, Galanter, and Pribram (1960, pp. 43–44) identified several that we have not discussed. They noted the concept of "retroflex" introduced by L.T. Troland (1928) to name sensory feedback, E.B. Holt's revival of "circular reflexes" (Holt 1931), and Edward Tolman's "schematic sowbug" (Tolman 1939), which the authors termed "a perfectly respectable feedback mechanism." Particularly important to Miller, Galanter, and Pribram was Dewey's critique of the reflex arc, which contained insights that the authors found still powerful even after six intervening decades of theory-building and research (Dewey 1896; see section 2.5).

Miller, Galanter, and Pribram had been schooled in twentieth-century behaviorism. They were attentive and sensitive, however, to criticism of the stimulus-response-reinforcement paradigm of behavioral psychology, from Dewey's early article to Chomsky's soon-to-be-classic review of Skinner's *Verbal Behavior* (Chomsky 1959). The central theoretical problem with behaviorism has always been the difficulty of building up complex cognitive capabilities and behavior from simple stimulus-response-reinforcement units. Miller, Galanter, and Pribram sought an alternative building block. They found it in the form of a blend of two ideas: the feedback loop, as it appeared in the work of Wiener, Ashby, and the other cybernetics writers, and the recursive DO-loop from digital computer programming, as it appeared in the then recent work on problem solving carried out by Allen Newell, J.C. Shaw, and Herbert Simon.

As we have noted (section 3.3), a central theme running through the work of Herbert Simon has been how people, individually and collectively, solve problems. One line of this work, dealing with problem solving in organizations and represented by March and Simon (1958), flows naturally out of Simon's work described in section 3.3 relating to feedback and mutual causality. In another line, investigating the processes by which individuals solve problems, feedback as "circular causality" is not present. The important organizing concept is instead the individual as an "information processor" (Newell, Shaw, and Simon 1958). The model for "man as problem solver" is a computer possessing a number of memories, a number of primitive information processes, and a definite set of rules for combining processes into whole programs of processing. Analysis is focused discretely on individual mental moves, like moves in a game or steps in a mathematical proof. The loops that appear in the theory are the recursive loops that appear in computer algorithms.

A direct ancestor of the TOTE unit, the Logic Theorist (Newell and Simon 1956) is a computer program designed to simulate the behavior of a person trying to find proofs in symbolic logic. It was written in IPL, a LISP-like computer language capable of handling non-numerical symbols, words, and lists. In the Logic Theorist there are numerous loops of the form

generate theorems:
test if theorem of the form A implies B,
if false continue generation;
if succeed stop and report proof

(see Newell and Simon 1972, p. 115). Miller, Galanter, and Pribram had access to much of the published and unpublished work of Newell, Shaw, and Simon up to 1959. In their development of the TOTE unit, they based much of their thinking on it.

The TOTE Unit

TOTE stands for Test-Operate-Test-Exit. In a TOTE unit, one Tests to see if a goal is met, Operates to approach the goal, Tests again, and Exits from the loop when the goal is reached. The authors' motivating description of the TOTE unit placed it in the context of a reflex, in which thresholds of neural stimulation must be exceeded all along the neural path to trigger action. They concluded:

The general pattern of reflex action, therefore, is to test the input energies against some criteria established in the organism, to respond if the result of the test is to show an incongruity, and to continue to respond until the incongruity vanishes, at which time the reflex is terminated. Thus, there is "feedback" from the result of the action to the testing phase, and we are confronted by a recursive loop. The simplest kind of diagram to represent this conception of reflex action—an alternative to the classical reflex arc—would have to look something like Figure [4.7].

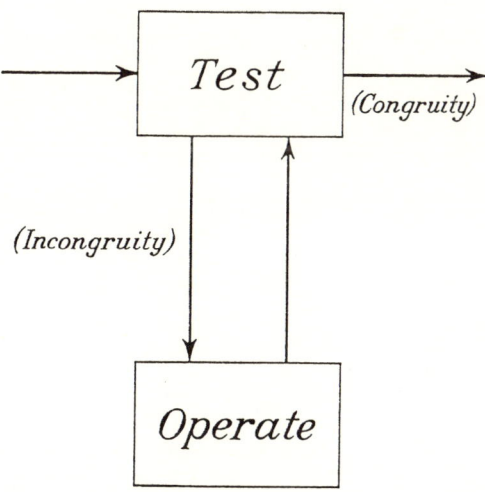

FIGURE 4.7: The TOTE unit. Source: Miller, Galanter, and Pribram (1960, p. 26).

The interpretation toward which the argument moves is one that has been called the "cybernetic hypothesis," namely that the fundamental building block of the nervous system is the feedback loop (Miller, Galanter, and Pribram 1960, pp. 26–27).

Miller, Galanter, and Pribram's concern was not particularly with neural activity or reflexes, however. Their interest was much more general:

> The reflex is not the unit we should use as the element of behavior: the unit should be the feedback loop itself. . . . The reflex should be recognized as only one of many possible actualizations of a TOTE pattern. The next task is to generalize the TOTE unit so that it will be useful in a majority—hopefully, in all—of the behavioral descriptions we shall have to make (p. 27).

They hoped to make the feedback loop the basic unit for all psychological theory. Following Dewey (1896), they concluded that the concepts of the reflex arc, stimulus, response, and reinforcement all had to be reinterpreted as aspects of psychological feedback loops (p. 30).

The authors saw several possible meanings for the arrows in a TOTE unit: flows of energy, the transmission of information, or the flow of "an intangible something called control" (p. 28). The latter could be thought of simply as "succession," the sort of passage of control that takes place in a computer program when the control of the machine's operation passes from one statement to the next. Energy is the appropriate interpretation at the level of neural activity. In complex psychological processes, it made the most sense to the authors to interpret the arrows as transfers of information or control.

A TOTE is a fundamental organizing or coordinating unit. Miller, Galanter, and Pribram saw it as the building block for all Plans linking Images to Action. The great advantage of their conception of the TOTE unit is that it builds swiftly into hierarchies of TOTEs that can represent plans of any complexity. Any operation (O) in a TOTE can be expanded and replaced by an entire TOTE of its own. The authors' classic example, shown in Figure 4.8, was hammering a nail.

The basic idea in hammering a nail is $T_1 O_1 T_1 E_1$, where

T_1 represents "Is the head of the nail flush?"
O_1 represents "Hammer!"
and E_1 represents "Exit from the 'Hammer' TOTE."

The plan calls for repeating the sequence T_1, O_1 until the test shows the job is done. The operation O_1 of striking with the hammer could be broken down, as it is in Figure 4.8, to show the test and operation required for the hammer to be up and ready to strike. The more complete plan for hammering becomes $T_1(T_2O_2T_2E_2)O_1T_1E_1$, where

T_2 represents "Is hammer up?"
O_2 represents "Lift hammer."
and E_2 represents "Exit from the 'Lift Hammer' TOTE."

The authors observed that this example, although exceedingly simple, suggests clearly how "circles within circles" can yield a hierarchical tree, and how such a tree of feedback loops, though not sequential in itself, can organize and coordinate a sequence of behaviors (pp. 34–37).

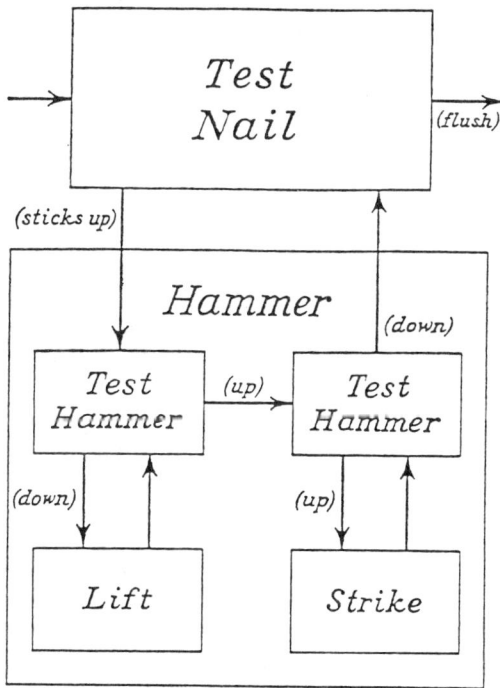

FIGURE 4.8: The hierarchical Plan for hammering a nail. Source: Miller, Galanter, and Pribram (1960, p. 36).

Simulation

Miller, Galanter, and Pribram raised the natural question of how to test a theory based on a TOTE hierarchy—indeed, how to approach tests of the TOTE framework itself. In their discussion of problem solving, for example, they exposed the enormously complex maze of all the possible heuristics people might use to guide their thinking and behavior. Any problem-solving theory must be very complex to come close to matching reality. And worse, such a theory must deal with internal, cognitive behavior that is all but unobservable. How can such a theory be tested?

They suggested that the solution lay in computer simulation:

> No benign or parsimonious deity has issued us an insurance policy against complexity. However, there is no need to become discouraged on that account, for within the past decade or so electronic engineers have begun to develop computing machines that are big enough and fast enough to serve as models for testing complicated theories. Describe the theory carefully, translate the description into a computer program, run the program on a computer, and see if it reacts the same way organisms do. Now that we know how to write such programs—especially since the work of Newell, Shaw, and Simon—we can begin to test ideas that would probably have seemed impossibly complicated to an earlier generation of psychologists (p. 182).

The test for such complex theories is the behavior they generate:

> If the heuristic devices that thinkers say they are using can be translated into programs that reproduce the results obtained by the person, then we have every reason to believe that the heuristic device was a true description of his procedure in solving the problem. If the heuristic method is ambiguous, the program simply will not work. With this test available, therefore, heuristic rules of thumb can indeed be proposed as elements of a serious theory of thinking (p. 183).

At least part of the solution, then, to the problem of evaluating such complex feedback theories of psychological behavior is "re-enactment," that is, simulation.

Thus, like Forrester, Beer, and others, the authors linked their feedback view to computer simulation. They noted that the concept and symbolism of a TOTE enables hierarchical structures underlying behavior to be described in the computer language developed by Newell, Shaw, and Simon. Theories of human behavior expressed in TOTEs are amenable to computer simulation. That new possibility was exciting to these authors, who saw in it "a new license to conjecture." The speculations of "mentalists" could at last be made somewhat testable and empirically verifiable (p. 56). There was now the possibility of a scientifically respectable "subjective behaviorism" founded on the simulation of TOTE hierarchies (pp. 211–213).

It should be clear, however, that the computer simulations of feedback processes that Miller, Galanter, and Pribram had in mind are significantly different from those initiated by Forrester and used by Beer (see sections 3.3 and 4.2). The simulations of Forrester and Beer are numerical. They trace over time the dynamics of quantities. Feedback structure is described in terms of quantities and influences that cause those quantities to increase or decrease. The mathematics involved is essentially that of differential (or integral) equations. In contrast, the simulations of TOTE hierarchies that Miller, Galanter, and Pribram envisioned involve symbols, not quantities. The elements in a TOTE loop represent discrete events or operations, not numbers. The mathematics that is most akin to TOTE simulation programs is symbolic logic, not differential equations. The output of such simulations would be symbolic sequences representing specific behaviors, like moves in a game, utterances in a conversation, or physical actions.

There are also differences in how these two types of simulation programs "control themselves." In the simulation of a TOTE hierarchy, the program would be "under the control of" only one statement at a time. Only one loop would be active at any given time (or, if you prefer, only one set of nested loops in the hierarchy). In a system of differential equations representing a complex mutual causal/information feedback system, many loops (conceivably, all of them) are simultaneously active.

Analog and Digital Feedback

In addition there are differences in the way Miller, Galanter, and Pribram use the word "feedback." In their formulation it is closer to "knowledge of results." As they noted, the arrows in TOTE loop can represent "succession"—in tracing around the loop one can read the arrows as "and then . . ." The arrows in the feedback diagrams we have previously discussed cannot be read that way. They represent "influence," and they have a polarity, positive or negative, associated with them. The arrows in a TOTE unit have no such polarity. A TOTE loop itself seems like a negative feedback loop because it represents a controlling process: the action in the loop repeats until the Test shows that some "incongruity" is eliminated. It is difficult to translate into TOTEs the circular causal feedback loops of Mill or Myrdal.

At the heart of the differences between TOTEs and the feedback loops in the servomechanisms thread is the discreteness of a TOTE unit.

Miller, Galanter, and Pribram discussed the issue in terms of "analogue" and "digital" processes (pp. 90–93). The context for their discussion was the learning of motor skills and habits, such as learning to fly an airplane:

> The input to an aviator, for example, is usually of a continuously varying sort, and the response he is supposed to make is often proportional to the magnitude of the input. It would seem that the good flier must function as an analogue device, a servomechanism. The beginner cannot do so, of course, because his Plans are formulated verbally, symbolically, digitally, and he has not yet learned how to translate these into the continuous, proportionate movements he is required to make. Once the subplan is mastered and turned over to his muscles, however, it can operate as if it were a subprogram in an analogue computer. But note that this program, which looks so continuous and appropriately analogue at the lower levels in the hierarchy, is itself a relatively stable unit that can be represented by a single symbol at the higher levels in the hierarchy. That is to say, planning at the higher levels looks like the sort of information-processing we see in digital computers, whereas the execution of the Plan at the lowest levels looks like the sort of process we see in analogue computers (pp. 90–91).

Thus, to these authors the TOTE unit fits the discrete, language-like characteristics of planning at higher cognitive levels. Continuous psycho-motor processes would apparently be under the control of feedback loops that look more like those of the servomechanisms thread. For the moment we merely observe that there are differences here; we shall later return to the question of their significance for the evolution of the feedback concept.

Psychology in Terms of TOTE Hierarchies

The authors proceeded to push the notion of TOTE hierarchies as the "structure of behavior" as far as they could. They discussed values, intentions, instincts, habits, roles, the integration of plans, remembering and forgetting, speaking, problem solving, and even the question of where plans come from. The theory-building is reminiscent of Deutsch (1948) (see section 3.2), but in book length it is naturally developed in much greater depth and goes further in seeking empirical support and proposing critical tests. Throughout, the role of the TOTE hierarchy is to form the connection between Images of behaviors and Plans for their execution.

Plans are executed, the authors asserted, simply because humans are alive. Behavior of some sort is inescapable. The question is which Plans to execute (p. 62). Why do we intend to do what we do? The authors hypothesized that values associated with the Images we have must determine the behavior we exhibit. They assumed therefore that the test phase of any Plan may draw on Images. The resulting evaluation determines whether to initiate the Plan, whether to continue it once it has begun, and whether to abandon it in favor of other Plans that are linked to more highly valued Images. A hierarchy of TOTE feedback units may therefore reflect a hierarchy of values (p. 63). A complex Plan associated with a highly valued Image may call for some operations that are by themselves negatively valued, and the Plan may yet be executed. It might be abandoned in midcourse, however, if the negatives associated with some subplans come to outweigh the value of the overall Plan.

The notion of an uncompleted Plan raises the issues of remembering and forgetting. It suggests that we have (at least) two kinds of storage capabilities for Plans, a "dead storage" for Plans not in use and a "quick-access, working memory" for Plans being executed. It is only upon transferring a Plan to the working memory that uncompleted parts of a Plan become intentions. Lewin had postulated a "tension system" to explain the recall and resumption of interrupted tasks. The authors suggested (p. 68) that resumption would depend upon whether the uncompleted Plan remained in the working memory. They found support for their hypotheses in Lewin's own observation that writing down an intention often leads to forgetting it. The forgetting, they hypothesized, is the result of freeing the working memory for other Plans.

But why do I forget to look at my written reminders? The authors proposed several mechanisms in terms of their TOTE framework that could account in general for forgetting. Most obviously, something is forgotten when its Plan is left incomplete. The Plan may have been abandoned because others came to have higher immediate value; it may have encountered an impossible step; the planner's Image might have changed, thereby calling for a new Plan to be placed in the working memory; the partial execution of the Plan may itself have changed the value associated with the Image; and if a Plan is "flexible," meaning that its parts can be executed in any order, it would be more easily interrupted and less easily recalled. For our purposes it should be observed that the role of TOTE units and the feedback concept in these explanations is minimal. Much more significant is the notion of the human being as an information processor with limited working memory that is carefully husbanded. The TOTE perspective is strongly influenced by images of the computer.

Instincts fit neatly into the TOTE framework. Miller, Galanter, and Pribram characterized them as "involuntary, inflexible, inherited Plans" (p. 74). They were encouraged by the appearance of hierarchies of behaviors in Nicholas Tinbergen's classic work on instincts. They concluded that the known empirical results could be explained by chaining or concatenating inflexible TOTE hierarchies. Konrad Lorenz's discoveries on imprinting and innate releasing mechanisms of instinctive behavior could also be interpreted in terms of TOTEs. The newly hatched gosling that learns to follow its mother (or Lorenz) could have an instinctive "follow that" Plan that comes to define the gosling's image of "mother."

Habits and skills can be thought of as originally voluntary Plans that become involuntary and inflexible. The higher mammals, particularly humans, are seen as being able to build up their own "instincts" in terms of TOTE hierarchies. The authors found support for their point of view here in the acknowledged role of language in the learning of skills.

"Metaplans" and the phenomenon of self-reference

In passing to higher level cognitive functions, the authors drew perhaps their most significant conclusion about a theory of psychology based upon TOTEs. They had observed that people function with many ongoing Plans in various stages of completion.[6] Plans have to be integrated into a single stream of behavior. Conflicting Plans, Plans that cannot be integrated together, could be the source of indecision, emotional conflict, frustration, and perhaps serious psychological disturbances.[7] They concluded that there must be a mechanism for integrating Plans, and that it must be of the same form as the Plans themselves—a feedback TOTE hierarchy (p. 98). Here, however, the Tests and Operations in the TOTEs deal not with units of behavior but with other TOTE hierarchies themselves. To use one of Stafford Beer's favorite notions, there must be "metaplans," Plans that talk about Plans.

The authors encountered the need for metaplans in remembering (pp. 125–138), speaking (pp. 139–158), and problem solving (pp. 159–194). In their analysis of spoken language, the simple skill level of the pronunciation of sounds that combine to form an utterance falls in the category of a "motor Plan." Although complex, the hierarchical TOTE structure that is capable of "parsing" a given sentence, comparing sounds with desired sounds, and controlling the speech muscles seems reasonably understandable. The difficult problem is the Plan behind the motor Plan. Where do sentences come from? What assures they are grammatical? How is sentence structure learned? The authors approached the problem from the point of view of Chomsky's notion of transformational grammar (Chomsky 1956, 1957). They concluded that a "grammar Plan" could in principle be formed from TOTE hierarchies. For English the grammar Plan could comprise fewer than 100 rules of phrase formation, 100 transformations, and perhaps 100,000 rules for vocabulary and pronunciation—not so complex that it could not be mastered by a child "after ten or fifteen years of constant practice." Significantly, such a transformational view requires that TOTEs act upon TOTEs. There must be metaplans for speech. And since humans seem to be the only animals capable of speech, the authors wondered if what separates humans from other animals on the evolutionary scale is the capacity to use Plans to construct Plans to guide behavior. Is the ability to work with metaplans uniquely human?

The concept of using Plans to talk about Plans arose again as the authors addressed the psychology of problem solving. They were interested in the Plans that help us define problems, guide our efforts, recognize solutions, and stop the problem solving behavior when a solution is found or we transfer our attention elsewhere. For some problems it is clearly possible to have an algorithm, or a systematic approach, for searching for a solution. Such an algorithm could undoubtedly be expressed in terms of a hierarchy of TOTE units. But in the majority of problem-solving efforts people do not use systematic approaches. They make use of heuristic rules of thumb—Plans for the discovery of Plans. Here the theory requires TOTEs for the creation of TOTEs.

In highlighting the need for such metaplans, Miller, Galanter, and Pribram exposed a set of problems that is still stimulating research today. Complex TOTE hierarchies involving metaplans have the property of "self-reference."[8] Included in this notion are the concepts of self-organization—using Plans to manipulate Plans; self-transformation—using Plans to alter (rewrite) Plans; and autopoesis—using Plans to invent Plans. These are clearly loop ideas. Information emanating from the system returns to affect the structure and behavior of that same system. In the cybernetics literature, which we should emphasize again is distinct from the feedback thread in the social sciences that we are now tracing, these concepts have been integrated into the concept of feedback.

A comprehensive treatment of the literature associated with these ideas is beyond the scope of this book. Suffice to say, a distinction should be made between two kinds of loop ideas. One is the self-reinforcing and self-correcting feedback loop whose history we have traced in the social sciences from the mid-1700s. There is polarity attached to this feedback concept. The other loop idea is the transmission and return of discrete messages, like successive moves in a game. The concept of loop polarity essentially does not exist in such a situation. One would be hard pressed to draw a diagram of a positive and negative loop structure for a chess game, or, as we have observed, for any TOTE hierarchy. The primary focus of this study is on the loop concept united with the notion of loop polarity. The real significance of the emergence of the ideas of self-organization, self-transformation, and autopoesis for that focus is how those ideas brought about a shift in the evolution of the feedback concept. We shall observe, as the work of Miller, Galanter, and Pribram has already suggested, that the concept of feedback tied to self-reinforcing and self-correcting mutual causal loops gradually merges with the more encompassing and more diffuse notion of "knowledge of results."

4.4 Magoroh Maruyama and Deviation-Amplifying Feedback

We have observed that writers in the cybernetics thread tended to emphasize negative feedback processes, almost to the complete exclusion of positive feedback loops. One of the first in the cybernetics thread to try to right the balance was Magoroh Maruyama. He pointed out that positive, self-reinforcing, or "deviation-amplifying" processes are information-feedback phenomena and have an important role to play in mutual causal systems. The study of such things would constitute such a significant development for cyberneticians that he suggested it be labeled "the second cybernetics" (Maruyama 1963).

To Maruyama deviation-amplifying mutual causal systems were ubiquitous:

accumulations of capital in industry, evolution of living organisms, the rise of cultures of various types, interpersonal processes which produce mental illness, international conflicts, and the processes that are loosely termed as "vicious circles" and "compound interests"; in short, all processes of mutual causal relationships that amplify an insignificant or accidental initial kick, build up deviation and diverge from the initial condition (Maruyama 1963, p. 304).

The list here sounds familiar. One can easily imagine that Maruyama was acquainted with writings we have previously noted, for example Keynes and Samuelson on the capital investment multiplier, Bateson on culture contact and on schizophrenia, Richardson on arms races, and perhaps even Wallace on evolution (see section 2.5). In fact, Maruyama traced some of the history of the idea of positive feedback, citing in particular the econometric work of Tinbergen and Myrdal's writings on race relations and on economic development. But Maruyama attached another label to the concept of positive feedback that added a new dimension to it. He associated "the first cybernetics," the study of deviation-counteracting processes, with the concept of "morphostasis"—literally, "static form or structure." He linked the deviation-amplifying processes of his "second cybernetics" with "morphogenesis"—creation of structure or change in form.

Maruyama emphasized the property of positive loops to amplify an initial "kick," which might be a random event and could push the self-reinforcing process in any direction. As an example, he cited Myrdal's discussion of a developing economy:

> Thus, in economically underdeveloped countries it is necessary not only to plan the economy, but also to give the initial kick and reinforce it for a while in such a direction and with such an intensity as to maximize the efficiency of development per initial investment (p. 305).

He found the same amplification of an initial "kick" in the weathering of a rock. A small crack collects a bit of water, which freezes and splits off more of the rock, widening it to the point that organic matter collects, providing a toe-hold for the roots of a tree that further widen the crack. Cities, he suggested, follow a similar self-reinforcing pattern. The region in which the city begins may be essentially homogeneous, but the initial settlement starts a process that leads to the concentration of activity in what eventually becomes the city:

> The secret of the growth of the city is in the process of deviation-amplifying mutual positive feedback networks rather than the initial condition or in the initial kick. This process, rather than the initial condition, has generated the complexly structured city. It is in this sense that the deviation-amplifying mutual causal process is called "morphogenesis" (p. 305).

In exposing the role of positive feedback in societies and organisms, Maruyama's 1963 article made four specific contributions to the understanding of feedback in the cybernetics thread. First was a philosophical or theoretical point. He argued that positive feedback helped to explain some of the puzzles attending the application of the second law of thermodynamics to biological systems. For Maruyama the increase in structuredness in living systems that grow and evolve was not completely explained by Wiener's characterization of information as "negative entropy." For Maruyama the important kind of information supporting the increase in complexity and structure in such systems is positive feedback. Wallace had seen evolution in terms of negative feedback, as organisms and their environment mutually adapt to each other. In contrast, Maruyama emphasized a number of positive loops inherent in the evolutionary process.

Second, in the course of his discussion of the deviation-amplifying characteristics of organic growth, Maruyama emphasized what he saw as the role of positive feedback loops in structure generation. He showed how a small number of rules could govern the growth of a complex pattern of colored cells in an array. He concluded:

> Biologists have been puzzled by the fact that the amount of information stored in the genes is much smaller than the amount of information needed to describe the structure of the adult individual. The puzzle is now solved by noticing that it is not necessary for the genes to carry all the information regarding the adult structure, but it suffices for the genes to carry a set of rules to generate the information (p. 308).

The information contained in the genes, or in the rules governing growth in any system, is thus amplified by the growth process itself. For that reason Maruyama identified morphogenesis, usually thought of as the development of an embryo into an adult individual, with deviation-amplifying positive feedback.

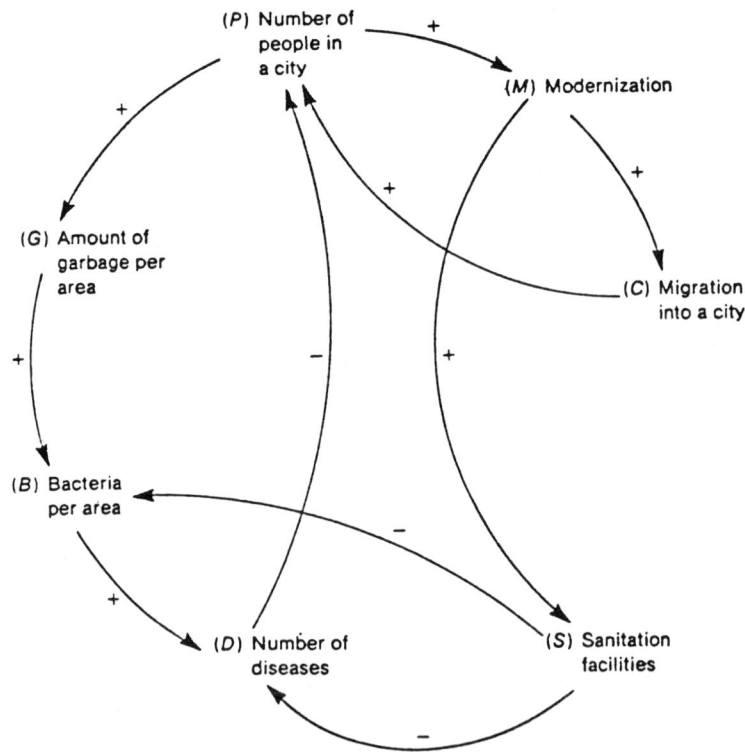

FIGURE 4.9: Maruyama's signed causal-loop diagram. Source: Maruyama (1963, p. 311).

Maruyama's third contribution to the understanding of positive feedback in the cybernetics thread was his introduction of signed causal-loop diagrams and the characterization of feedback loops in their terms. Figure 4.9 shows the example on which he based his discussion. It purports to describe forces influential in the growth of a city. Maruyama defined the + sign to mean that the variable at the head of the arrow changes in the same direction as the variable at the tail. In the figure, an increase in garbage brings about an increase in bacteria; similarly, a decrease in garbage causes a decrease in bacteria. Negative signs on the causal links have the opposite property: a change in the variable at the tail (ceteris paribus) causes a change in the opposite direction in the variable at the head.

Maruyama noted several positive, deviation-amplifying loops in this figure, one in particular involving population, modernization, and migration, and another linking population, modernization, sanitation facilities, and disease. The latter is self-reinforcing because of the double negative it contains. He generalized to the following characterization of positive and negative feedback loops:

In general, a loop with an even number of negative influences is deviation-amplifying, and a loop with an odd number of negative influences is deviation-counteracting (p. 312).

Maruyama concluded his discussion with the following observation:

The system shown in the diagram contains several loops, some of which are deviation-amplifying and some of which are deviation-counteracting. Whether the system as a whole is deviation-amplifying or deviation-counteracting depends on the strength of each loop. A society or an organism contains many deviation-amplifying loops as well as deviation-counteracting loops, and an understanding of a society or an organism cannot be attained without studying both types of loops and the relationships between them. It is in this sense that our second cybernetics is essential to our further study of societies and organisms (p. 312).

The Significance of the Maruyama Article

The "Second Cybernetics" clearly marked a new line of investigation in the cybernetics thread. At the same time it reflected a number of characteristics of that thread that we have previously observed. In the tradition of Wiener and Rosenblueth, it contains numerous references to important issues in the philosophy of science, such as the processes of evolution and the applicability of the laws of thermodynamics to living systems. Like the previous writings in the cybernetics thread, it made little attempt to use the concept of feedback to analyze dynamics.

At most, the dynamic conclusions were limited to broad generalizations of whether a system as a whole is deviation-amplifying or -counteracting. The comments along these lines were akin to those of previous writers in the cybernetics thread on stability versus instability.[9]

But Maruyama's article accomplished several significant things in the evolution of the feedback concept in American social science. First and foremost, it linked the feedback concept of cybernetics with the notion of mutual causality familiar to virtually all social scientists. It established that information-feedback processes and loops of mutual causality were fundamentally the same concept. The loops of Mill, Myrdal, Bateson, Richardson, and others were merely the other half of the feedback story told by Wiener and the cybernetics writers. Second, it promulgated a simple tool, the causal-loop diagram, for the representation of mutual causal feedback structures. The use of such diagrams has since become common in both feedback threads. While not solely responsible for the development, Maruyama's article no doubt contributed to it. The article was widely read and continues to be cited and discovered anew by authors in diverse fields. Notable examples are Wender (1968) and Ashton (1976).

Although solidly in the cybernetics thread in the evolution of the feedback concept in the social sciences, the "Second Cybernetics" shows evidence of sharing some of the characteristics of the servomechanisms thread. To be sure, the link between feedback structure and patterns of behavior is not yet in evidence in Maruyama's article, but there are hints of developments to come. Harking back to the input-output emphasis of the Macy conferences, Maruyama talked about exogenous, unpredictable "kicks" that set deviation-amplifying feedback processes in motion. However, he placed even stronger emphasis on the dominating influence of the deviation-amplifying internal structure of the system. In that sense he shared one of the characteristic points of view of authors in the servomechanisms thread. His use of a closed loop diagram with no exogenous influences to address dynamic phenomena associated with cities is further evidence of a blurring of the distinctions between the feedback threads. We have seen related tendencies develop over time in the work of Beer. The hints of an unraveling of these threads and a reweaving into a common strand became stronger later, particularly in the work of Powers (1973) and Weick (1979).

4.5 Political Science: Karl Deutsch and David Easton

Two of the strongest proponents of the feedback perspective in political science have been Karl Deutsch and David Easton. As we have noted (section 3.2), Deutsch was present at the creation of the cybernetics thread. His article (Deutsch 1948) was one of the first by a social scientist to make use of the engineer's concept of feedback to build social theory. In *The Nerves of Government* (1963) Deutsch returned to the themes he had introduced in that early article, amplifying and extending them considerably. Easton's feedback writings (Easton 1957, 1965a, 1965b) interleave with Deutsch's, with the result that each refers at times to the other's work. Both use the feedback concept in the ways we have come to associate with the cybernetics thread, although Easton increasingly cites work done in the servomechanisms thread.

Easton's Political Systems Analysis

Easton saw his task as the construction of "empirically oriented general theory in political science." (Easton 1965a, p. 3). Others, he noted, had been moving in similar directions, but with different basic units of analysis. Herbert Simon had focused on the administrative decision. Deutsch was centering on the message and its networks. Easton chose the "system." Four notions were central to his version of political systems analysis: political life as a system of behavior, the physical and social environment of the political system, the response of the political system to internally and environmentally generated stress, and feedback. For Easton, it is the last two notions, response and feedback, that distinguish his approach from other systems perspectives in political theory. He summarized his basic position as follows:

It is helpful to interpret political phenomena as constituting an open system, one that must cope with the problems generated by its exposure to influences from these environmental systems. If a system of this kind is to persist through time, it must obtain adequate feedback about its past performances, and it must be able to take measures that regulate its future behavior (1965a, pp. 24–25).

The central question in Easton's work is what keeps a political system in power. Feedback plays a vital role. In Easton's view, political survival cannot be adequately understood without the concepts of feedback and response. We will observe here, however, that although he traces his understanding of feedback to Wiener and the cyberneticians, Easton's feedback is tantamount to the comparatively simpler notion of "knowledge of results." As we shall see, the notion of feedback he uses is largely devoid of loop polarity.

Easton had sketched his ideas in an article in the late 1950s (Easton 1957). Over the next eight years he expanded his framework for political analysis to book length (1965a) and endeavored to fill out the framework in a second, still more detailed and expansive book (1965b).

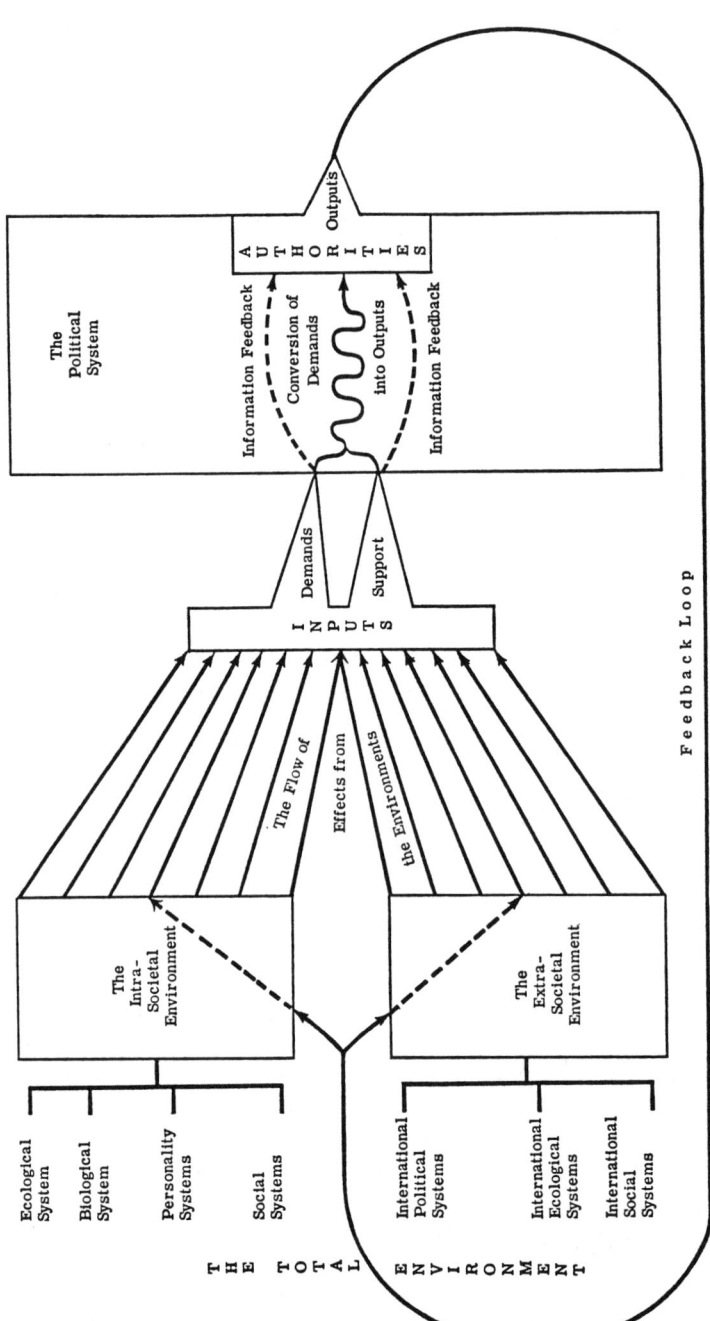

FIGURE 4.10: Easton's "dynamic response model of a political system." Source: Easton (1965a, p. 110; 1965b, p. 30).

Figure 4.10 appeared in both books and contains Easton's basic idea of the role of feedback as it operates between the political system and the society it serves. Of this figure, he wrote:

> Broadly, this diagrammatic representation of the functioning of a political system suggests that what is happening in the environment affects the political system through the kinds of influences that flow into the system. Through its structures and processes the system then acts on these intakes in such a way that they are converted into outputs. These are the authoritative decisions and their implementation. The outputs return to the systems in the environment, or, in many cases they may turn directly and without intermediaries back upon the system itself. . . . There is a continuous flow of influences or outputs from the political system into and through the environments. By modifying the environments, political outputs thereby influence the next round of effects that move from the environment back to the political system. In this way we can identify a continuous feedback loop (1965a, p. 111).

To reduce complexities to manageable proportions, Easton collapsed the enormity of influences on the political system to two types: demands and support. Somehow some information about demands and support is communicated through the political system to the authorities, whose actions have effects back on the environment, thereby creating Easton's fundamental feedback loop. The influence of the early cybernetics writings here is unmistakable. We have seen that the Macy meetings came to define feedback as the alteration of input by output. Easton used the same input/output viewpoint and terminology. In fact, he emphasized it in his next figure, which "stripped the rich and complex political processes down to their bare bones" to show that "after all, in its elemental form a political system is just a means whereby certain kinds of inputs are converted into outputs" (p. 112). That simplified diagram, shown as Figure 4.11, is a direct descendant of a feedback figure in Wiener's *Cybernetics*.[10]

Political systems in Easton's view are put under stress by demands and lack of support. The outputs of the political system—its laws, policies, administrative decisions, statements of consensus, and even its political favors—are the responses it makes to satisfy or mollify demands and to kindle support. They are the result of its efforts to cope with the stresses placed upon it. In Easton's view the capacity of the political system to respond to those stresses depends upon the quality of the information it receives and its ability and willingness to act on it:

> Information about the state of the system and its environment can be communicated back to the authorities; through their actions the system is able to act so as to attempt to change or maintain any given conditions in which the system may find itself. That is to say, a political system is endowed with feedback and the capacity to respond to it. It is through the combination of these prop-

erties—feedback and response—which until recent years had gone virtually unrecognized, that a system is able to make some effort to regulate stress by modifying or redirecting its own behavior (p. 128).

The information required by authorities to stay in power ("be able to respond and thereby seek to cope with stress") includes the conditions prevailing in the environment and the political system itself, information about the supporting state of mind of the members of the system, and information about the effects produced by previous outputs. Without such good information, the political system would fall:

> Only on the basis of knowledge about what has taken place or about the current state of affairs with respect to demand and support would the authorities be able to respond by adjusting, modifying, or correcting previous decisions. . . . Without such feedback, behavior would be erratic or random, unrelated in any causal way to what had previously occurred. . . . It is apparent that feedback plays a prominent role in the way in which the members of a system meet stress. . . . Possessed of feedback information, the members of a system are able to infuse their efforts with direction and purpose. It is for reasons such as these that feedback has been recognized as a central phenomenon in human behavior, both individual and collective (pp. 129–130).[11]

It is evident here that Easton believes that feedback is a vital concept in political analysis. Unfortunately, it is also evident that Easton's feed-

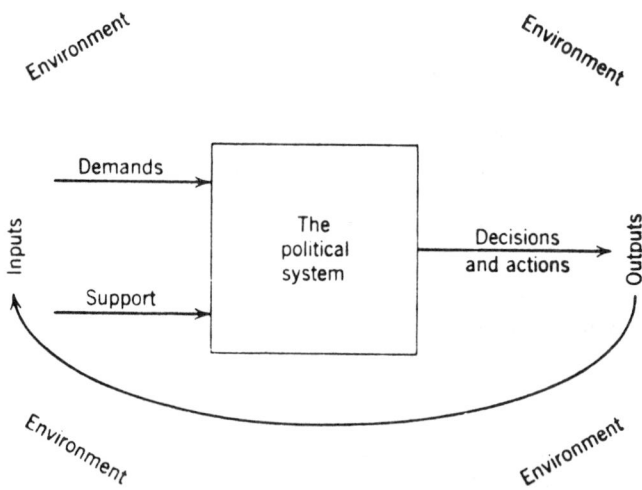

FIGURE 4.11: Easton's "simplified model of a political system," showing the input/output emphasis in his use of the feedback concept. Source: Easton (1965a, p. 112; 1965b, p. 32).

back concept is a mere shadow of what others have meant by the same term. It contains none of the mutual causality of Myrdal or Merton, or for that matter Maruyama. Easton's feedback loops are not self-reinforcing or self-correcting—they are merely sequences of action and information that close back upon themselves. Loops are created, but the loops do not imply dynamic behavior. Easton's feedback is most akin to knowledge of results of a move in a game, like chess. The mutual causal structure embedded in the sequence of moves in such a game is hidden, and in some sense a good deal less important to the players than detailed analysis of each successive current situation.

A Systems Analysis of Political Life

The tendency to treat feedback as "knowledge of results" continued in Easton's most significant feedback work (Easton 1965b). In it he devoted six full chapters and numerous other paragraphs and sections to feedback in political systems. The principal difference between this work and his previous efforts is the level of detail it contains. The spirit and main message about the role of feedback in the survival of political systems is unchanged. In view of that, the following discussion does not attempt to review all of Easton's thoughts on the subject of feedback in this work, but instead focuses on those ideas in the book most relevant to our characterization of the two conceptual feedback threads.

Easton defined a feedback loop in terms of action as well as information:

In brief, the feedback loop will identify a set of processes, composed of information and related outputs and their consequences, that enables a system to control and regulate the disturbances as they impress themselves on the system (1965b, p. 367).

Like Deutsch he distinguished various kinds of feedback. "Simple feedback" refers to goal-seeking behavior. "Complex feedback" is involved in learning from past experience and in "purpose regulated behavior."[12] Both of the latter involve the capability of the system to change its own goals. To Easton, such feedback is of a different, and in some sense higher, order (pp. 368–371).

Expanding on his earlier simplified diagrams of feedback in political systems, Easton (1965b) presented a multi-loop version of Figures 4.10 and 4.11, shown in Figure 4.12. Loop I in that figure represents a feedback process within a sector of the political system: "One member (or group of members) may express his political views to another; the other may respond critically or otherwise; in light of the response, the first member may change his opinions or behavior" (p. 373). Loop II

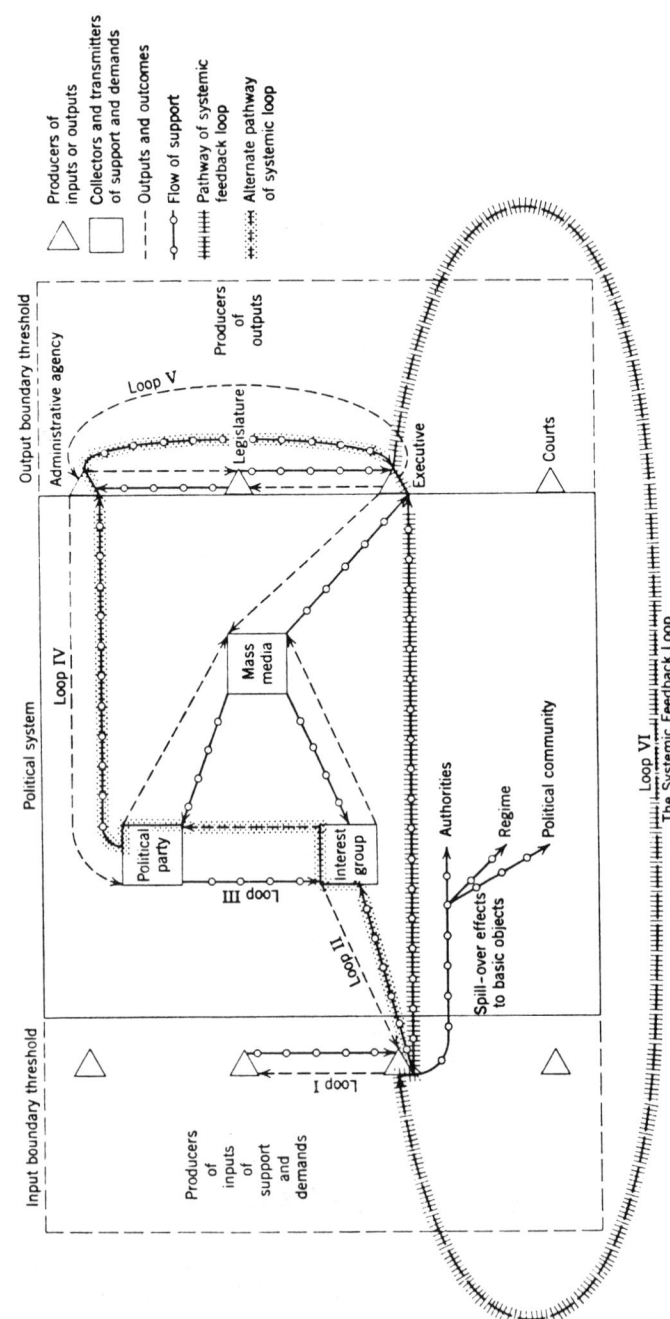

FIGURE 4.12: Multiple feedback loops of a political system. Source: Easton (1965b, p. 374).

shows an interchange of information and benefits between a member and an interest group. Easton asserted that this loop could control the behavior of the interest group, "in the next round." Loop III represents the mutual exchange of information, benefits, and support between an interest group and a political party. Loop IV moves to the next level of aggregation, representing the feedback processes involving a political party and unit of government. Loop V represents feedback loop connections among such "producers of [political] outputs." Finally, through Loop VI, the "systemic feedback loop" emphasized in Easton's earlier diagrams, the outputs of the political system return to influence the behavior of the individuals or groups that initiate political demands.

Dynamics

Easton himself saw some shortcomings in such a description, devoid as it is of any notion of loop polarity or dynamics. He felt it necessary to observe that Loop I, for example, "does not add much to an understanding of the relationship when stated so simply, without including the properties that a dynamic analysis will later bring out as associated with the feedback processes" (p. 373). His ultimate aim was such a dynamic analysis, but it was beyond what he viewed to be the state of the feedback arts of the time, at least in political science. Of loop polarity he observed that to assume each output leads always to an increase or decrease in the flow of support implies the assumption of linearity. Like many other feedback authors we have encountered, he believed that such linearity in social systems is "utterly unlikely" (p. 375).

Nonetheless, he was convinced that nonlinear dynamic analyses of political behavior would eventually be possible:

The practicality of a more detailed analysis is beyond doubt, however. Although even the small chain of loops we have here seems very complex to the unaided mind, modern computer technology reduces the solution of the decision functions involved to a simpler matter. The vital task would be to construct an initial verbal and then mathematical model that took into account at least the central variables that influence decisions for each unit in a loop and to settle on plausible rules governing behavior (footnote, p. 376).

It is probably no accident that the approach Easton described sounds like that pursued by Forrester in *Industrial Dynamics*. Easton was familiar with the work and cited it approvingly several times.[13] Yet it is interesting that he felt the need for two large volumes of preparatory spadework before such feedback simulation models could be attempted in political systems analysis. The citation evidence here indicates a natural tendency for the two feedback threads we have identified to become aware of each other over time. The actual use of the concept in Easton's work, however, serves to emphasize further the differences between the two threads.[14]

Perceptions

Easton stressed the importance of perceptions in the analysis of political systems. It did not matter, he asserted, how observers outside a social system view the intent or success of a given output of the political system. What matters is only how the output is perceived by the members of the social system. Accidental or "autonomous" misperception is significant, as are deliberate attempts by the government to deceive the public through "symbolic gratification" (1965b, 387). This emphasis on perceptions is part of the lore of political science: Easton observed, for example, that the importance of perceptions in political systems is considerably magnified because causality is so muddied in social systems. It is probably impossible for people to know for certain whether a political move has actually caused an observed outcome. But the importance of perceptions is further amplified by Easton's information feedback perspective. Perception is, after all, the receipt and possible translation or distortion of information. Noise, delays, and other transmission problems were central concerns of information theory, which contributed so much to the cybernetics perspective. Any feedback theorist would consequently place great weight on the significance of perceptions in social systems.

Nonlinearity of feedback response

Easton was concerned about the problem of determining the direction and intensity of feedback response to what he called a "stimulus"— an output from the political system that becomes an input to the social system (1965b, pp. 404–409). He was convinced that such response would be nonlinear: a drop or increase in support for the government is probably not directly proportional to the "apparent intensity of the stimulus" (p. 409). He suggested two reasons for the nonlinearity of response: lags and chain reactions. Of the two, however, only the notion of chain reactions has the capacity to generate truly nonlinear response. A lag or delay merely postpones and perhaps dampens the response: if the response would have been a proportional rise in support without the lag, it would still be (eventually) proportional with it.

However, a chain reaction, as Easton intended the phrase, refers to a self-reinforcing, positive feedback process. Such a loop can dramatically alter the behavior of the system:

> A small stimulus may trigger a massive response. A shot fired by the military during a rally to protest continued indifference of the authorities to the demands of members, may be the beginning of a revolution (p. 409).

Such a change would be a dramatic shift in loop dominance which, in a formal model, could only be the result of nonlinearities. The problem of determining the degree of feedback response in such a situation troubled Easton. He noted, however, that "judgments about the effect of feedback stimuli are constantly being made in practical political situations" (p. 409). People make such determinations in their mental models all the time. In principle, then, it ought to be possible to make such determinations in formal models. Here Easton sounds substantially like Forrester, but he stopped far short of a formal simulation model of a political system.

Discreteness and continuity

Easton's emphasis on "stimuli," "responses," and "the next round" of influences implies a discrete view of the behavior of social and political systems. The focus is on events, much like moves in a game, as noted above. The modeling approach in the servomechanisms thread tends to emphasize continuity and a focus on persistent feedback structure underlying events. Several times in *A Systems Analysis of Political Life* Easton gave evidence of being aware of the distinction. In a footnote to the paragraph cited above on revolution as a chain reaction started by an event at a rally, he commented:

> Of course, a small action that does trigger a large revolt is an indication of underlying smoldering discontent. In the same way, a delayed reaction in personal emotions may be set off by a trivial incident, revealing a past history of dissatisfactions and grievances. The triggering incident is only the last in a series of stimuli, and the underlying factors contribute to the political "elasticity" of the situation (1965b, footnote, p. 409).

The note implicitly raises a classic question about the nature of causality in social systems. What, for example, is the "cause" of the revolution—the shot fired at the rally or the "underlying smoldering discontent"? One is an unpredictable, almost random event; the other is persistent underlying structure. Easton's insistence on couching political and social system behavior in discrete, event-oriented terms (stimulus, response) suggests he sees events as the causal agents. Yet he raised the question here of whether processes underlying events are not more important.

He returned to that question in a slightly different guise at the end of the book in a chapter on "The Goals of Systems Analysis." One such goal is a "dynamic view" of the behavior of social and political systems. For Easton, a dynamic view would reveal

> the kind of work, not that individual members or groups in their particularity get done, but that the system as a network of interrelated actions itself manages to perform (477).

Individuals and their actions—the "events" of a behavioral system—would presumably be portrayed in aggregate form. Aggregation smooths over the particular characteristics of individual people and events, creating a more continuous, less discrete picture of the behavior of a social system. Easton suggested here that the goal of his approach is such an aggregate, structure-oriented view of political systems. He was striving for a feedback view of the dynamics of social systems, but dynamic patterns are nowhere in evidence in his work. He saw himself to be laying the foundations for such a perspective for others to pick up and develop.

The Nerves of Government

In 1963 Karl Deutsch returned to the themes he had set out much earlier (Deutsch 1948) and expanded them to book length in *The Nerves of Government*. This later work focused on the models used by people to address questions about government and society. The book contains, in fact, an eloquent statement of the essential role that models play in human thinking. For Deutsch, the word is not restricted to explicit representations such as a scale model of an airplane, an architect's drawings, or a list of equations. Representations or structurings of reality that exist solely in the mind also qualify. "We have seen," asserted Deutsch, "that men think in terms of models" (Deutsch 1963, p. 19). It is thus not a question of whether to use models, but only what kind to use.

The Nerves of Government argued for a new model for thinking about social systems and politics. The old models of "mechanism," "organism," and "historical process" had had their day and each had ultimately proved wanting:

> Mechanism and the equilibrium concept cannot represent growth and evolution. Organisms are incapable of both accurate analysis and internal re-

arrangement; and models of historical processes lacked inner structure and quantitative predictability (1963, p. 79).

As in his 1948 article, Deutsch advocated a "cybernetic model" of people and society. At the heart of that model is his concept of a "self-modifying communications network," or more simply a "learning net," which he defined as

> any system characterized by a relevant degree of organization, communication, and control, regardless of the particular processes by which its messages are transmitted and its functions carried out—whether by words between individuals in a social organization, or by nerve cells and hormones in a living body, or by electric signals in an electric device (p. 80).

As we have noted previously, this "learning net" concept is an offspring of the neural nets of McCulloch and Pitts from the early days of the cybernetics movement.

Much of the book reiterates and amplifies the ideas Deutsch had outlined in his 1948 article, treated in section 3.2. In particular, his definition and characterizations of feedback are repeated verbatim, as is much of his discussion of his feedback view of consciousness, will, autonomy, and so on. A principal difference between the two works is the book's continued stress on the importance of quantifying and empirically justifying cybernetic models of social systems. The book repeatedly outlines potentially promising areas for empirical research in political science that stem from Deutsch's cybernetics perspective. Three particular discussions in the book are of additional interest here: his treatment of Talcott Parson's "general interchange model"; his comments on positive feedback, and his emphasis on autonomous structural change.

The "general interchange model"

One of the models Deutsch proffers as a promising point of departure for social systems research is the "general interchange model," which he attributes to Talcott Parsons. The model is a closed-loop view of transactions in a social system. It grows out of the premise that there are four functional prerequisites for any such system: pattern maintenance, adaptation, goal attainment, and integration. For human society Parsons and Deutsch postulated households as the maintainers of societal patterns; the economy, they suggested, can be viewed as the mechanism of societal adaptation; goal attainment is the purview of the polity; and the whole mass of functions and subsystems is integrated by culture, which includes education, religion, and mass communication. Deutsch painted a verbal picture of these four subsystems as vertices of a square connected by six "flows" or "transactions" which could be represented by the sides and diagonals of the square, much as I have shown in Figure 4.13.

The model raises questions about the nature of the interchanges between each pair of subsystems. Deutsch described several such flows. Households contribute labor to the economy, for example, receiving in return currency. The economy contributes resources and skills to the polity, from which it receives "dependable expectations" for the continuance and evolution of the society. Between the household and the polity pass support, loyalty, demands, and responses to demands, just as Easton extensively described. Deutsch stopped short, however, of describing all the transactions between subsystems, and instead outlined a number of quantitative studies suggested by the model. His purpose was to move toward quantifying the interchange model and developing its predictive capabilities:

A part of this development would be the application of cybernetic concepts to the system, making large and more explicit use of time variables as well as of

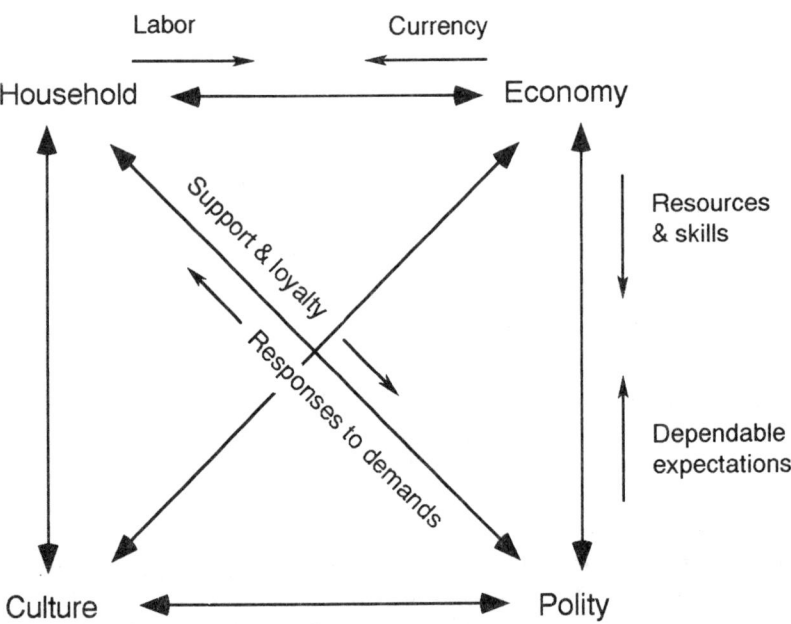

FIGURE 4.13: A representation of Parsons's "General Interchange Model" for a social system, adapted from Deutsch (1963, pp. 116–127).

probabilistic and statistical considerations. This would mean, among other things, the measurement or estimation of the extent and probable distribution of imbalances in the transaction flows; of the corresponding loads upon the equilibrating or adjusting mechanisms in the subsystems; of the lags, gains, and leads in their responses; and hence of the probable stability and future states of the entire system and its parts (1963, p. 128).

Deutsch did not describe further what such a quantitative model would look like. It is clear, however, that cybernetics is meant to contribute much more to this view of society than just the concept of feedback. Indeed, Deutsch characterized cybernetics much as Beer did, in terms of stochastic variables:

Cybernetics is based on full dynamics including changes of state; and it combines these full dynamics with statistics. Cybernetics is the study of the full dynamics of a system under a statistically varying input (p. 90).

Positive feedback in Nerves of Government

Another departure from his early article was a treatment of positive feedback processes in social and political systems. He used the term "amplifying feedback" and found examples of it in panics in crowds and economic markets, runaway inflation, armament races, and "the growth of bitterness in an extremely divided community." The latter two examples trace to the work of L.F. Richardson and Gregory Bateson (see section 2.5). Believing that amplifying feedback processes can get out of control and "damage or wreck the system in which they arise," he asserted that they should be of great interest to social scientists.

His comments on conflicts are a particularly interesting application of the concept of the gain of a positive feedback loop.

If each side is convinced that it must have a margin of superiority over the other, amplifying feedback may result for the whole system.... The course of the resulting amplifying feedback sequence could be forecast and perhaps even controlled, however, by observing and, if possible, controlling the growth or decline of the amount of gain at each cycle, that is the size of successive increments in performance, of which it is composed. If gestures are answered by more vigorous gestures, and threats with more vigorous counterthreats, but if care is taken to keep the competitive increase at each stage somewhat below the increase during the stage that preceded it, then it should be possible for both contending sides to "keep themselves covered" at each step, and, without ever accepting inferiority to the rival, to reduce the sequence of wage-price rises, military threats, and so on, to some foreseeable and perhaps tolerant maximum level (p. 194).

The implication is that if the other side behaves similarly a policy of deliberate "underretaliation" could safely move conflicts or arms races from escalation to stability. In this regard, Deutsch cited President Kennedy's announcement in 1962 that the United States would respond to Russian nuclear tests with a smaller series of similar tests, unless a test ban agreement were reached. The purpose of such a move is to reduce the gain of the self-reinforcing process to the point that it converges rather than grows without bound.[15]

Also present in Deutsch's discussion of arms races is a suggestion for a potential source of negative loop constraints on positive feedback processes:

> Many amplifying feedback processes depend on some external supplies or facilities in maintaining their rate of gain, such as forest fires that require fuel, or arms races that require economic resources. In such situations we may find a phase of constant or increasing gain, and thus accelerating performance, followed by a phase of slowly or rapidly decreasing gain, as the limits of available supplies or facilities are approached (p. 194).

The result would be a shift in dominance from positive loops to negative ones, producing something akin to logistic behavior. Although Deutsch does not say so, the shift in dominance he describes is an example of the nonlinearity of social systems insisted upon by many of the authors in both feedback threads.

Goal-changing, learning, and structural change

In *Nerves of Government* Deutsch repeated his earlier characterizations of several hierarchical levels of feedback, from simple goal-directed feedback to goal-changing feedback. In his view all levels are essential for truly autonomous "self-steering":

> All self-steering networks have three basic elements: receptors, effectors, and feedback controls. What "selfhood," or autonomy, such simple networks have is in their feedback controls.... More complex systems can change their goals, or "reset" their feedbacks, by interaction with information from their past, stored in particular memory devices.... Still more complex networks may include processes of "consciousness," of internal monitoring of certain states of the net.... One test of "functioning" on all these levels would be the capacity to learn, that is, to produce internal rearrangements in the system so as to bring about changes in its behavior (pp. 128–131).

Autonomy, or "self-steering" as Deutsch called it, is a recurring theme in the book. The phenomenon requires feedback relating current conditions to goals, as he repeatedly asserted, but in addition it requires memory, the action of which is also governed by feedback:

To be truly autonomous, such extended self-steering organizations must also carry with them a wide range of stored data from the past, and they must govern the recall of such items from memory by means of feedback processes similar in essence to those by which they obtain information from the outside world (p. 220).

Furthermore, truly autonomous organizations must be able to learn, which Deutsch characterized as "an observable change in [an organization's] behavior in response to the repetition of some unchanged stimulus." Such a change in behavior, he asserted, "could only be accomplished by a change in inner structure" (p. 220). The possible structural changes he envisioned included changes in attention, preferences, goals, the location and sequence of decision points, and the content and organization of memory.

Presumably, the healthy individual or organization strives for changes that enhance its ability to cope and thrive. Yet changes could be dysfunctional. Deutsch saw six types of "failures or pathologies" that might result from inner structural change in an organization: the loss of power; the loss or impairment of information channels to the outside world; the loss of "steering capacity" or the ability to modify behavior quickly enough; the loss of "depth of memory" including recall, screening, and recombination of stored data; the loss of "capacity for partial inner rearrangement" (loss of flexibility, becoming rigid); and the most extreme case, the loss of capacity for "fundamental rearrangement of inner structure" (p. 222). This last failure refers to the loss of the ability to experience such dramatic restructurings as "personality change," "reformation," "rebirth," and "conversion." All but the first of these potential failures in autonomous systems, the loss of power, relate directly to Deutsch's cybernetic feedback perspective. And all refer, in greater or lesser degrees, to changes in the structure of the system.

Deutsch's discussion of the dimensions of growth of a political system is closely related to these structural change ideas. Again he saw six such dimensions: growth in manpower; economic growth; growth in "operational reserves" available for the pursuit of new goals; growth in autonomy or self-determination; growth and change in the political system's own patterns of organization and communication ("strategic simplifications"); and finally, growth in goal-changing ability. The last two involve inner structural change. The last one, the deepest dimension of growth, involving increases in the system's ability to learn and develop, includes the ability to produce genuine novelty (pp. 251–253).[16]

Thus once again in the cybernetics thread we see an emphasis on self-reference, self-organization, and self-change. It appeared earlier in Deutsch's discussion of the limitations of game theory in the analysis of social systems and conflicts.[17] His principal problem with the game theory approach was that it is a "static theory":

> Generally, present-day game theory assumes no change in the performance characteristics of the game's elements during the time that the game is in progress. . . . [Likewise,] it does not provide ordinarily for changes in the rules of the game. Taken together these two restrictions seem to cut it off from the description of much of the process of learning (p. 57).

Thus to Deutsch, analyses of social systems must allow for structural change, including the substitution of different goals, learning, and deep changes in memory and information flow.

Deutsch's treatment of the subject of self-reference and autonomous structural change, however, does not appear to raise the questions that troubled Beer and Miller, Galanter, and Pribram and others and prompted them to appeal to Russell's theory of logical types (see sections 2.4, 4.2, and 4.3). Perhaps a more accurate interpretation, however, is that the potential for contradiction that some saw in such self-referencing systems is eliminated by the hierarchy of levels or types of feedback that Deutsch worked from. The hierarchy, from minor goal-directed loop structures to pervasive restructuring loops, helps to define for Deutsch the nature of the "self" of the individual or organization:

> In self-determination, the location of the "self" can be sought at the location of the feedback circuits of the relatively highest hierarchical type. . . . When the channels of the highest type are lost or disrupted, the self-controlling behavior of the system is lowered to the next lower level, and its more primitive remaining "self" must now be sought at the location of the relatively highest type of feedback circuits that continue to function (p. 130).

We shall see the idea of a noncontradictory hierarchy of feedback levels developed further, with a slant somewhat more akin to the servomechanisms thread, in the psychological theory-building of William Powers (1973) (see section 4.7).

Dynamics and diagrams in Nerves of Government

We have observed that one of the differences between the cybernetics and servomechanisms threads in the social sciences is their treatment of dynamic behavior. *Nerves of Government* continues the pattern we have seen previously in the cybernetics thread, displaying no graphs of dynamic behavior. However, Deutsch did associate the feedback concept somewhat with dynamic phenomena, if only verbally, contrasting it significantly with the "equilibrium" approach he saw prevalent in various places in the social sciences. The equilibrium approach can only deal with "a very restricted field of science, called 'steady state dynamics'." In contrast, the feedback concept promises to be able to deal with transients, sudden changes, growth, evolution, and "friction" in a system inhibiting or distorting change (1963, pp. 89–90).

Deutsch saw at least four advantages of the feedback view over the equilibrium approach.

First of all, in feedback processes, the goal situation sought is outside, not inside, the goal-seeking system. Second, the system itself is not isolated from its environment but, on the contrary, depends for its functioning upon a constant stream of information from the environment, as well as upon a constant stream of information concerning its own performance. Third, the goal may be a changing goal.... In the fourth place, a goal may be approached indirectly by a course, or a number of possible courses, around a set of obstacles (pp. 186–187).

But in addition, a feedback focus would, in principle at least, enable social scientists to identify and measure some new elements in social systems. In particular they could evaluate a given system, by focusing on "the number and the size of its mistakes, that is, the under- and overcorrections it makes in reaching the goal" (p. 187).

Important feedback concepts that determine such dynamic behavior, and presumably aid the analysis of it in social systems, are in Deutsch's view the notions of "load," "lag," "lead," and "gain" (pp. 90, 187–191). Load refers to the amount of information impinging on the system, much like Easton's notion of demands on a political system, but includes the notion of the "extent and speed of changes in the position" of goals the system is trying to reach (p. 187). Lag refers to delays in the system; lead, according to Deutsch, refers to the extent to which future system states can be accurately predicted. Gain, in Deutsch's interpretation, is the amount of corrective action taken for a given distance from a goal. In political systems, these concepts lead to questions about rates of change in the international or domestic situation, perception and action times on the part of policy makers or participative democracies, speed and size of response of different types of regimes, and the efforts of governments to predict and anticipate problems (pp. 189–190). "Considerations of this kind," Deutsch suggested, "may be of some help in the long and seemingly unpromising debate concerning the 'superiority' of this or that political system" (p. 190).

The advantage of the cybernetics approach, Deutsch asserted, is that it "is based on full dynamics including changes of state; and it combines these full dynamics with statistics" (p. 90). Whatever dynamics appear in the book are treated verbally, however. Like others in the cybernetics thread, Deutsch sketched no graphs over time, nor did he use any feedback structure to account for observed patterns of behavior.

With one exception, Deutsch's entire treatment of feedback in *Nerves of Government* is verbal. The one exception is in an appendix, where he displays a diagram that he calls "a crude model" of information flow in foreign policy decisions. The picture is reproduced in Figure 4.14. The arrows in the diagram are information paths along which messages travel. In this sense the arrows in Deutsch's diagram are analogous to those in a TOTE hierarchy, along which travel messages of the outcome of tests and operations. Deutsch's diagram is not a picture of causal influences. Consequently, its links and loops do not have polarities attached to them. One might infer, however, that most of the efforts in the system are to establish appropriate goals and strive to hold the system near them, so the loops would have the character of goal-seeking negative loops.

Yet much more important to Deutsch than loop polarity is the nature of the information, the messages, that are sent along the information paths he identifies. Link and loop polarities do not appear to be particularly important to him. The behaviors of political systems that interest Deutsch are not tied as much to positive and negative loops as to the content of specific political messages and the nature of specific political moves.

Other references to the feedback concept in Deutsch's works

Some corroboration and some contrary evidence for these conclusions can be found in Deutsch's other major publications. In *Nationalism and Social Communication* (Deutsch 1953/1966) we find, for example, a number of graphs over time and an explicit differential equations model in an appendix. Their presence suggests that Deutsch shares more of the characteristics of the servomechanism thread than *Nerves of Government* would indicate. However, a detailed investigation of the role that feedback plays in *Nationalism and Social Communication* shows that conclusion is not justified. Deutsch did not use the concept of feedback to understand or model the dynamic patterns he displayed. The references to feedback in *Nationalism* are focused on communication, not mutual causality.[18]

Perhaps the most direct justification of these claims begins at the end of the book, with the differential equations model developed by Robert Solow to capture the structure of Deutsch's arguments. Deutsch saw two processes as fundamental in the understanding of the growth of nationalism in a society: the rate of assimilation of different peoples in the society, and the rate of mobilization of people in the society. The

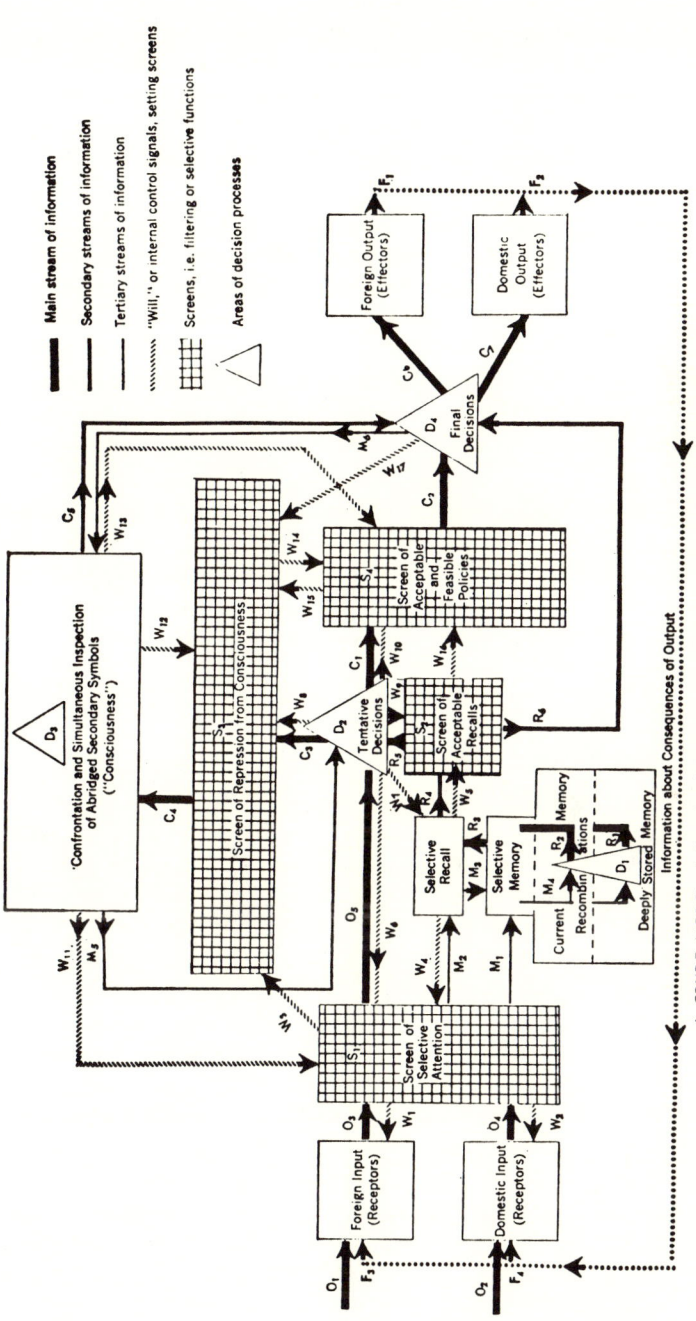

FIGURE 4.14: Deutsch's diagram of information flow in foreign policy decisions. Source: Deutsch (1963, p. 258).

latter refers to such processes as growth in literacy, political awareness, and urbanization. Solow tried to capture the essence of Deutsch's verbal presentation in the following equations (Deutsch 1966, Appendix V, pp. 235–239):

$$\frac{dP}{dt} = pP,$$

$$\frac{dA}{dt} = (a + c)A \qquad [\text{or:} = kA(M - A)],[20]$$

$$\frac{dD}{dt} = (d)D - cA,$$

$$\frac{dM}{dt} = (b + m)M,$$

where

P = total population,
    p = natural rate of increase of P;
A = assimilated population,
    a = natural rate of increase of A;
    c = rate of assimilation of net entry of outsiders into A;
M = mobilized population,
    b = natural rate of increase of M,
    m = rate of entry of outsiders into M;
D = differentiated population,
    d = rate of natural increase of D.

The "differentiated" population is the unassimilated part of the total population. Thus $P(t) = A(t) + D(t)$, for any time t. Similarly, Deutsch identified an "underlying" population U as the portion that is not mobilized, so $U(t) = P(t) - M(t)$. The loop structure of this model consists of four largely independent positive feedback loops, as shown in Figure 4.15.[20]

Solow exhibited the solutions of these equations and showed that A, the assimilated population, can only exhibit exponential growth. D, the remaining "differentiated" population, can either show growth or increasingly steep decline (eventually becoming negative) or a combination of the two. He noted that there is nothing inherent in the differential equations that keeps $P = A + D$, or $M < P$, and that the assumption of constant fractional growth rates of A and D actually contradicts the assumption of a constant fractional growth rate of P. He concluded that the most realistic solution to such flaws would involve abandoning the assumption of constant values for a and c:

> It will be seen that this model is crude indeed. Refinements would have to include changes in the rate of assimilation a as a function of changes in the mobilization rate, and particularly as functions of the rise of the mobilized but unassimilated group H (not specified in the formal model).... More complex models involving changing rates of growth, and particularly changes in a, d, p, h, etc., could be constructed. There seems to be no reason why the lengthy calculations required by such models could not be handled by existing types of computers, electronic or mechanical, but any calculations and models of this kind would clearly go beyond the limits of this study (p. 239).

It is important to realize that such changes in the coefficients, if represented endogenously in the model, would come from nonlinearities incorporated in the model's structure. In spite of Deutsch's repeated concern for the changing structure of social systems, the model is linear and has none of the shifting loop dominance characteristic of nonlinear models. Furthermore, as Solow pointed out, its linearity actually produces a contradiction: the coefficient p cannot be constant if $P = A + D$, and A and D are changing at constant exponential growth rates. Why were nonlinearities not incorporated?

The answer is that Solow's model is a translation of merely the

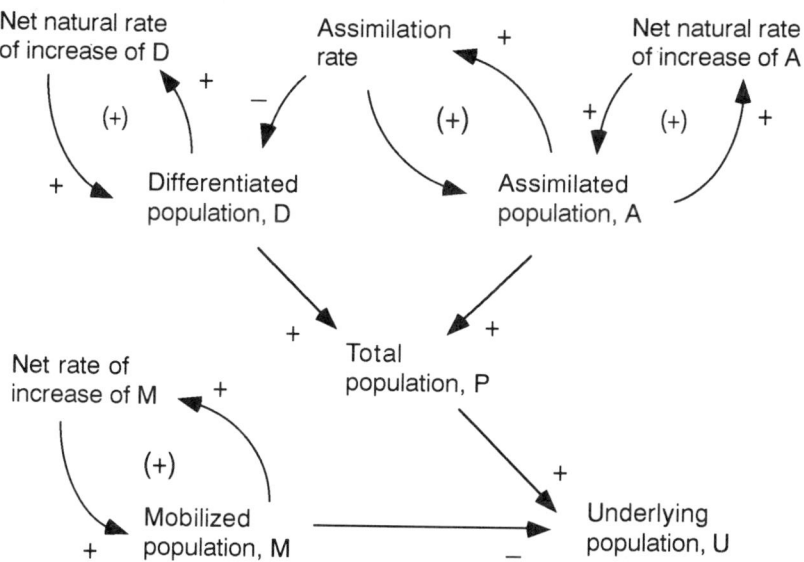

FIGURE 4.15: Loop structure of the Deutsch/Solow model of assimilation and mobilization in a society, abstracted from Deutsch (1966, pp. 235–239).

framework of Deutsch's verbal analyses. Neither Deutsch nor Solow ever referred to the positive feedback loops implicit in the mathematical model. In his verbal treatment Deutsch focused exclusively on the "rates" of growth or decline of various subpopulations in the countries he investigates, meaning the fractional growth rates. Each such rate that he estimated from empirical data is "per capita the initial value" of the subpopulation. These estimates correspond to Solow's p, a, c, and so on. Deutsch did, in fact, hold those initial fractional change rates constant as he projected population changes into the future. However, neither Deutsch nor Solow talked about the positive feedback structure of the system that results from this focus on fractional rates of change. That structure is an artifact of the focus on fractional rates, not a tool for conceptualizing the feedback structure of the system. The behavior of the system was not traced to this positive feedback structure, but rather to the values of the fractional rates drawn out of the empirical data and projected forward in time.

A deeper attempt to derive the dynamic behavior of the system from a structure of feedback loops would include the feedback structure that alters endogenously over time the fractional growth rates. However, both Deutsch and Solow put such an effort beyond the limits of the current study, although they envisioned the possibility:

Any more general quantitative comparison between the relatively high assimilation speeds among mobilized persons and the considerably lower assimilation speeds among the underlying population remains to be worked out. Likewise to be worked out would have to be the mutual interaction of the different rates of change which thus far have been discussed as if they were wholly independent of each other. Actually, an increase in the mobilized and assimilated population N amplifies under certain conditions the demand for new recruits for N, just as growth in the mobilized but differentiated population H may create under certain conditions a secondary demand for more recruits for H. In mathematical terms this could perhaps be expressed by a multiplier, or by models based on "feedback" patterns rather than on simply equilibrium. But such refinements will have to wait until a good deal more data are accumulated (p. 163).

Thus a lack of good empirical data is seen to stand in the way of building a formal model based on feedback to capture mutual interactions and endogenous changes in fractional rates.

Yet Deutsch discussed in qualitative terms a number of factors influencing the rates of mobilization and assimilation. And remarkably, instead of describing these mutual influences in feedback terms, he couched them all in terms of the balancing metaphor reminiscent of David Hume and Adam Smith. After discussing the rate of mobilization in terms of "the outcome of the push and the pull of a demand and supply balance involving several distinct processes" (pp. 153–156), Deutsch analyzed the factors influencing assimilation:

> The rate of assimilation measures the second great process which, together with the process of mobilization, determines largely the outcome of national development. The rate of assimilation in its turn depends on a considerable number of elements. We may perhaps best visualize these as acting on each other in six overall balances. The first two of these balances are largely technical and linguistic; the third and fourth balances are largely economic. The fifth and the sixth are largely cultural and political in nature. Together, in the interplay, they determine the nature and the speed of the process of assimilation (p. 156).

He then proceeded to discuss these balancing mechanisms in detail. Nowhere does he observe that they could be viewed as negative feedback loop systems.

The choice of metaphor here is significant in our story of the evolution of the feedback concept. Deutsch was at the forefront in bringing the engineer's concept of feedback into the social sciences. Why did he choose here the balancing metaphor instead of a perspective based on feedback and mutual causality? There are two possible answers. First, it is conceivable that he may not have seen the mutually interacting processes in feedback terms. We have seen that his use of the feedback concept tended to focus on the feedback of particular messages, not on mutual causal loops. He repeatedly referred to the concept of feedback as a notion of "communications engineers," not, significantly, servomechanisms engineers. The second possibility is that he himself made all the links between balances and feedback/mutual causal mechanisms, but that he wrote his analyses in terms of the balancing metaphor because it was familiar to his audience. The balance of supply and demand had almost 200 years of social science tradition attached to it. Lewin's field theory (Lewin 1951) and Festinger's cognitive dissonance (Festinger 1957) are both phrased in terms of the balancing of opposing forces. Certainly, the balancing metaphor was prevalent in the social sciences at the time and would help to communicate Deutsch's understandings of the processes he saw operating in the dynamics of nationalism.

Whatever the reason for Deutsch's choice of the balancing metaphor instead of feedback and mutual causality in this discussion, we are left with the fact that what Deutsch wrote did not explicitly link the feedback structure of the system to the dynamics he focused upon. The inspiration he drew from cybernetics enlightened his understanding of the role of communication in society, not the role of positive and negative loops of feedback and circular causality.

4.6 Karl Weick and the Study of Organizations

The feedback perspective on complex systems gained a strong advocate in Karl Weick, a significant organizational studies scholar. Feedback plays a prominent role in Weick's widely used introductory text in the field, *The Social Psychology of Organizing* (Weick 1969/1979). The literature on organizations, he observed, is replete with words such as "connection," "relation," "link," "network," "interdependence," and "reciprocal" (p. 65). At the heart of such notions he placed the concept of feedback, viewing it as reciprocal or mutual causality.

Weick's feedback sources are largely authors in the cybernetics thread, notably, Ashton (1976), Reusch and Bateson (1951), Bateson (1972), Beer (1975), Boulding (1971), Buckley (1967, 1968), D.T. Campbell (numerous works), Maruyama (1963, 1974), McCulloch (1965), J.G. Miller (1971), Slack (1955), Vickers (1967), von Foerster (1967), and Wender (1968). He also cited Simon (1957, 1962) and March and Simon (1958), authors associated initially at least with the servomechanisms thread (Weick 1979, pp. 265–283). Unlike Deutsch, Miller, Galanter, and Pribram and others we have recently encountered in the cybernetics thread, however, Weick does not take his cue from communication theory and does not focus on the feedback of messages. Instead, he illuminated what he regarded to be the ubiquity and significance of positive and negative causal loops in complex organizations.

Weick's initial reference to feedback comes in the first five pages of the work, introduced with a newspaper quotation about farm machinery. Farmers, it seems, had been buying heavier and more sophisticated machinery to save labor costs. But the heavier machines compact the soil making it difficult for roots and water to reach each other. "The subsoil then must be tilled with an even larger, deeper plow," the article asserted, "which, of course, requires a more powerful tractor to pull it" (Weick 1979, p. 2). Weick used the observation to introduce what he saw as one of the characteristic problems of organizing:

> The problem created when heavy machinery packed the soil and necessitated heavier machinery to break it up is an eloquent example of feedback loops that sometimes turn into vicious circles. The cause-effect relationships that exist in organizations are dense and often circular. Sometimes these causal circuits cancel the influences of one variable on another, and sometimes they amplify the effects of one variable on another. It is the network of these causal relationships that imposes many of the controls in organizations and that stabilize or disrupt the organization. It is the patterns of these causal links that account for much of what happens in organizations. Though not directly visible, these causal patterns account for more of what happens in organizations than do some of the more visible elements such as machinery, time clocks, and pollution equipment (p. 7).

Developments in the Cybernetics Thread 229

Weick involved his reader in a feedback view of organizations with an exercise focused on a meeting he or she has recently attended. Weick gave the diagram shown in Figure 4.16 and asked the reader to connect the phrases in the diagram with causal arrows showing the effect of one variable in the meeting on another. He described the now familiar conventions for determining the polarity of causal influences in such a diagram. If an increase in a variable tended to cause another variable to increase, the reader was to draw an arrow and label it with a plus sign. If an increase in one produced a decrease in another, the arrow was to be labeled with a minus sign. (Weick's source for these prescriptions was Maruyama (1963).)

Weick then used the presumed results of this exercise to motivate and define interdependence and positive and negative loops. He defined an independent variable as one which has arrows emanating from it but none pointing to it. Similarly, a dependent variable is one that has arrows coming into it but none leaving. Any variable that has arrows both coming in and going out he termed interdependent. Some

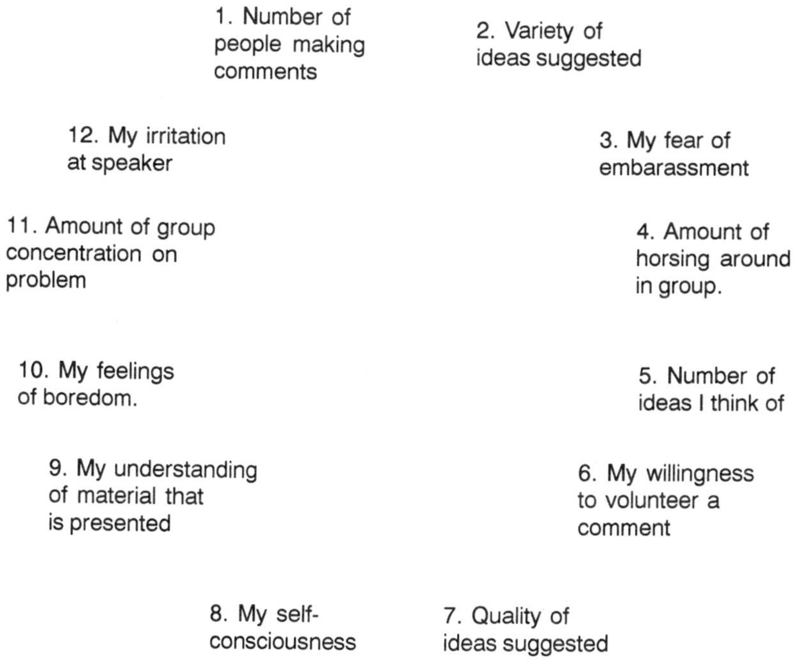

1. Number of people making comments

2. Variety of ideas suggested

12. My irritation at speaker

3. My fear of embarassment

11. Amount of group concentration on problem

4. Amount of horsing around in group.

10. My feelings of boredom.

5. Number of ideas I think of

9. My understanding of material that is presented

6. My willingness to volunteer a comment

8. My self-consciousness

7. Quality of ideas suggested

FIGURE 4.16: Diagram for an exercise for the reader to analyze the events of a discussion or meeting; used by Weick to motivate a discussion of positive and negative feedback. Source: Weick (1979, p. 70).

of the interdependent variables in the reader's diagram are likely to have formed loops. Citing Maruyama (1963), Wender (1968), and Bateson (1972) respectively, Weick termed a loop with an even number of negative signs to be "deviation-amplifying," "vicious," or "regenerative." Similarly, a loop with an odd number of negative signs "imposes stabilities," "counteracts deviations," and is "self-regulating." Thus in Weick's view there is apparently complete interchangeability among the notions of interdependence, mutual causality, and feedback.

Of positive feedback loops, Weick asserted:

> In a causal loop with an even number of negative signs there is no regulation or control. Once a variable begins to move in a particular direction, either up or down, the variable will continue to move in that same direction until the system is destroyed or until some dramatic change occurs (p. 72).

Presumably, he meant that an isolated positive loop by itself has this property, not a positive loop embedded in a complex interdependent structure with other, potentially stabilizing loops.

Like some of his predecessors in the cybernetics thread, Weick used his feedback perspective to focus on stability versus instability:

> When people examine relations in organizations they look for interdependent variables, causal loops, and the presence or absence of control. Loops that are deviation-counteracting mean that the system is basically stable; loops that are deviation-amplifying mean that the system is basically unstable. Whether this instability leads to constructive or destructive growth is of importance (p. 74).

His subsequent attempts to deduce the stability or instability of an interdependent system are interesting, partly because they are clever verbal interpretations of some quantitative ideas we have encountered before, and partly because they illustrate rather dramatically some of the negative implications of the separation of the two feedback threads we have been tracing.

Inferring the Implications of Loop Structure

One way to try to deduce the fate of a system containing more than one loop, Weick suggested, is to try to determine its "most important loop":

> If the most important loop is deviation-counteracting, then the system in which the loop is embedded will be deviation-counteracting. If the most important loop is deviation-amplifying, then the system will be deviation-amplifying. The difficulty with this method of predicting is that the judgement of a loop's importance may often be purely arbitrary (pp. 74–75).

One could eliminate some of the arbitrariness, he suggested, by concentrating on the loop(s) containing the variables with the greatest number of inputs and outputs. Those variables, being in some sense the "most interdependent," were seen by Weick to be plausible indicators of the behavior of the entire system. He applied his criterion to the causal-loop diagram contained in Maruyama (1963) describing interdependencies in a city (see Figure 4.9), finding that[21]

> The most important loop within this system is deviation-amplifying, and therefore the entire system is deviation-amplifying; we predict that it will eventually destroy itself unless one of the relationships changes in sign, another relationship is added, or some relationship is deleted (p. 76).

Here Weick was trying to deduce one aspect of behavior—stability or instability—from the system's feedback structure expressed as a causal-loop diagram.[22] His suggestion was to search for what others call dominant loops (Richardson 1984). In the servomechanisms thread, the question of behavior would be answered by a quantitative model, either solved analytically or simulated to determine its behavior. Dominant loops would emerge from analyzing simulation experiments (Richardson and Pugh 1981) or by sophisticated mathematical methods (N. Forrester 1982; Richardson 1986).

One could try to deduce the stability or instability of the system another way, Weick suggested, if one preferred to assume that all the loops in the system are essentially of equal importance. We know the fate of a single causal loop by the number of negative links it contains. So, by analogy, Weick suggested that the stability or instability of a system of interdependent loops could be determined by counting the number of negative *loops* it contains:

> We would predict that any system will survive as a system only if it contains an odd number of these negative loops. If the system contains an even number of negative loops, then their effects will cancel one another, and the remaining positive cycles will amplify whatever deviations may occur (p. 76).

Or, alternatively he suggested, one might count the total number of negative relationships in the entire system, weighing each by the number of loops it enters into. Again by analogy to the situation in a single loop, Weick suggested that if the total number of negatives is even the system is deviation-amplifying, and if it is odd the system is deviation-counteracting (p. 76).

Weick applied these procedures to Maruyama's urban causal-loop diagram (Figure 4.9) and found, in contradiction to his previous conclusion, that both imply that the Maruyama system is deviation-counteracting. Rather than conclude that his criteria are faulty, he suggested:

> The prediction differs from the earlier prediction based on the assumption that the loops were of unequal strength, thereby illustrating the point that assumptions make a difference. It is conceivable that both conclusions about this system are correct. The system will continue for some period of time, but due to such things as the differential speed with which the cycles are completed, the magnitude of the changes at each variable, the tightness of the couplings among variables, the number of times each loop is activated, and the effects of exogenous variables, increasing amounts of instability could be introduced (p. 76).

Unfortunately, Weick is very far off base here. Of course assumptions make a difference, but that is not the reason for the contradictory conclusions from Maruyama's diagram. There is simply no foundation for Weick's inferences based on the number of negative loops or the total number of negative relationships in a system. It is easy to construct examples of stable systems with an even number of negative loops, or unstable systems with an odd number of negative loops.[23] Furthermore, it is hard to imagine that every system of, say, 1,000 interconnected causal feedback loops always switches from stable to unstable, or vice versa, with the addition of just one more negative link or loop. How a multi-loop system behaves over time depends not just on the structure and polarity of its feedback links and loops, but also on the number of delays and accumulations (stocks, averages, levels, or integrations) in the system, and also on the nonlinearities or potential shifts in loop dominance in the system.

Weick made a serious error here, one that substantially invalidates much of the feedback thinking in his book. But the significance for us is not in the error itself, but rather in what it implies about the separation of the feedback threads we have been tracing. It is hard to imagine a social scientist familiar with the use of the feedback concept in the servomechanisms thread who would have proposed the analogy Weick made. People in that feedback thread build formal models whose primary purpose is to trace out the dynamic implications of a system's feedback structure. If it were as simple as Weick's rules suggest, there would be little need of formal models. Furthermore, experimenting with formal models of feedback systems would show quite quickly that their behavior has little to do with the simple number of negative feedback loops.

I believe that Weick was pushed toward his mistake by two forces that do not fit together very well. One is his own interest in deriving substantive conclusions about the behavior of complex organizations from his certainty that they contain causal feedback loops. The other is the pattern of feedback thinking previously established and perpetuated in the cybernetics thread from which Weick draws his feedback understandings. Weick wanted to deduce behavioral implications, but what he had read about the feedback concept did not adequately prepare him for the task. The focus on stability versus instability in the cybernetics thread allows only yes or no answers. The behavior that Weick is interested in is far richer. The error he made here reinforces what a number of our feedback authors have emphasized: couching feedback analyses in verbal terms or causal-loop diagrams does not provide the reliability that mathematical models or simulation offer.

Weick's work thus provides additional telling evidence for the separation of the two feedback threads we have been tracing. But *The Social Psychology of Organizing* is important in the evolution of the feedback concept in the social sciences for more fundamental reasons. Weick unequivocally asserted the importance of feedback and circular causality in organizations, and he introduced clever verbal equivalents for some of the quantitative feedback concepts we have encountered.

Nonlinearity and Structural Change

The notion of nonlinearity and loop dominance appeared indirectly in Weick's discussions of "uptight variables," "reversing a variable," and "discrediting a cause map." He drew the phrase "uptight variable" from Bateson (1972, pp. 496–497). It signifies a quantity close to its upper or lower limit, a variable that is severely constrained in its flexibility to move in a given direction:

> If this happens, then all other variables related to that single uptight variable will now also become frozen, and that's true even if they are not near the extremes of their limits. . . . When variables freeze up, danger results because they are then unavailable to absorb changes occasioned by normal fluctuations (Weick 1979, p. 78).

The idea is related to one of Ashby's early observations, that controls can make a stable system unstable (see section 3.2). It is also clearly a verbal representation of the quantitative notion of nonlinearity. In nonlinear systems, like the Verhulst population equation, for example (section 2.2), loop dominance can shift to controlling negative feedback loops that prevent the movement of certain variables in particular directions. Verhulst's population near its maximum would be, in Weick's terminology, "uptight." Weick suggested that uptight variables could be responsible for some of the difficulties the reader may have experienced in the meeting or discussion that he or she had diagrammed in loop terms earlier.

Besides becoming "uptight," a variable can "reverse its direction." Weick noted two senses for this latter phrase, describing both in terms of the causal-loop diagram reproduced in Figure 4.17. The structure consists of a number of positive feedback loops associated by Wender (1968) with depression. Once present, depression is reinforced through each of these positive loops and, according to the diagram at least, can become increasingly severe. However, Weick asserted, at least one of the variables could be induced to change its direction, thus "reversing the variable" and transforming a vicious circle into a virtuous one:

If coping ability [in Figure 4.17] can be increased by training and/or a success experience, then the downward spiral of depression can be halted and self-evaluation can increase (p. 84).

In organizations, he noted, such reversal of direction "may require social action and redefinition," and it may be difficult to accomplish because "inertia and prior understandings work against the redirection." The causal mechanism for the change in direction is left outside the diagram, a chance happening or otherwise unspecified causal structure.

Weick's other meaning of "reversing a variable" involves more than a change in direction of a variable. Causal feedback systems, he suggested, are vulnerable to changes in structure, through reversals of the polarity of causal influences:

The causal network of depression portrayed in [Figure 4.17] is vulnerable, and one point of vulnerability is "aid from others." As a person's self-evaluation drops, depression increases, coping ability decreases, one has to depend on others for aid, which lowers even further one's self-evaluation, and so on. This vicious circle is fueled by the interpretation that getting aid from others is a sign of weakness and dependence. Upon reflection, however, the meaning of aid can shift from one of weakness to the more positive interpretation that it's okay to ask for help and it's great to have others who care enough to come to one's aid. If, after reflection, that interpretation becomes salient, then the negative sign between aid and self-evaluation in [Figure 4.17] becomes a positive sign, which turns amplification into stability (p. 84).

Thus a variable may be "reversed" if the sign of an influence on it reverses. Again, however, Weick did not diagram the underlying processes or influences that determine the sign of the causal influence from "aid" to "self-evaluation" or cause it to change. Such changes result from "reflection and contemplation"—activities, Weick asserted, that "frequently result in reversing of variables and a reversal of one's fate" (p. 83).

The ideas presented here focus on structural change of a causal feedback system and echo concerns we have seen in both feedback threads. In the cybernetics thread authors have argued for metaplans and the abilities of systems to "rewrite" their own structure. In the servomechanisms thread structural change is portrayed in shifts in loop dominance caused by nonlinearities. As we would expect, Weick's presentation is more akin to the "rewriting" notions of the cybernetics thread.

Weick's most interesting assertions about structural change in a feedback system fall under the label of "discrediting a cause map." He took as an example the assumption of a simple linear relationship between the quality of a performance and the number of criticisms (in rehearsals perhaps): the greater the number of criticisms, the higher the

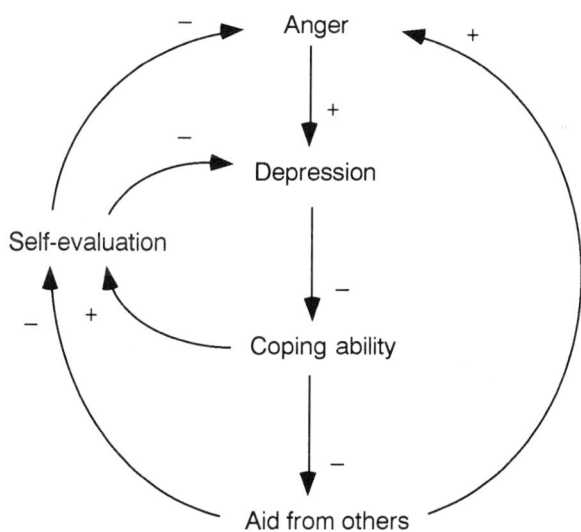

FIGURE 4.17: A set of connected variables thought to be associated with depression. Source: Wender (1968), as discussed in Weick (1979, pp. 83–84).

quality of the performance. He then listed the following ways in which such a causal influence could be changed:

1 The causal direction could be reversed: the quality of performance causes the number of criticisms received.
2 The sign of the linkage could be changed: as criticisms increase the quality of performance decreases.
3 Two variables could be decoupled: criticism has no effect on quality.
4 The direction of the relationship could be removed: criticisms affect quality, but there is no regularity to the direction in which this relationship moves.
5 Dissolve the variables: there is no such thing as criticisms or quality of performance.
6 The coupling could be tightened: criticisms always, immediately, and in direct proportion affect the quality of performance, an assertion that is stronger than the one originally formed. This introduces change in the speed with which the cycle is completed and in the amount of influence exerted on other variables that are related to these two.
7 The coupling can be loosened: criticisms have a very modest regular effect on the quality of performance.
8 The effects of the originating variable are canceled by another pathway: even though criticisms affect quality directly, they also affect the patience of one's peers, and as this patience wears thin the quality of performance goes down. Therefore, criticisms may improve quality, but they have even more effect on peers and peer pressure decreases quality, which undercuts any direct effect from criticisms.
9 The variables are related in a curvilinear manner: criticisms affect quality of performance directly up to a point, but beyond that criticisms have just the opposite effect—they degrade quality (pp. 85–86).

I find this list interesting for two reasons. First, it is a rather complete compendium of the ways that a given bit of causal feedback structure in a system can change. As a system dynamicist, I see the list as a neat verbal statement of the potential changes that nonlinearities can endogenously create in a quantitative model of a feedback system. In that line of thinking, it is interesting to note that the last two entries in Weick's list can account for all the rest: other causal pathways, together with nonlinearities (curvilinear relationships), could produce any of these other ways a cause map can be "discredited." Presumably, Weick listed these other ways because he viewed them differently, as the results of some processes of change other than the shifts in dominance that his numbers 8 and 9 can produce—randomness, perhaps, or unspecified exogenous influences.

Weick's list is also interesting because it seems to suggest that the causal feedback structure of a system is very changeable. He acknowledged that some changes are harder to accomplish than others, but asserted that "anything that is done to create a cause map can be undone to change it" (p. 86). The apparent malleability of causality would suggest to most readers that the feedback structure underlying events at one moment may be completely different from that underlying subsequent events. In contrast to some of the feedback authors we have previously encountered, Weick does not seem to be focusing on causal structure that persists through time.[24]

In spite of the changeability of feedback structure he sees, Weick unequivocally asserted the importance of a circular causal view.

Most managers get into trouble because they forget to think in circles. I mean this literally. Managerial problems persist because managers continue to believe that there are such things as unilateral causation, independent and dependent variables, origins, and terminations. . . . If you become obsessed with interdependence and causal loops, then lots of issues take on a new look. If you take loops seriously, for example, then you realize that some very sacred ideas such as self-power and self-determination are fictions. If most of our behavior is embedded in causal circuits, then whatever we do will come back to haunt and control us. The revenge of the environment is perhaps the most obvious example (pp. 86–87).

Erroneous Implications of Feedback Loops in Organizations

In the remainder of the work, Weick made use of causal-loop diagrams and arguments to communicate his views on the nature of organizations and the fruitful ways of looking at them. He adopted an evolutionary point of view that he attributed largely to Campbell (1970, 1972, 1974), which focuses on change in organizations as a process of natural selection. The central structure in Weick's presentation is shown in Figure 4.18. He devoted an entire chapter to each of the terms shown in the diagram, as well as another one addressing the structure they form together. We shall have to be content here with brief definitions.[25]

In Weick's view enactment is what happens in organizations, and thus represents a mix of environmental and internal variation. Selection is the process of perceiving what is happening, as in "selective attention." What is perceived is a function of what has been enacted and what has been the traditional or persistent way of viewing things in the organization. Retention is simply the process of storing what has been perceived.

Circular causality is rampant in this view of organizations. What happens (enactment) in an organization depends a great deal on what the organization remembers as its nature and purpose (retention); what it retains depends on what it perceives to have been happening (selection); and what it selects to attend to depends both on what is happening and on what the organization has become accustomed to perceiving. Weick suggests that the basic theme of this view of organizations is captured entirely in the neatly circular question, "How can I know what I think until I see what I say?" (p. 133). An organization, in Weick's view, does not know what it is about until it looks backward and imposes structure on what has been enacted in its name.

Weick gave numerous specific examples of the general structural pattern shown in Figure 4.18. In each the structure has the following properties:

(1) ecological change and enactment are linked causally in a deviation-amplifying circuit, (2) enactment is linked to selection by a direct causal relationship, indicating that the volume of enactment will have a direct effect on the volume of selection activity that occurs, (3) likewise selection has a direct effect on retention, meaning that an increase in the amount of selection activity will trigger a corresponding increase in the amount of retention activity, (4) retention affects both selection and enactment and these effects can be either direct or inverse, depending on whether the person decides to trust his past experience (+) or disbelieve it (−) (p. 132).

The structure therefore consists of sometimes positive and sometimes negative loops, depending on the specifics of the situation.

Weick added a generic negative feedback loop to complete his view of the way natural selection operates in organizations. The loop underlies how organizational processes are built up. In Weick's view such processes are composed of "cycles" or "double interacts" between people, and they are assembled in organizations to reduce the "equivocality" in inputs to any part of the organization. The greater the equivocality, the fewer rules exist in the organization for dealing with

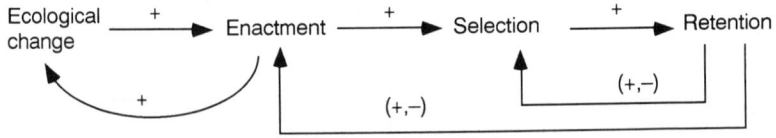

FIGURE 4.18: The four processes of organizing. Source: Weick (1979, p. 132).

the input, so the more "interlocked cycles" between people will naturally develop to deal with the disturbing equivocality.

The result of this reasoning, which Weick presented in much greater detail (pp. 110–118), is the negative loop shown in Figure 4.19. Weick superimposed this loop twice in Figure 4.18, as a part of both the selection and retention processes. He noted that if the signs on the links from retention to enactment and selection are the same, the resulting structure contains an even number of negative links and loops. He concluded:

> it should be noticed that unless members of the organization are ambivalent toward their past experience, these organizing processes will be unstable, and the entire system will be deviation amplifying. If members completely trust their past experiences there are six negative signs, and if they completely distrust their past experiences there are eight negative signs—both being even numbers. Only simultaneous trust and disbelief will stabilize the system. (p. 133).

Weick's conclusions here are again based on the erroneous analogical connection he sees between negative links in loops and negative loops in systems. As we have observed, there is no basis to conclude that a feedback structure with an even number of negative signs is unstable. Weick's insights about organizations may be correct, but they are in no sense justified by the loop structure he presents. The error here is unfortunate, because Weick has a potentially significant feedback view of organizational structure and dynamics. It would be interesting to see what understandings could be produced by a marriage of Weick's views with the servomechanisms thread.

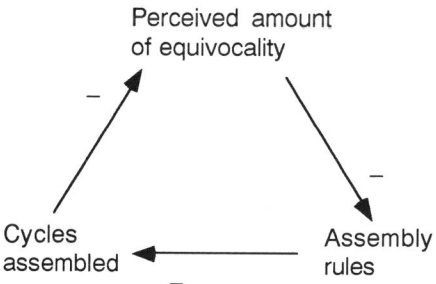

FIGURE 4.19: Weick's generic negative feedback loop governing the development of "cycles" or "double interacts" at any level in an organization (Weick 1979, p. 133).

4.7 The Psychology of William Powers

The final figure whose work we will investigate in detail in this chapter is something of an anomaly in our story. William T. Powers is a theory-builder in psychology. In the tradition of McCulloch and Pitts, he is fundamentally interested in the question of how the human brain could be organized to produce the human behavior we observe. By the feedback scholars he cites,[26] he places himself squarely in the cybernetics thread. But in several important ways, as we shall see, his use of the feedback concept has more in common with the servomechanisms thread. In fact, in Powers (1978) he identifies a number of "blunders" in social scientists' use of the feedback concept and traces them back to tendencies he perceives to have been inadvertently initiated by Wiener and others in the earliest days of cybernetics. Powers's work is evidence of a reweaving of the two feedback threads.

Powers's first article dealing with the feedback concept (Powers, Clark, and McFarland 1960) appeared in the year Miller, Galanter, and Pribram's *Plans and the Structure of Behavior* was published. The two works shared some of the same purposes. In particular, both sought to place the feedback concept at the foundation of psychology, and both attempted to do so through a hierarchy of feedback units. The differences between the two approaches highlight important ideas in the evolution of the feedback concept in the social sciences. The ideas Powers expressed in that first article developed into his major work, published in 1973 with the intriguing title *Behavior: The Control of Perception*. The discussion that follows is based upon that book and Powers (1978).

Background

Even more than Miller, Galanter, and Pribram, Powers is fundamentally critical of behaviorism. More precisely, he objects to what he sees as the particular concept of cause and effect that underlies "scientific method" in behaviorist psychology. As he sees it, the prevailing view holds that the proximate cause of what an organism does lies outside that organism: "the best the organism can do is to modulate the connection from the stimulus that is the cause to the behavior that is the effect" (Powers 1973, p. 1). That point of view can be traced, he suggests, to a basis in the known structure of the nervous system, consisting of receptors (sense organs), inner neural connections, and effectors (muscles).

But, Powers asserts, the conclusion that there is an "orderly march of cause and effect from stimulus object to sensory receptor, and from muscle to the eventual behavioral result" is not justified. A stimulus that manages to result in a regular behavioral response does not act on the nervous system the same way every time; a stimulus that does act on the nervous system in a consistent way does not always produce the same observable behavior; a given set of muscle tensions does not always produce the same observable behavior; and likewise, when a given observable behavior repeats, the muscle tensions responsible for it are not necessarily the same every time (p. 5).

Powers rejects what he sees as the traditional solution to such dilemmas—the invocation of the notion of "natural variability," and the application of statistics to tease out behavioral regularities.

"Significance" in experimental results [has] come to mean something other than "importance." It now means a little triumph over nature's noise level (pp. 5–6).

His alternative approach is to build a conceptual model of the human brain, based on the concept of feedback, that is consistent with what is known about neural physiology, and that is indeed capable of producing or explaining observed human behavior.

We have seen similar or related efforts before, notably the neural nets of McCulloch and Pitts (1943), the TOTE unit of Miller, Galanter, and Pribram (1960), and the problem-solving work of Newell, Shaw, and Simon (e.g., 1958, 1959). Powers was familiar with these and had praise for them all, particularly the work of Miller, Galanter, and Pribram.

But Powers found two fundamental problems with these efforts that he sought to correct in his own approach. First, TOTE formulations and problem-solving simulations had not been, in his view, models of the brain, but rather models of the behavior that is caused by the brain. Powers sought a behaving model of the brain that rests not "on descriptions of what the brain does," but rather "on concepts of the functions in the brain that make it capable of doing these things" (p. 17):

A model in the sense I intend is a description of subsystems within the system being studied, each having its own properties and all—interacting together according to their individual properties—being responsible for observing appearances (p. 15).

Furthermore

... Most of modern cybernetics, including related computer-simulation studies of thought processes, is in the class of externalized theories rather than models in the sense intended here (p. 17).

Powers's background had been in physics and astronomy. He sought a psychological model with an inner structural integrity and an explantory power analogous to the molecular theory of matter.

The second problem he saw with previous attempts to formulate models of the brain was their discrete character. Earlier theorists had made a mistake, he believed, in using the digital computer as their underlying metaphor. He deliberately chose a continuous representation based on an analog model, and his reasons stemmed partly from his physical science background:

For purely practical reasons, engineers treat electrical current as a smoothly variable continuous quality and ignore what they know to be too detailed a representation of reality for their purposes. The level of detail one accepts as basic must be consistent with the level of detail in the phenomena to be described in these basic terms. ... No one neural impulse has any discernible relationship to observations (objective or subjective) of behavior. Even if we knew where all neural impulses were at any given instant, the listing of their locations would convey only meaningless detail, like a half-tone photograph viewed under a microscope. If we want understanding of relationships, we must keep the level of detail consistent and comprehensible, inside and outside of the organism (p. 21).

The argument is similar to Forrester's justification for a continuous point of view of social system structure and dynamics (see section 3.3). Powers concluded that neural current, defined as a number of neural impulses per unit time, is the appropriate basic measure of neural-system activity, not the neural impulses themselves.

He then proceeded to develop at some length a picture of the basic neural structures in the brain, including micro-structures that add, subtract, multiply, branch, amplify, integrate and differentiate with respect to time, and perform logical operations. Armed with these micro-structures, he proceeded to define a perception or a perceptual signal as "a neural current in a single fiber or bundle of redundant fibers." The magnitude of every such perceptual signal, he assumed, is related to some set of "primary sensory-nerve stimulations." These continuous perceptual signals are the foundation of the feedback theory of human behavior that he developed:

The analogue or continuous variable model provides a way to describe perceptions as the outcome of a process whereby an external state of affairs is continually represented inside the brain as one or more continuous neural signals. These signals—perceptual signals—are reality as far as the brain experiences reality (p. 39).

Feedback and Human Behavior

Like other feedback authors we have encountered in the discussion in the preceding sections, Powers finds feedback to be an inherent, essential fact of life:

What an organism senses affects what it does, and what it does affects what it senses. Only the first half of that commonplace observation has been incorporated into most psychological concepts of nervous system organization (p. 41).

While he is understandably critical of those who have not taken feedback into account, he is even more critical of the way the concept has been, in his opinion, misunderstood or misapplied.

He notes, for example, that D.D. Hebb used the word "feedback" to describe sequences of stimuli and responses:

"Any behavioral response to a single stimulation thus produces a sensory feedback which can act as the initiator of a second response, whose feedback initiates a third response, and so on" (Hebb 1964, p. 58; quoted in Powers 1973, pp. 41–42).

Powers finds this a first approximation to a "closed loop of causes and effects," but asserts that it is incorrect. Hebb's model does not behave like the organism:

The model behaves in a series of discrete, well-separated responses alternating with discrete stimuli. The real organism behaves in a smoothly continuous manner, with both responses and stimuli continually changing and continually interacting (p. 42).

Again, the objection to a discrete view rather than a continuous one, but here the focus is not on the underlying neural model but on circular causality in observable behavior. Powers's objection here is precisely what concerned John Dewey about the reflex arc more than 75 years before (see section 2.5), although Powers does not indicate he knows of the august company he keeps.

Annett's treatment of the feedback concept (Annett 1969), drawing a distinction between action feedback (which occurs during a response) and learning feedback (which occurs later) fares somewhat better in Powers's view:

> Action feedback occurs as soon as a response begins and can affect the response while it is happening. Learning feedback, which Annett prefers to call by an older terminology, knowledge of results (KR), is generally delayed enough to prevent the response from being directly affected by it. Annett correctly recognizes behavior with action feedback as an example of servomechanism behavior and illustrates this type of device with a diagram essentially like the one Norbert Wiener used in 1948 as a model for a kinesthetic feedback control system (p. 43).

The diagram Annett showed is reproduced in this volume as Figure 3.2b. Powers notes that the way it is drawn allows the feedback unit as a whole to be thought of in terms of stimulus (input) and response (output), just as behaviorists are wont to do. But in Powers's view, that diagram has significantly misled a generation of social scientists. He observed, correctly, that the signal at the "input" location in Wiener's diagram is more properly thought of as the reference signal against which some existing condition in the loop is compared. In stark contrast to the behaviorist's external point of view, it is not "input" from the external environment but rather a reference condition internal to the organism.

Powers concluded:

> The general importance of feedback is well recognized, but only a small percentage of behavioral scientists have dealt with it correctly even in restricted areas of behavior. It is my purpose here to expand somewhat on the ideas of these other workers and to try to show that feedback, when correctly analyzed, is the central and determining factor in all observed behavior (p. 44).

The Tracking Experiment

Powers (1973) develops the details of his feedback view of behavior in terms of a classical experiment in which a subject attempts to track a moving spot on a screen with a cursor that he or she controls with a joystick. In this context Powers's notion of inner "reference signals" becomes vivid. At first glance the subject appears to be motivated to eliminate any error between the perceived positions of the target spot and the cursor. However, suppose the experimenter whispers in the subject's ear to keep the cursor two inches above the spot. The subject can easily do that, and observers would see a steady-state error between cursor and target:

What is important is that the subject is not simply responding to "error" as we naively saw the situation at first. He is responding to an error, but the error is of a totally different kind. It is the difference between some condition of the situation as the subject sees it, and what we might call a reference condition, as he understands it. . . . The reference condition determines where the spot of light will be; the target does not. The motions of the target simply tend to cause a disturbance of the actual state of affairs away from the reference condition, and the subject moves the stick in any way that is required to cancel the effects of those disturbances before any large errors result (pp. 45, 46).

Powers notes that the reference condition is not directly observable. It is easy to define, however, as the condition that calls forth no response from the subject (p. 45). It functions as the goal, or more precisely the perception of the goal, of the situation. The subject has two perceptual signals: one for the actual position and one for the desired, or reference, position. Not being directly observable, the reference perception is a hypothetical construct. But seen as a part of a feedback structure it has real behavioral meaning:

Since behavioral feedback of any significance is always negative (because it is always in opposition to disturbances from a goal conditions), it follows that there will always be a tendency to move toward a zero-error condition calling for no effort, and thus one will always be able to discover the reference condition (if he is clever enough). By the same token, one will always be able to discover what the subject is controlling, for if disturbances are applied which do not in fact disturb the controlled aspect of the environment, the subject's behavior will not oppose the disturbance (p. 47).

The reference condition—the internal perceptual signal—is the condition of zero perceptual error. Hence the meaning of the title of Powers's book: Behavior is the organism's attempt to control its perceptual signals, to bring them into a state of zero error.

Powers's view alters very significantly the common interpretations of "stimulus" and "response" and changes how we view their causal connections. In this feedback interpretation of a tracking experiment, the "stimulus" is not the target spot. The stimulus is the disturbance that creates a discrepancy between reference conditions and actual conditions, as the subject perceives them.

This means that when an event occurs that tends to disturb a controlled quantity, the organism will respond. The response, however, will be related to the cause of the disturbance only indirectly, through a commonly affected quantity—the one under control by the organism (pp. 48–49).

Why does a given action follow a given disturbance? Why are different disturbances sometimes followed by the same response? Why does the same disturbance sometimes result in different responses? Behaviorism has trouble with such questions. In Powers's feedback view, however, the reasons are clear. The "stimulus," seen as an error between the perception of some condition and its reference perception, is a function of two things. A change in either can bring a change in behavior—a response. And changes in either that bring about a condition of "zero error" result in no behavioral response.

In Powers's view, these observations show that the classical stimulus-response view of behavior had to fail. In its place should be a psychology of human behavior founded on the feedback concept and the control of reference perceptions. Though born in the comparative simplicity of a tracking experiment, in Powers's view the theory covers all human behavior:

> The main proposition in this book is that all behavior is oriented all of the time around the control of certain quantities with respect to specific reference conditions. The only reason for which any higher organism acts is to counteract the effects of disturbances (constant or varying) on controlled quantities it senses. When the nature of these controlled quantities is known together with the corresponding reference conditions, variability all but disappears from behavior (pp. 47–48).

To give his feedback view the potential to be so universally applicable, Powers postulates a hierarchical organization of neuro-muscular and psychological control systems.

Powers's Hierarchy of Psychological Control Systems

In the tracking experiment, a whisper in the subject's ear can change the observed behavior: the cursor stays an inch above the target, or it leaves the moving target altogether and sticks to an edge, and so on. Clearly, the reference condition for the cursor can change. Thus some psychological process acts to set that reference condition. It is even conceivable that the subject could get bored and leave: some other set of reference conditions could supersede all those associated with performing the experiment. Powers concludes that there is a hierarchy of negative feedback control organizations visible in the subject's behavior (1973, p. 52). In this hierarchy, higher level systems set the reference conditions for lower level systems and in return receive information about errors between controlled conditions and their reference values.

Powers pursued in considerable detail the possible structure of such a hierarchy of psychological control systems within the individual (1973, Chapters 6–14). The details are not particularly of interest in this investigation. Powers himself acknowledges that the specifics of the hierarchy he sketches may very well be wrong. But the general characteristics of his conceptualization are worth noting here.

The hierarchy is built from replications of a generic "control unit of organization," shown in Figure 4.20. This unit has a clear negative feedback loop structure: the goal of the control unit is to reduce the perceptual error signal to zero and to maintain it there in the face of environmental disturbances. To illustrate the hierarchical structures that can be built from such control units, Powers described a model of a person in the tracking experiment. Shown in Figure 4.21, the model contains three levels of control, as may be seen by noting the three "comparators" in the figure.

In view of the similarity of purpose between Powers's work and that of Miller, Galanter, and Pribram (1960), it is interesting to compare Powers's general model (Figure 4.20) with the TOTE unit (section 4.3). There are two principal differences, both of which Powers specifically intended. First, although I cannot vouch for its biological accuracy, Powers's structure has a clear biological flavor. The TOTE unit does not. As Powers said, the TOTE is a description of behavior, not an explanation of how the organism actually functions neuro-physiologically. Second, Powers's structure is phrased in continuous terms: neural signals are treated as current flows rather than as the discrete firings of synapses. In contrast, the TOTE unit is based fundamentally on a discrete point of view: operations and tests follow one after the other in orderly sequences.

Furthermore, in applications of Powers's general control unit, one can assign link and loop polarities, since the items in the loop can be viewed as quantities (representing the strength or amount of neural current). The elements in a TOTE loop are not quantities, and its links represent sequencing rather than positive or negative influences. As we have observed, one infers the negative polarity of the TOTE unit not from its links but from its operation as a control device. Thus Powers's generic control loop model is much more like what authors in the servomechanisms thread would have produced, had any of them addressed the issues that concerned Powers and Miller, Galanter, and Pribram. A system dynamicist in the Forrester tradition would feel right at home with the structure and dynamics reflected by Figure 4.21. Indeed, in its complete form Powers expects his hierarchical feedback control model to aim toward "a literal block diagram of the functions of the human nervous system" (p. 78).

Powers's entire theoretical hierarchy of feedback control structures purportedly governing all human behavior contains at least eight or nine levels of control. Acknowledging the complete structure to be "preliminary," Powers labeled its levels as shown in Table 4.4. In Powers's words:

The model consists of a hierarchical structure of feedback control organizations in which higher-order systems perceive and control an environment composed of lower-order systems; only first-order systems interact with the world.

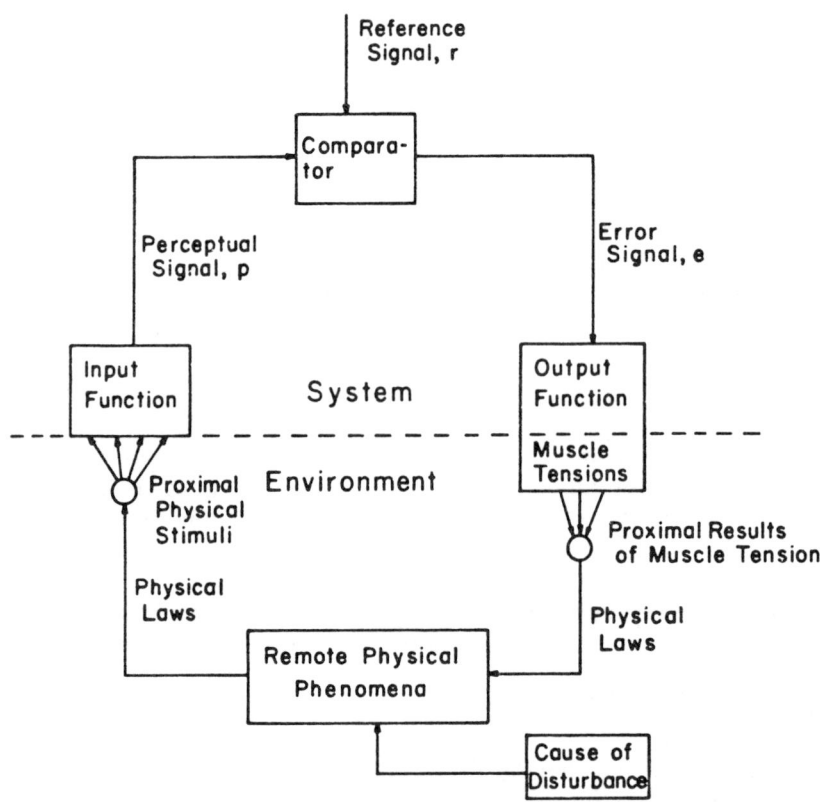

FIGURE 4.20: General model of a feedback control system and its local environment. Source: Powers (1973, p. 61).

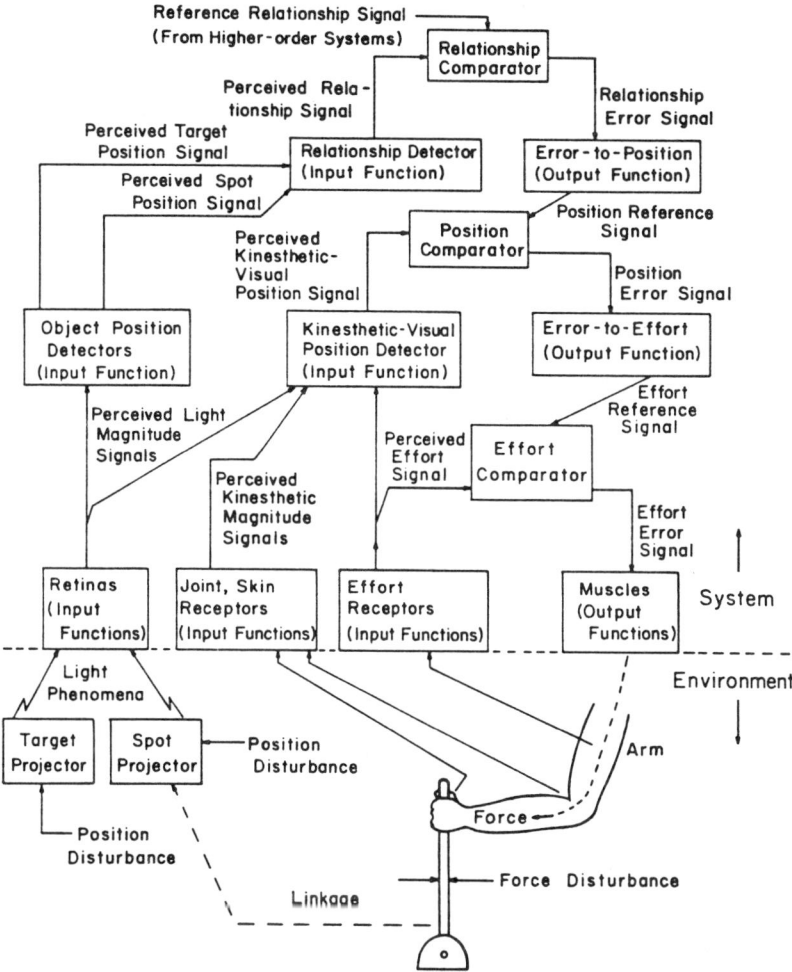

FIGURE 4.21: Three-level control system model of a person in the tracking situation, illustrating Powers's conception of a hierarchy of psychological control systems composed of negative feedback control units. Source: Powers (1973, p. 71).

TABLE 4.4 Hierarchy of psychological control systems, abstracted from Powers (1973, pp. 82–176)

Level	Focus of control	Examples
First-order	Intensity	Muscle tension; spinal reflexes
Second-order	Sensation	Kinesthetic perception
Third-order	Configuration	Posture; grasping; phonemes
Fourth-order	Transitions	Motion; time; change; warming
Fifth-order	Sequence	Walking; word sequences
Sixth-order	Relationships	Cause-and-effect; categorization
Seventh-order	Programs	Looking for glasses to read with
Eighth-order	Principles	Problem-solving heuristics
Ninth-order	System concepts	Perception of unities in abstraction
Higher levels	Spiritual phenomena?	

The entire hierarchy is organized around a single concept: control by means of adjusting reference-signals for lower-order systems. This is almost certainly an oversimplification, if only because the control systems are all nonlinear to some extent. There may well be other models of control such as adjustment of parameters of control systems (p. 78).

Up to the fourth or fifth levels, Powers finds neurological evidence for the control systems he postulates and suggests where in the body's neurophysiology they reside. Above the fifth level, he is less certain of the accuracy of the hierarchy, although he is absolutely convinced of the need for some such hierarchy in psychological functioning. (It is interesting to compare Powers's hierarchy with the one proposed by Beer [see section 4.2] for the components of a viable system. There are clear correspondences, with Beer's being more aggregated and more linked to the physical parts of the nervous system and Powers's being more detailed and, at the higher levels, more speculative.)

Awareness and learning

For what we call perception or awareness, Powers concludes that the brain must have its own "model" structured according to much the same hierarchy (pp. 147–153). Ashby and Deutsch (recall section 3.2) made very similar arguments. Powers's treatment of learning is also similar to that of other writers in the cybernetics thread. Learning is fundamental reorganization—"not a change in the way existing components of a system are employed . . . but a change in the properties or even the number of components" (p. 180). Powers's ideas here were strongly influenced by Ashby (1952), Pask (1960), and Yovitz and Cameron (1960). In particular he cited Ashby's homeostat as an example of a device based upon the feedback concept that is capable of self-reorganization. Furthermore, he identified his entire view of learning as "an extension of Ashby's concept of 'ultrastability'" (p. 183).

In his model for learning Powers hypothesizes that

there is a separate inherited organization responsible for changes in organization of the malleable part of the nervous system—the part that eventually becomes the hierarchy of perception and control (p. 182).

The essential character of this reorganizing system is that

It senses the states of physical quantities intrinsic to the organism and . . . controls those quantities with respect to genetically given reference levels (p. 183).

Elements of this reorganizing system are thus genetically determined and genetically stored "deep structure."

The structure and functioning of the reorganizing system is almost identical to the control unit and the control system hierarchy that the reorganizing system creates and modifies. "Intrinsic reference signals" are compared with signals representing the state of the hierarchy and its environment, producing "intrinsic error signals" that drive reorganization. The essential differences are that the inputs to the reorganizing system are "intrinsic signals" rather than "perceptual signals," and the outputs are reorganizations of the psychological control system hierarchy.

Powers argues that the reorganizing system is characterized by negative feedback. There is feedback present, he says, because the intrinsic state affects behavior (by setting reference conditions for all lower order processes), and conversely, changes in the behavioral organization affect the organism's intrinsic state. The feedback must be negative, he argues, because intrinsic error will stimulate reorganization of the control hierarchy unless the error is zero (p. 186).

Intrinsic error is self-correcting simply because reorganization is in principal capable of altering any behavior pattern, and these alterations are terminated by the behavior pattern (if one exists) that succeeds in restoring intrinsic error to zero. . . . The behavior pattern that reduces intrinsic error to zero stops the process of reorganization, and therefore that behavior pattern will persist (p. 186).

He emphasizes that the reorganization process is independent of the kind of behavior being reorganized.

The criteria for terminating some reorganizing do not depend on a control system's achieving some goal state for its perception, not at any level in the hierarchy including those levels associated with problem-solving behaviors. They depend only on physiological effects of carrying out any given behavior. Reorganization may terminate when one fails to remember or fails to solve a problem, if those failures result in restoration of intrinsic error to zero. One will then have become organized to fail in those situations. The reorganizing system has no pride (p. 187).

The lack of explicit goal states within the reorganizing system for the behavior that it controls is reminiscent of von Bertalanffy's distinctions in his concept of the open system (section 3.2).

The great distinction between Powers's view of learning and the stimulus-response approach is related to what is learned. In the stimulus-response view, the organism learns a specific response to a specific event. In Powers's view, learning is a continual process of reorganization that persists as long as there is intrinsic error (p. 192).

Conflict and control

Behavior: The Control of Perception concludes with an heroic extension and application of these ideas to the phenomena of interpersonal conflict, defense, and control. In Powers's feedback point of view, the usual methods of interpersonal control reduce to two choices: interference—disturbing the other's controlled quantities; or deception—altering the other's perceptions. But attempts to control others do not always work, and Powers traces the difficulties to the feedback concept:

We have tried to deal with behaving systems as we deal with inanimate objects, because until the advent of feedback theory the precise difference between animate and inanimate objects could not be correctly understood.

The behavior of an animate object—an organism—is governed by internal reference signals. The ultimate determinant of the organism's choice of reference signals is its set of intrinsic reference levels, which are not only internal to the organism but are inaccessible to external influences. The behavior of an organism can be influenced . . . , but the behavior of organisms is not organized around the control of overt actions or any randomly noticed effects they produce. It is organized around control of perceptions. That is why the two methods of control we have just discussed seem to work if not carried too far. They do not in fact control perceptions, but only actions. Organisms do not care how they act as long as the actions do not disturb the perceptions they do care about (pp. 264–265).

To Powers, this is the key point of his entire theory. Many of the details of the hierarchies he describes could ultimately prove to be misguided, but he would be surprised if future research falsifies the basic idea:

the concept of behavior as a feedback process organized around the control of perception, and reorganized as a way of maintaining ourselves in a peculiarly human condition defined by intrinsic reference levels (p. 272).

Application to Psychotherapy

Powers's essential notion—the concept of behavior as a feedback process organized around the control of perception—found its way to William Glasser, a noted psychiatrist, author, and developer of the approach to counseling known as reality therapy (Glasser 1965). In Powers's work, Glasser found the theoretical foundation for his own approach to psychotherapy. In *Stations of the Mind: New Directions for Reality Therapy* (1981), Glasser detailed his new view of "BCP psychology," psychology based upon the notion of Behavior as the Control of Perception, and sketched how this feedback view of behavior can work in understanding one's own stresses and in helping others.

Glasser's feedback view is captured in Figure 4.22, which appears on the inside front cover and repeatedly in various forms throughout the book. Translating some of Powers's theory into a more colloquial vocabulary, Glasser focuses on "comparing stations" in the human brain. A comparing station is a locus of brain activity at which a perception and its comparable internal reference perception come together. If they match, the perception is "controlled"; if they do not match, a "perceptual error" results and the person must do something intended to reduce it. All behavior, in Glasser's view, is traceable to such perceptual errors:

For all practical purposes *all my behavior is initiated by the error signal caused by the detection of an error in an open comparing station. When there is perceptual error there is always an error signal, and I must do something. It is a neurological fact of life; it cannot be disregarded.* The only way that I can avoid dealing with this error is to close down the station and thus cut off the error (Glasser 1981, pp. 52–53, emphasis in original).

In response to a mismatch between a perception and its comparable reference perception, a person can *redirect* his or her behavior, *reorient* toward new information that eliminates the perceptual error, or, in the most extreme cases, *reorganize* to select random new behaviors generated in an effort to reduce the perceptual error. Reorganization is the one tool available at birth; it is the source of learning. Redirection and reorientation to new information are more efficient processes for reducing perceptual error that emerge as a baby matures.

There is a marked resemblance between these views and Festinger's concept of cognitive dissonance (Festinger 1957). Indeed, cognitive

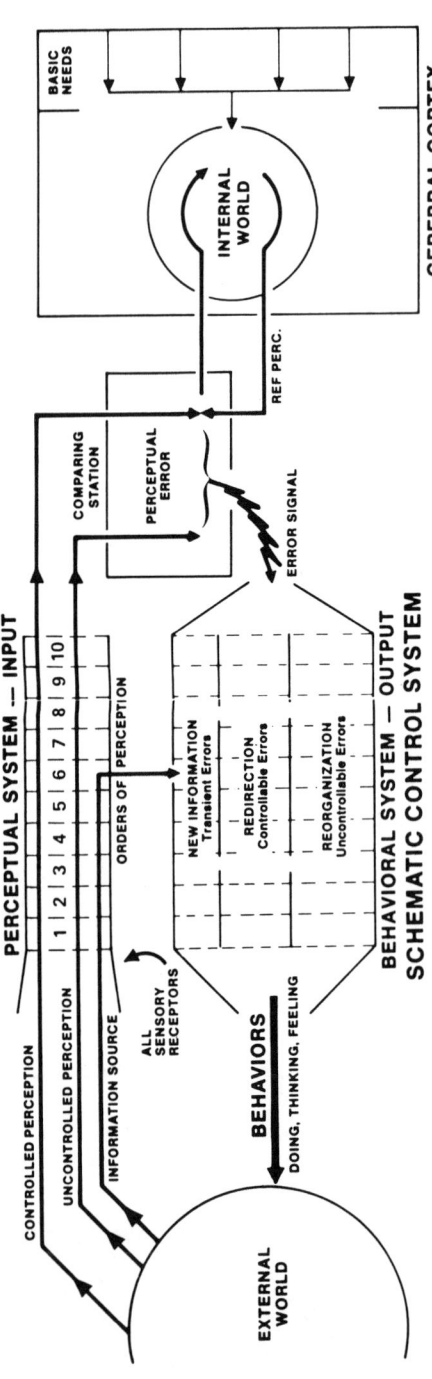

Figure 4.22: Schematic overview of BCP psychology. Source: Glasser (1984).

dissonance must be a kind of perceptual error. Festinger, we have noted (sections 1.1 and 4.5), did not phrase his theory in feedback terms, but it was replete with implicit negative feedback loops, as Figure 4.23 illustrates. It is especially interesting to note that Festinger predicted that people experiencing cognitive dissonance would seek out different kinds of information in an effort to reduce their dissonance. Powers and Glasser posit exactly the same process, and raise it to the level of one of three principal psychological tools underlying all human behavior. Given the fame of Festinger and the principle of cognitive dissonance, it is reasonable to conclude that both Powers and Glasser were aware of the earlier idea. Whether they consciously transformed Festinger's thinking into feedback terms, or did so unconsciously, automatically, as they developed their feedback view of behavior, would be impossible to say without asking them.

Counseling from a feedback control theory point of view

Powers's hierarchy of psychological control systems appears in Glasser's view as "orders of perception," ranging from intensity and sensation at the low end to principles and systems concepts at the high end. This perceptual hierarchy provides Glasser with a focus in counseling based on BCP psychology. In Glasser's feedback view, psychological distress and debilitating behaviors such as acute headaches or depression are the result of large perceptual errors. A person, consciously or more likely unconsciously, selects various behaviors—"headaching" or "depressing"—in an effort to reduce those perceptual errors. The higher a perceptual error is on Powers's scale of orders of perception, the greater the psychological distress. Thus one effort in counseling is to help a client move a perceptual error down the scale, toward a "comparing station" of less psychological significance. The final goal of BCP counseling is the same goal each of us has, to reduce our "total perceptual error."

Glasser's counseling approach, reality therapy, suggests that a counselor work through eight steps with a client. The first four—Making friends, What are you doing now?, Is this behavior helping you?, and Making a plan—all involve explicit references to feedback and the control of perceptual errors.

Like many other therapists, Glasser finds Making friends to be crucial, and he traces its significance to the basic need of belonging: ". . . when belonging reduces their total error, they drive their behavioral system much less frantically" (1981, p. 268). A sense of belonging—acceptance—by the counselor can reduce the total perceptual error, and when that happens, "better behaviors immediately become

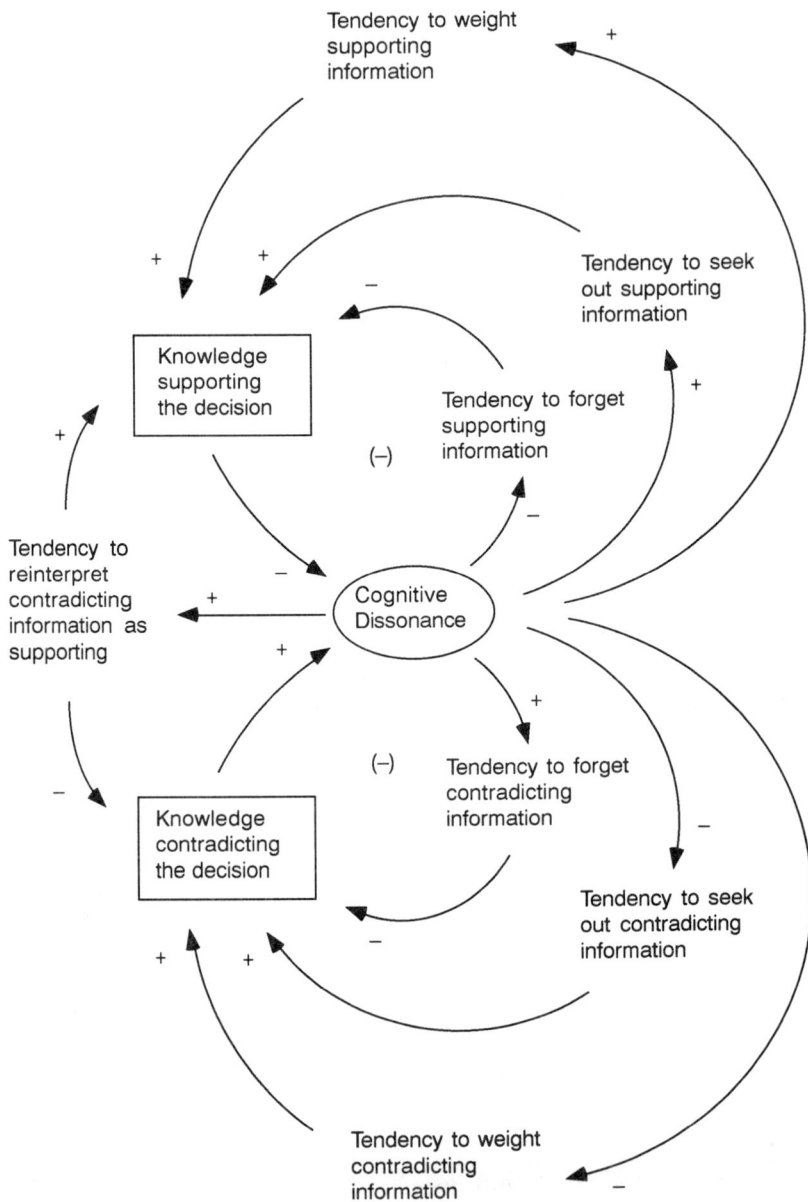

FIGURE 4.23: Festinger's theory (1957) of cognitive dissonance represented in terms of feedback loops.

possible." Glasser eventually has his client look at the BCP chart (Figure 4.21) to become familiar with its feedback control theory perspective. That enables therapist and client to "take a hard look at the uncontrolled reference perceptions that are coming from the client's internal world."

The next three steps form a unit. In asking What are you doing now?, Glasser pushes toward the notion that even thinking and feeling are behaviors. For example, he substitutes the term "depressing" for depression. That change allows therapist and client to investigate the extent to which a given behavior is helping or hurting. This evaluative step is uncomfortable—total perceptual error will increase momentarily, until the client devises a substitute behavior that can reduce his or her total perceptual error. Thus it must be tied closely to the fourth step, Making a plan to do better. In helping the client stick to a plan, the therapist should be dogged, Glasser asserts, but should never use punishment. Criticism simply increases total perceptual error; criticism from the therapist-now-friend is likely to result in large errors at comparing stations rather high up on the scale of orders of perception. In general, the counselor is cautioned to do nothing that raises the client's total perceptual error.

Eventually, acting on plans devised to reduce the feedback discrepancy between the client's internal reference conditions and her perceptions of reality will, in Glasser's view, result in less total perceptual error and more choices for healthy behaviors.

Addiction fits neatly into such a feedback view of behavior (Glasser 1981, 201–226). Some drugs, such as heroin, have their addicting effect because they increase pleasurable sensations by stimulating or mimicking neuro-transmitters that are felt as pleasure. In Glasser's BCP perspective, their positive pleasurable effects can swamp the negative effects of perceptual errors. Other drugs, such as alcohol, are addicting not because they increase pleasure but because they can hide pain. Glasser concludes,

> It seems that alcohol addiction can be almost completely explained by one simple but profound effect which occurs at all of our open comparing stations, that is, it reduces our ability to sense a perceptual error (p. 209).

Tranquilizers have the same effect and can be similarly addicting. Glasser uses this feedback control point of view to explain both physical and psychological withdrawal symptoms.

The significance of these ideas for us in this study is not their correctness—there are surely many approaches to therapy and counseling that have strong devotees and theoretical bases. Here the significance of Glasser's views, and the work of Powers on which he bases them, is the explicit foundation of BCP psychology on the feedback concept. The concept of "perceptual error" is the engineering control theorist's "error signal." Behavior that reduces such error signals will be selected over behavior that sustains or increases them. A hierarchy of control systems is postulated that selects what perceptual errors to attend to. Human behavior is thus a consequence of a profoundly intricate feedback control system organized, according to Powers and Glasser, to control what we perceive.

Powers's Relationship to the Two Feedback Threads

As we have noted, by the feedback authors he cites Powers places himself squarely in the cybernetics thread. However, he is extremely selective in what he takes from that line of feedback thinking. He takes the concept of feedback as his fundamental viewpoint, but he explicitly rejects the digital or discrete view that characterizes much of the use of the feedback concept in the cybernetics thread. He takes a number of Ashby's ideas, including the homeostat and ultrastability, but he completely ignores requisite variety and the restriction to "input manipulation/output classification." He makes use of Wiener's block diagram of a basic feedback control unit, but he significantly (and correctly) alters the usual interpretation of the input to the loop, labeling Wiener's input as the "reference signal" and noting that the input as "disturbance" is not shown in Wiener's figure (recall Figures 2.7 and 3.2). He shares the interest in brain functioning that stimulated the earliest cyberneticians, and he shares their exclusive focus on negative feedback processes, but in what I regard to be a remarkable article he eventually turns to the mathematics of servomechanisms to explain some of his thoughts.

Powers's "Quantitative Analysis of Purposive Systems"

Powers's most extensive use of mathematics appears in a later work (Powers 1978). The article is fascinating for what it says about the use of the feedback concept in the social sciences, but it is also of considerable interest to us for its citation evidence, which shows that Powers by this time had direct links to the servomechanisms literature.[27]

The four blunders

In this article Powers exposed four serious mistakes he perceives in the use of the feedback concept by social scientists, which he labels the machine analogy blunder, the objectification blunder, the input blunder, and the man-machine blunder. There were suggestions of these ideas in Powers (1973), but in this later article his thinking had clearly crystallized.

In the *machine analogy* blunder, Powers accuses other social scientists of mistaking the fundamental importance of feedback in human behavior:

> There are and have been for some time scientists who think of control system models of behavioral organization as a mere analogy of human behavior to the behavior of a technological invention.

But that point of view is grossly mistaken, according to Powers. Feedback is a fact of living systems, not a mere analogy.

> Servomechanisms have always been designed to take over a kind of task that had previously been done by human beings and higher animals and by no other kind of natural system.... The servomechanism has always been only an imitation of the real thing, a living organism, and the engineers who invented it first had to be, however unwittingly, psychologists. The analogy developed from man to machine—not the other way (1978, p. 418).

One is reminded of the primitive person washing clothes in a river by beating them with a stick. Imagine the difficulty of designing a washing machine that could sense, as the primitive person can with ease, when the clothes are sufficiently clean. In the machine analogy blunder, Powers correctly identified a serious stumbling block to the deep and accurate appreciation of feedback in human systems.

In the *objectification blunder,* Powers finds fault with the prevailing "scientific" study of behavior that insists on trying to understand behavior in terms of its objective appearance. In the process he suggests a significant reinterpretation of Skinnerian psychology:

> A natural control system can be organized only around the effects that its actions (or independent events) have on its inputs (broadly defined), for its inputs contain all consequences of its actions that can conceivably matter to the control system.
> This was Skinner's [*The Behavior of Organisms,* 1938] momentous discovery. He concluded that behavior is controlled by its consequences, unfortunately expressing the discovery from the observer's or user's point of view. From the behaving system's point of view, however, Skinner's discovery is better stated in the following way: Behavior exists only to control consequences that affect the organism. From the viewpoint of the behaving system, behavior itself, as output, is of no importance. To deal with behavior under any model strictly in terms of its objective appearance, therefore, is to miss the reason for its existence. Cybernetics and especially engineering psychology simply took over this erroneous point of view from behaviorism (p. 419).

From Powers's feedback point of view, behavior exists to control perceptual signals. Unless one tries to look at what is being controlled, one misses the significance of any behavior.

The *input blunder* is the misinterpretation of the "input" in Wiener's feedback block diagram, which we have previously described. The blunder is that Wiener's input is the reference signal or goal state for the quantity being controlled in the feedback loop. The overwhelming tendency is for social scientists dealing with natural (living) systems to treat that reference signal erroneously as sensory input or stimulus.

> In natural control systems, there are no externally manipulatable reference inputs. There are only sensory inputs. Reference signals for natural control systems are set by processes inside the organism and are not accessible from the outside. Another name for a natural reference signal is purpose (p. 419).

The fourth mistake Powers perceives—the *man-machine blunder*—underlies several of the preceding errors. It is the tendency, however unwitting, to treat people as machines. In Powers's view, this mistake manifests itself in the behavioral psychologist's persistent tendency to miss the existence of people's internal reference signals.

Mathematics and a fundamental behavioral illusion

Powers turns to the mathematics of servomechanisms engineers to draw out the significance of these blunders for the study of psychological phenomena. The structural setting for his equations is shown in Figure 4.24 (together with the original figure caption cautioning the reader against misinterpretation). He begins with two simple equations:

$$q_o = f(q_i), \tag{1}$$

$$q_i = g(q_o) + h(q_d), \tag{2}$$

reflecting the assumptions that output q_o is a function of the input q_i, and input is a function of both a disturbance q_d and a feedback signal $g(q_o)$. The function f represents characteristics of the behaving organism; the function g represents environmental feedback effects; the function h represents perception or sensation.

If there were no feedback (the "open-loop" situation), simple substitution shows that the output quantity would be

$$q_o = f[h(q_d)]. \tag{3}$$

That is, the organism's behavior would be a direct function of the disturbance. However, with the negative feedback link present and sufficiently influential (a "high loop gain"), Powers is able to show that the system approaches the following "ideal" system:

$$g(q_o) = q_i^* - h(q_d), \qquad (4)$$
$$q_i = q_i^*. \qquad (5)$$

In such an ideal system, any disturbance q_d is immediately counteracted, and the system holds its input quantity q_i at the no-disturbance value q_i^* at all times. (It is an "ideal" system because it requires an infinite loop gain to accomplish this feat. Compare to the discussion of Maxwell's analyses of proportional control in section 2.1.)

Armed with this mathematical machinery, Powers then exposes what he calls "a behavioral illusion of such significance that one hesitates to believe it could exist" (p. 425). Solving equation (4) for the output q_o, one finds

$$q_o = g^{-1}[q_i^* - h(q_d)], \qquad (6)$$

which bears a remarkable resemblance to the open-loop situation captured in equation (3). Both express the output q_o as a function of the disturbance q_d. The principal difference is that in the open-loop situation the organism function f appears, while in the ideal closed-loop situation the environmental feedback function g appears.

It is a crucial difference. Powers observes that experimenters igno-

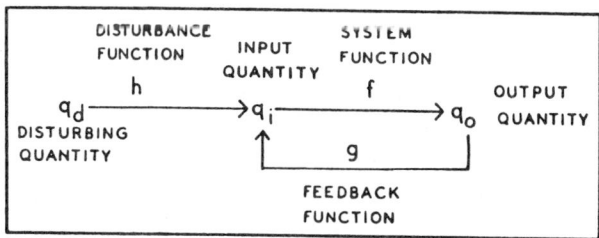

Figure 3. Relationships among variables and functions in the quasi-static analysis. (The topological similarity of Wiener's [1948] diagram, adapted in the present Figure 1, is of no significance because these variables and functions all pertain to observables outside the organism. This is not a model of the organism; it is a model of the organism's relationships to the external world.)

FIGURE 4.24: Feedback structure that Powers analyzes quantitatively. Source: Powers (1978).

rant of the feedback view are very likely to misunderstand what they see:

> If one varies a distal stimulus q_d and observes that a measure of behavior q_o shows a strong regular dependence on q_d, there is certainly a temptation to assume that the form of the dependence reveals something about the organism. Yet, the comparison we have just seen indicates that the form of the dependence may reflect only properties of the local environment [the function g^{-1}, not the function f]. The nightmare of any experimenter is to realize too late that his results were forced by his experimental design and do not actually pertain to behavior (p. 426).

Powers illustrates the significance of this behavioral illusion in several experimental settings, and notes in each case that what we observe is not the function f that describes the organism, but rather the function g that describes the "physics of the feedback effects."

> This property of [negative feedback] systems is well known to control engineers and to those who work with analog computers. It is time behavioral scientists became aware of it, whatever the consequences (p. 427).

To Powers, the consequences would be nothing short of a revolution in behavioral science.

To illustrate, Powers concludes this remarkable article with reported results from six empirical experiments that are difficult to explain from a behaviorist point of view but fit neatly into Powers's feedback framework. In the most complex experimental settings, a subject in a tracking experiment is given four cursors influenced by different random disturbances and, in some nonlinear fashion, by the position of the joystick. The subject may pick which cursor or cursors to attend to, and even pick what aspect of the display to hold constant (position, difference in positions between two cursors, etc.).

Powers reports that in these experiments the statistical evidence an experimenter would collect is very misleading. Since the joy-stick affects all the cursors, a typical correlation between the handle position and one of the noncontrolled cursors is frequently as high as 0.8. Furthermore, the correlation between a controlled quantity and its disturbance is normally lower than 0.1 and for a practiced subject is frequently zero to two significant digits. Finally, the correlation between handle position and the magnitude of the disturbance is normally higher than 0.99. Yet the subject cannot sense any of the disturbances except through their effects on the inputs, the cursor movements which he or she also affects directly. Powers concludes:

> If the controlled input quantity shows a correlation of essentially zero with the behavior, any standard experimental design would reject it as contributing

nothing to the variance of behavior. But the disturbance that contributes essentially 100% of the variance of the behavior can act on the organism only via the variable that shows no significant correlation with behavior. Not only the old cause-effect model breaks down when one is dealing with a [negative feedback] system, the very basis of experimental psychology breaks down also (p. 434).

Summary

These are not shy ideas. Powers's interpretations and uses of the feedback concept suggest the need for fundamental changes in the ways social scientists do business. In particular, Powers's work correctly identifies significant flaws in the way some have interpreted and made use of the feedback concept. In this study that is an important development. But Powers's controversial views are potentially more far reaching. He himself concludes that his feedback approach forms the basis for "a scientific revolution in psychology and biology, the revolution promised by cybernetics 30 years ago but delayed by difficulties in breaking free of older points of view" (Powers 1978, p. 434).

In light of the evolution of the feedback concept we are tracing, it is evident that Powers himself has broken free of some of the newer points of view of the cybernetic thread as well. He has adhered to a continuous view based on quantitative notions, while others in the cybernetics thread emphasized discrete, sequential processes described and analyzed linguistically. As a consequence, he has by-passed the issues associated with meta-languages that so influenced Beer, Miller, Galanter, and Pribram, and others. The logic thread is entirely absent in Powers's work. Sociologically, in terms of citation influence, he is in the cybernetics thread. Methodologically, his use of the feedback concept has a great deal in common with the servomechanisms thread. He is almost alone in the cybernetics thread in making use of the mathematics of servomechanisms engineers to argue serious social science conclusions.[28] One is prompted to wonder if his physical science background predisposed him early on to think somewhat like a servomechanisms engineer before he ever encountered the cybernetic ideas that so influenced the direction of his researches.

4.8 Summary and Directions

In Chapter 3, we observed that the cybernetics thread—the chain of feedback thinkers and thoughts in the social sciences that can be traced to Wiener and other writers in the field of cybernetics—began with the following associations:

- Feedback was defined in terms of input and output. It was seen as the influence of output back on input.
- The use of the feedback concept was limited to loops of negative polarity.
- Feedback was viewed as the mechanism of homeostasis and control.
- The negative feedback concept was not associated with the general concept of mutual or circular causal processes.
- The concept of feedback was used to address philosophical and theoretical questions relating to control.
- The stability of a feedback system, and the conditions producing instability, were a central concern.
- Feedback analyses are usually verbal rather than mathematical or pictorial.
- Feedback was viewed as the information transmitted in messages. The concept was associated with communication networks and information theory.
- Feedback was associated with the creation of intelligent machines and the automation of human functions.

In this chapter we have seen how these associations have developed in the work of a selected number of social scientists. The feedback concept continued to be viewed in terms of input to and output from a system, with particular emphasis, especially evident in the work of Beer and Deutsch, on the stochastic nature of inputs. Treatments continued to be largely verbal, although increasingly augmented by diagrams. Issues of philosophical or theoretical significance associated with the feedback concept continued to dominate. The dynamic focus continued to be on essentially two polar possibilities—stability or instability.

Another characteristic of the early work in cybernetics is evident in the work of several of these later writers: an emphasis on the brain and neurophysiology, either as a direct focus of study, as in Powers, or as an analogy, as in the work of Beer and Deutsch. The latter is particularly interesting for our feedback focus, because the concept of feedback becomes interpreted in terms of neural pathways. For Deutsch the early cybernetics notion of a neural net becomes the communication network of a complex intercommunicating system, literally the nerves of government. Feedback as transmission of information takes place in such networks, but the concept of self-reinforcing and self-correcting feedback loops of positive and negative polarity is only minimally in evidence.

In Beer's work, which is the closest that we have investigated to ongoing work in the actual field of cybernetics, the brain serves as both a source of analogies about management systems and as a master monitor and controller. Systems as complexly interconnected as brains, economies, and corporations are not to be understood, but they can be managed by cybernetic principles. The computer, as an electronic brain, emerges as the substitute for human understanding in the control of corporations and economies.

I do not know what to make of Beer's later use of system dynamics models. Their presence in his work appears to contradict his insistence that exceedingly complex systems cannot be understood. There are two reasonable possibilities. His view of the potential for understanding complex feedback interactions in an economy may have evolved over time, or he may have seen system dynamics models as another attempt to manage complex systems by computer technique.

In the work of several of these authors we have noted that feedback loops were interpreted as the transmission and return of discrete messages. In that form the concept is captured as a linguistic rather than quantitative phenomenon, as in the TOTE unit and Deutsch's communication nets. There are two results of this translation of the feedback concept into linguistic terms. One trivializes the notion of feedback to the level of chit-chat in a conversation. At this trivial extreme, without the concept of loop polarity, the idea reduces to the common terminology of "giving feedback," as in giving constructive criticism. At that level, feedback as "knowledge of results" is vastly weakened as a scholarly tool, a mere shadow of the powerful circular causal concept that excited Myrdal and guided the functionalists. Feedback at that level is about as useful to social scientists as learning that the "key to playing great chess" is to know that players alternate moves.

The other result of interpreting the feedback concept in linguistic terms is to associate it with the phenomena of self-reference, self-organization, and autopoesis, concepts that are certainly at the opposite end of the scale from trivial. It is in this direction that Deutsch's work and the TOTE unit of Miller, Galanter, and Pribram point. However, in spite of the potential significance of this direction, it is also not the powerful circular causal concept that Myrdal and others saw. What remains of feedback as circular causality in the cybernetics thread in the social sciences is largely exemplified by the work of Muruyama and Weick. It is characterized by the use of causal-loop diagrams and largely intuitive inferences about their implications for the system or problem under investigation.

Repeatedly we see in these works an emphasis on change in the feedback structure of a system. Weick's nine ways of "discounting a cause map," Maruyama's morphogenesis, TOTEs for changing TOTEs in Miller, Galanter, and Pribram, and other ideas in Chapter 4 suggest

266 Developments in the Cybernetics Thread

that structural change is an important concern to social scientists linked to the feedback concept. The concern matches the insistence on nonlinearity of structure in the servomechanisms thread, but in the cybernetics thread it is phrased differently. The principal mechanism of structural change in this thread is largely a linguistic idea or a visual one: structure is "rewritten" or "redrawn," and a new structure results. In formal models the idea connects most closely to modern computer programs in LISP that can rewrite parts of their own computer code.

Directions

We can discern in these writings four basic directions of development of ideas associated with the feedback concept in the cybernetics thread. Each of the following are currently active lines of thinking in the social sciences that have origins or direct connections to the feedback authors and works we have considered in section 3.2 and Chapter 4:
- a general systems thread—linking feedback with stability/instability and morphostasis/morphogenesis in living systems
- a communications thread—centering on linguistic loops, conversational planning, reflection, self-organization, and structural change
- an artificial intelligence thread—centering on the creation of intelligent machines, the automation of human functions and control, and computer simulation of human cognitive capabilities
- a causal loop thread—striving to derive dynamic implications from diagrams of feedback structure

There is also an ongoing line of work in the field of cybernetics itself that continues to emanate messages to the social sciences.

The feedback knowledge that a social scientist obtains from the cybernetic thread falls, I believe, into one or more of these categories. Some ideas from these threads have found their way into social science texts (e.g., Buckley 1967, Kuhn 1975).

Of central significance is still the lack of a fundamental association of the feedback concept with patterns of dynamic behavior—one of the primary notions that tends to characterize the servomechanisms thread. There are indications, however, in the late work of Beer and in the work of Maruyama, Weick, Easton, and Powers, of a tendency to move in directions more similar to the servomechanisms thread. And there are significant synoptic works such as James G. Miller's *Living Systems* (Miller 1978) that contain representatives of both threads of feedback thinking. I find Powers's work particularly intriguing, not only because of the content of his psychological theorizing but because of the depth of change in Powers's point of view that it seems to imply. Powers appears to move almost completely out of the conceptual thread from which he starts in 1960. Such a shift seems very rare.

Notes

1. A homomorphism is a mapping of one system onto another that preserves some structural relationships. An isomorphism is a special case of a homomorphism that loses no detail, that maps elements in one system one-to-one onto elements in the second system and preserves all their structural relationships. Beer uses the concept of homomorphism to represent a mapping that *reduces* variety. It is strictly a "many-to-one" transformation. In current common usage a "homomorphic image" of a system is a model.

2. See also Klir and Valach (1967), which develops in somewhat greater detail the binary/logical black box theory. Beer labels it "a 'must' for serious students of cybernetics" (Beer 1981, p. 409).

3. What happens in Beer's cybernetic control scheme is that information from the world situation at time t_1 (W1) passes through the models M1 to a management control point (labeled C1), from which emerges a forecast (M2) of the world situation at time t_2. M2 probably won't match W2, so information from the two are fed into the black box BB1, and ratios of corresponding terms emerge. Those ratios are used in the control point C2 to improve the predictions for the world situation at time t_3. If the forecast was only 70% of the actual, multiply by 1/.7 for the next forecast M3. Beer captured the resulting improvement by making the shape of the next forecast model M3 match more closely the world situation W2. Then there is a feedback interaction between the forecast M3 and the world situation W3 that is passed through the black box BB2 to improve still further the correction at C2, resulting in the marvelous prediction M4. A final black box BB3 and feedback loop helps to refine the original models M1a and M1b in case structural aspects of the system should change over time.

This description is a distillation of Beer (1966), pp. 313–330. I cannot claim to have captured the depth of Beer's discussion, in which the reader eventually learns that BB2 and BB3 are in fact the same black box, just as C1 and C2 are the same control point. See pp. 331–336 for a sketch of the statistical maneuvers designed into a black box control unit.

4. Echoing von Bertalanffy (section 3.2), the first claim confuses a "closed loop" of information and action with a "closed system," which exchanges no material or information with its environment. The second claim is more debatable, but it appears to be falsified in nonlinear systems, which can change dominant feedback (servo) structure (see section 2.2).

5. "Is there any science . . . which does not deliberate about some one particular thing in the city, but about the entire city itself, and in what way it may best orient itself towards other cities?" Quoted by Beer (1966, p. 401) from Plato, *The Republic* (Book IV).

6. In the coordination and integration of Plans in organizations and groups, Miller, Galanter, and Pribram saw a rich laboratory for the study of the processes of planning about Plans, citing in particular March and Simon (1958).

7. The ideas here are related to the "double-bind" theory of schizophrenia; see Bateson et al. (1956).

8. See Yovitz and Cameron (1960), von Foerster and Zopf (1962), von Neumann (1966), and Varela (1975, 1979).

9. At times, Maruyama's generalizations in this vein were extreme, as when he refers to "the study of other deviation-amplifying mutual causal processes such as history or mental illness" (p. 310). He meant that such phenomena contain both negative and positive mutual causal processes (personal communication), but the comment is jarring nonetheless.

10. Wiener (1948, p. 112), reproduced as Figure 3.2.

11. Here Easton referenced a variety of feedback publications, including Wiener (1948, 1954), Ashby (1952, 1956), Sluckin (1954), Vickers (1959a), Kuhn (1963), Deutsch (1948, 1963), and Forrester (1961). With the single exception of Forrester, his sources are thus authors in the cybernetics thread.

12. The phrase is from Vickers (1959a).

13. The citations are to Forrester (1961). Two are on the length and significance of delays in societal feedback systems (Easton 1965b, pp. 271, 410); one notes that Forrester's flow charts differ from Easton's (p. 71). Easton's first reference to Forrester's work singles out its treatment of the ability of social systems to "regulate their own behavior, transform their internal structure, and even go so far as to remodel their fundamental goals" (p. 19).

14. An example of an interesting twist of a connection between Easton's and Forrester's uses of the feedback concept concerns the distance from which each chooses to observe and analyze social feedback systems. Both stand back far enough to blur or aggregate details. Forrester, we have observed, advocated such a distancing because it appeared to him to yield the clearest view and the choicest insights. Easton, on the other hand, advocates a distant view because we do not know enough:

By standing so far off from a system that we do not see the multiple component units at the points of inputs and outputs, we shall greatly simplify the processes, it is true. It is as though we were initially reconciling ourselves to using a telescope rather than a microscope because we are not yet sufficiently confident of the units and processes that we want to lay open to detailed analysis (1965b, 377).

15. See section 3.2 for an example and discussion of such a loop from the work of Ashby.

16. Deutsch notes that his views about self-determination fit well with a suggestion of A.J. Toynbee's that "growth in organisms and civilizations should be measured by their increase in self-determination rather than by gains in size or complexity" (Deutsch 1963, p. 131). Deutsch's repeated references to Toynbee in *Nerves of Government* raises the possibility that one could find in Toynbee's writings intimations of the feedback loop.

17. Games have something of an information feedback structure, in the sense that one player's move constitutes "feedback" to another player on the consequences of the previous move. In that sense game theory could be said to be a feedback theory of behavior. The information loop that exists in games studied at this detailed level does not have polarity, however, so it is not central to this investigation.

18. Deutsch's statements about feedback in *Nationalism and Social Communication* are familiar from the previous discussion of *Nerves of Government* and are

consequently not repeated here. See Deutsch (1966, pp. 117, 120–122, 163, 166–167, 170).

19. Solow added this logistic expression for dA/dt to prevent D from going negative. It is not clear from the presentation whether the maximum M in this revised equation is the same as, or different from, the mobilized population M.

20. I have not represented dP/dt = pP in this diagram, since that equation contradicts the assumption that P = A + D.

21. The loop in Figure 4.9 that Weick determines to be the most important links population P, modernization M, sanitation S, bacteria B, and diseases D. It is a positive feedback loop because it contains an even number of negative signs.

22. There have been other, more quantitative attempts to answer the question of stability or instability from causal-loop diagrams. See, e.g., Levins (1974).

23. The Lotka-Volterra predator-prey system (section 2.2, Figure 2.12) contains three negative loops (and two positive ones) and is unstable, exhibiting sustained oscillations for any reasonable parameter values. The arms-race model of L.F. Richardson (section 2.2) contains two negative loops (and one positive one) and can be either stable or unstable, exhibiting exponential growth or goal-seeking behavior, depending on the values of the parameters chosen. A simple market structure containing two negative goal-seeking loops, one for supply and one for demand, could show very stable goal-seeking behavior. Ashby's example (section 3.2, Figure 3.4) that purports to be a goal-seeking positive loop is actually a system containing two negative loops and one positive one. In each of these cases and uncountable others, Weick's characterization fails.

24. There is some likelihood, however, that Weick intended the term "cause map" to refer to an image that a person has in mind—a cognitive map—not to a structural property of an external reality. Minds change notoriously quickly. Cognitive maps can too (see, e.g., Eden, Jones, and Sims 1979, 1983). This interpretation is consistent with much of what Weick says, but it is hard to reconcile with the importance he places on a circular causal view.

25. For the complete development, see Weick (1979, pp. 119–231).

26. The citations in Powers (1973) include Annett (1969), Ashby (1952), D.T. Campbell (numerous references), Gerard (1968), Gibbs (1970), McCulloch and Pitts (1943), Miller, Galanter, and Pribram (1960), Newell and Simon (1963), and Wiener (1948).

27. Powers (1978) cites Kelly (1968), Mayr (1970), and Starkey (1955).

28. The mathematics is familiar and important to Powers. He made further use of it in Powers (1978) to argue that a discrete "knowledge-of-results" or "stimulus-response-stimulus-response" view of behavior is fundamentally flawed. In his words "[such] analysis seems to succeed only because of the limitations of verbal or qualitative reasoning." To capture such a discrete view and prove his claim, he sets up a difference equation description of his basic input/output system:

$$q_o(t + 1) = F(q_i(t) - q_i^*), \tag{7}$$

$$q_i(t) = Gq_o(t) + Hq_d. \tag{8}$$

Equation (8) says that the organism changes its output q_o discretely at time t +

1 according to some function of the discrepancy between its input and its reference condition. Powers observes that this discrete system cannot be realistic as it is stable only for the range $-1 < FG < 1$. His solution is to modify equation (7) so that the organism's output closes only a fraction of the gap $F(q_i - q_i^*)$ between t and t + 1:

$$q_o(t + 1) = q_o(t) + K[F(q_i(t) - q_i^*) - q_o(t)], \quad (7')$$

where $0 < K < 1$. With this modification he is able to complete a mathematical analysis that shows that the system of equations [(7'),(8)] produces a picture of behavior more consistent with observation. He calls the adjustment in (8') a "dynamic constraint," and concludes:

Treating behavior as a succession of instantaneous events propagating around a closed loop will not yield a correct analysis, no matter how tiny the steps are made, unless this dynamic constraint is properly introduced (p. 428).

In Powers's view, feedback as knowledge-of-results or as sequences of stimulus-response pairs is thus incapable of accurately capturing behavior.

It is interesting to note that Powers's solution to the difficulties created by the discrete view of behavior involves the introduction of an accumulation into the equations. This is very like the solution suggested in section 3.2 for dealing with Ashby's concerns about the goal-seeking character of the discrete "positive" loop system $x_{t+1} = y_t/2$, $y_{t+1} = x_t/2$. Powers's revised equation (7') is also identical to one of Forrester's level equations, with K playing the role of the computation interval DT. We shall see similar problems with discrete difference equations in Culbertson's treatment of feedback in economic systems (section 5.3).

Chapter 5
Developments in the Servomechanisms Thread

> Everything an Indian does is in a circle, and that is because the power of the world always works in circles, and everything tries to be round.
>
> Black Elk, *Black Elk Speaks, Being the Life Story of a Holy Man of the Oglala Sioux*

5.1 Introduction

As we have defined it, the servomechanisms thread is an identifiable methodological and sociological line of feedback thought in the social sciences that differs from the cybernetics thread explored in Chapter 4. In section 3.3 we examined the early work in the servomechanisms thread in some detail, identifying and substantiating its differences from the cybernetics thread. We observed that the early writers in the servomechanisms thread all had direct links to the engineering control literature. The most important distinguishing characteristic of their work was its emphasis on deriving the dynamic behavior of a social system as a consequence of its feedback structure.

In this chapter, we shall trace the later directions taken by the early writers in the servomechanisms thread and the social scientists influenced by them. We shall find that the perceived power of feedback thinking in the servomechanisms thread stems from employing it from a particular conceptual distance. We shall also find that the tendencies of feedback thought that characterize this thread have not spread as widely in the social sciences as those we observed in the cybernetics thread in Chapter 4. Implications of these observations will be explored in detail in Chapter 6.

5.2 Feedback in the Later Work of Herbert Simon

We have observed that Simon was one of the early theorists who set in motion the servomechanisms thread in the social sciences. He was also one of the first social scientists to embrace the promise of computer simulation of human cognitive processes. His path-breaking studies of problem solving and thought (see Newell and Simon 1972) are based on a linguistic or information-processing model of cognition. Weizenbaum (1976, pp. 248–249) has observed connections between Newell, Simon, and Shaw's "means/end" analyses and Forrester's use of negative feedback. Forrester talks of a rate equation as a statement of a "discrepancy between the goal and the observed condition" (Forrester 1969, p. 13); Newell and Simon talk about "present objects" and "desired objects" and operators that reduce the differences. Nonetheless, there is an important difference. The emphasis in Simon's studies of how individuals solve problems is not on circular causality, but rather on the specific content and meaning of language-like problem-solving moves. The feedback concept of signed circular causal loops slipped into the background of Simon's thinking.

Just as he quickly adopted a feedback view of social system dynamics, Simon quickly came to regard feedback and circular causality as an obvious characteristic of social systems, one that seldom deserved special comment. Illustrative of this development is the offhand appearance of the feedback notion in March and Simon (1958). The book brings together a large number of propositions about the nature and behavior of organizations. Together the propositions form a complex picture of feedback effects and mutual causality. The authors were fully aware of the loop nature of the structures they described. Yet the word feedback does not appear in the index or the table of contents. In the few places it is used in the prose, it slips in without fanfare or apparent significance. The authors clearly expected that their readers would be familiar with the concept as a matter of course.

March and Simon observe early on, for example, that "there is a strong mutual interaction" between events and responses in an organization:

> The stimuli that are present at a given time are major determiners of what set [of responses] will be evoked or maintained; conversely, the set [of potential responses] at any given time will be a major determiner of what parts of the environment will be effective as stimuli. There is no circularity in this relation—just the usual kind of mutual interaction among variables of a dynamic system (March and Simon 1958, p. 10).

To these authors, feedback, mutual causality, and their connection to dynamics are just part of the social facts of life.

Causal-loop diagrams depicting feedback structures are interspersed throughout the book. The first ones, together with one of the authors' few explicit uses of the word feedback, appear in their discussion of Robert King Merton's "general bureaucracy model." Merton (1940) had argued that thinking of bureaucracies as "machines" could lead to

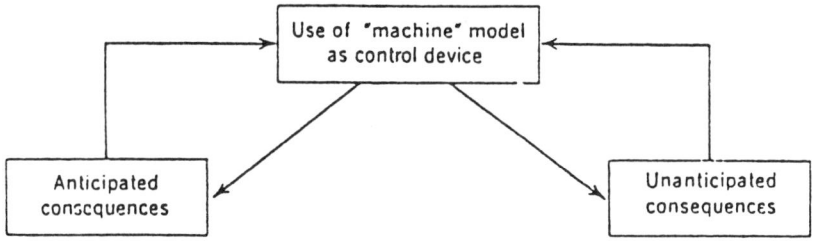

FIGURE 5.1: Merton's general bureaucracy model. Source: March and Simon (1958, p. 37).

control policies that had unanticipated consequences. March and Simon pictured Merton's argument in a closed-loop diagram, shown in Figure 5.1. (Once again, we have evidence of Merton as a feedback thinker.)

Merton's view was incomplete, however, from March and Simon's point of view. In an effort to explain why the members of an organization would learn to behave as Merton suggested, the authors felt it necessary to revise it as shown in Figure 5.2 to provide "at least one, and perhaps two, additional feedback loops in the system" (p. 40). The statement is, I believe, the first time the word is used in the book. No definitions are given; the reader is assumed to know what the idea means and not to be surprised by its use in this context.

Setting the stage for their later analyses of organizations, March and Simon sketched a number of other models of bureaucracies using the same sort of loop diagrams. They generalized these examples to produce a "general model of adaptive motivated behavior," which they represented both in causal-loop form (Figure 5.3) and in differential equations.

The formal model they associated with Figure 5.3 consisted of the following equations:

$$\frac{dA}{dt} = a(R - A + k) \qquad (a, k > 0), \qquad (1)$$

$$S = R - A, \qquad (2)$$

$$L = m(S^* - S) \qquad (S^*, m > 0), \qquad (3)$$

$$\frac{dR}{dt} = n(L - b - cR) \qquad (n, c > 0, b > 0), \qquad (4)$$

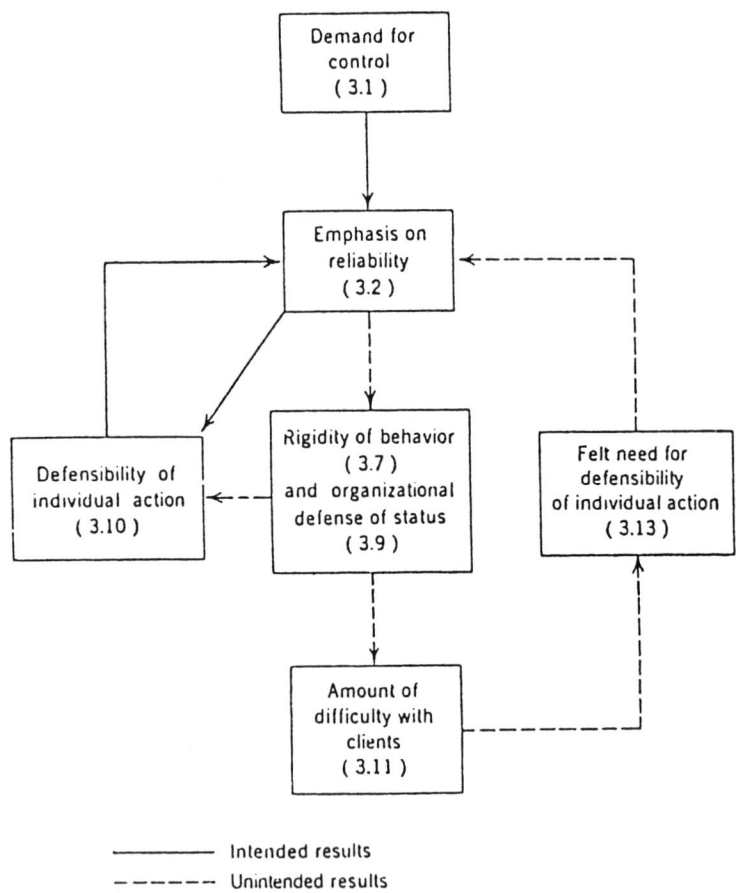

FIGURE 5.2: Simplified Merton model for organizational behavior. Source: March and Simon (1958, p. 49).

where

- A = level of aspiration,
- R = expected value of reward,
- S = satisfaction,
- L = search rate.

The authors made no mention of the dynamic behavior of this model. Their brief analysis focused on its steady state characteristics:

> Since a is positive, at equilibrium the aspiration level will exceed the reward. [Equation (3)] also postulates a "desired" level of satisfaction, S*, at which search for increased satisfaction would cease.... [Equation (4)] postulates that a certain amount of search, (b + cR), is required just to maintain the current level, R, of reward.... The system possesses a stable equilibrium (pp. 49–50).

In their use of loop diagrams and differential equations, March and Simon reflect tendencies we have been associating with the servo-

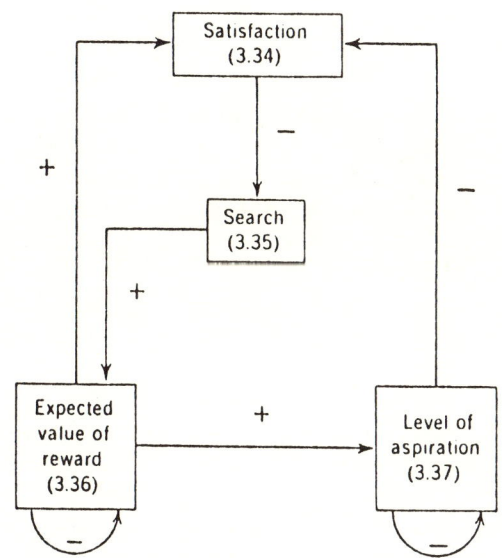

FIGURE 5.3: General model of adaptive motivated behavior. Source: March and Simon (1958, p. 49).

mechanisms thread, as we would expect from Simon. Their attention to the stability of their model rather than its dynamic behavior, however, makes them sound more like authors we have encountered in the cybernetics thread. The model was undoubtedly too simplistic for March and Simon to bother with its dynamic implications. Being linear, second order, and stable, the most it can do is show goal-seeking exponential patterns or sine waves that eventually damp out. Relating such model behavior to the behavior of real individuals in organizations would probably be difficult and would probably not produce the sorts of insights the authors were searching for.

Instead, in the remaining part of the book they developed an extensive set of propositions about the behavior of people in organizations, drawing on all the organizational literature that they could. Together, their propositions describe a complex circular causal feedback structure. Periodically, the authors stopped to sketch diagrams like Figures 5.2 and 5.3 showing the influence relations and interconnections contained in their propositions. The final result is a generic verbal and pictorial model of an organization, or rather of the behavior of people in organizations.

The authors stopped short of constructing a formal quantitative model for the complex structure they described. They noted, in fact, that the effort to quantify it would run into a thorny statistical problem. They considered the circular causal system they described to be "underidentified." Essentially, it contains relatively too many mutual causal processes and not enough causally independent variables to enable statisticians reliably to estimate parameters in the model from empirical data. Instead of a formal model connecting organizational structure and organizational behavior, the contribution of the book is the collection of knowledge about organizations and its codification into a system of mutual causal interactions.

The low key or almost implicit use of the feedback concept in *Organizations* characterizes, I believe, Simon's use of the idea from that point on. Information feedback and mutual causality are facts of life in complex systems, but they do not form for Simon a central focus. He does not attempt to derive patterns of dynamic behavior from the interaction of positive and negative loops in complex systems. His deeply significant work in the computer simulation of an individual's thought processes focuses on extremely detailed, disaggregate steps or problem-solving "moves." At that level of disaggregation the concept of loop polarity appears to be much less useful than the explicit nature of each successive move.

5.3 Use of the Feedback Concept in Economics

The Legacy of Tustin and Phillips

Initially, the work of Tustin (1953), Phillips (1954, 1957), and Allen (1956) brought the feedback concept and the mathematics of control engineers to the attention of economists. The writings of scholars in the cybernetics thread also became well known. Yet economists had their own very strong traditions of loop thinking, ranging from the earliest scholarship in the field—represented by Smith, Mill, Marx, and so on—to the most modern developments—econometric models in the pattern set by Koopmans, Tinbergen, and Klein. It also had at its core the General Equilibrium Theory of Walras and Pareto, with its elegant mathematics, its emphasis on optimization, and its comparatively static approach to economic dynamics.

Thus with the appearance of engineering notions in economics in the 1950s, several different scholarly cultures came in contact. Bateson's early feedback thinking on culture contact suggests the possible outcomes: fusion, destruction of one or both, or coexistence (with attendant schismogenetic processes maintaining their differences). While it is perhaps too harsh to say that the Phillips-Tustin approach has been destroyed by its contact with neo-classical economic theory, Cochrane and Graham (1976) conclude

> it did not penetrate the mainstream of macroeconomic thought. If the content of textbooks is an acceptable guide to penetration, one would have to argue that the control-theoretic analysis of Phillips and Tustin made a minimal impact on macroeconomic analysis. Although at times passing references to feedback were made, no explicit or rigorous development of the concept was pursued (Cochrane and Graham 1976, p. 244).

When it does appear in economics, the legacy of Tustin and Phillips tends to take one of two forms: the use of the mathematics of adaptive or optimal control to existing economic models, and the application of ideas from the cybernetics thread. In view of the distinctions raised in this study, the latter is more properly seen as a legacy of Wiener. A discussion of the former, the application of the mathematics of adaptive and optimal control to economic models, would carry us too far from our conceptual focus on the loop notion underlying feedback and circular causality. We will have to be content with the observation that the feedback concept in this optimal control thread hides behind the mathematics of control. It is not used as a tool for conceptualizing economic systems, nor for understanding economic behavior. Its attractions to economists are no doubt its perceived potential to automate (in a sense) the determination of desirable control policies for the management of a complex economy and its obvious kinship to the theme of optimization that grounds much of modern economics. (A start on the literature of adaptive and optimal control in economics is contained in the references.[1])

More appropriate for our purposes is an investigation of the conceptual uses of feedback in economics. Cochrane and Graham (1976) direct us to books by Axel Leijonhufvud (1968) and John Culbertson (1968). In Leijonhufvud's work, the feedback concept helps to illuminate the loop structure underlying economic theory, from the "Classics" to Keynes's *General Theory*. In contrast, Culbertson's use of feedback ideas tends more to obfuscate both economic theory and the value of a feedback view.

The Feedback Emphasis in the Work of Axel Leijonhufvud

A small article by Leijonhufvud (1970) published two years after his significant book on the economics of Keynes serves to introduce how the feedback concept and our two feedback threads have influenced his thinking. The purpose of the article was pedagogical, an attempt "to revive a soon 20-year-old paper of Goodwin's" (1951c) that had not, in Leijonhufvud's mind, seen the use it deserves.

> Goodwin dealt with the market as a servo-mechanism or homeostat. A homeostat is a control-device which regulates the behavior of a "machine" on the basis of information about, and evaluation of, its actual past performance, i.e., on the basis of feedback. The purpose of control is to bring actual performance into line with desired performance and this is done, in effect, through a process of trial and error.... We will regard, then, the market as a device for "solving" demand and supply equations by trial-and-error.
>
> ... We are going to regard the market as a servo-mechanism which continually feeds back the discrepancy between its actual and "desired" state and (hopefully) adjusts the actual state in the direction required to reduce the "error" towards the appropriate zero value (Leijonhufvud 1970, pp. 1, 2).

Leijonhufvud thus indicated his awareness of feedback notions from both of our conceptual threads, including the homeostat of Ashby from the cybernetics thread and the early work of Goodwin in the servomechanisms thread. By 1968 it would be a rare economist who was not familiar with elements from both threads. Predictably, however, Leijonhufvud selected elements of the two threads most appropriate for his economic focus, notably a dynamic perspective, phrased quantitatively, interpreted in terms of "deviation-amplifying" and "deviation-counteracting" loops, and incorporating a decidedly "endogenous" point of view. Some of the more uniquely cybernetic notions, notably requisite variety, self-reference, and the rewriting of structure do not appear.

Arguments of Walras and Marshall in feedback terms

Pursuing the idea he attributed to Goodwin, Leijonhufvud (1970) reinterpreted aspects of the economics of Walras and Marshall in feedback terms. As a market-clearing mechanism, Walras had postulated the iterative process of tâtonnement—literally, a "groping" toward the price at which supply and demand balance. In Leijonhufvud's difference equations representing tâtonnement, the change in price at any step in the goal-seeking process is a function of the difference between demand and supply:

$$\Delta p = f[D(p) - S(p)].$$

Price at time t is then computed from its price at the previous iteration plus the change:

$$p_t = f[D(p_{t-1}) - S(p_{t-1})] + p_{t-1}.$$

In these equations, $D(p)$ is the demand schedule, $S(p)$ is the supply schedule, and f is a function embodying the market's "rules of search":

$$f(x) \begin{cases} >0 \\ =0, \\ <0 \end{cases} \text{if } x \begin{cases} >0 \\ =0. \\ <0 \end{cases}$$

The result is a system consisting of two feedback loops, one involving supply and the other involving demand, which Leijonhufvud pictured as shown in Figure 5.4.

Leijonhufvud emphasized that such a formulation says that the market "does not know what it is looking for" (the equilibrating price), but "will know when it has found it" and will stop the search. What is required is merely that $D > S$ implies the price is "too low," and $D < S$ implies the price is "too high." With the rule of search expressed in f and with $x(p) = D(p) - S(p)$, the system will not be properly goal-seeking unless $x(p)$ has negative slope. The reason, in Leijonhufvud's reinterpretation of Walras, is that we need "unambiguously negative (error-reducing) feedback":

When ... the equations are such that this presumption [x'(p) < 0] is false, the servo-mechanism will not work; it will not "zero-in." Instead of having a *deviation-counteracting* control, we will find that the feedback effects are *deviation-amplifying*—the system departs further and further from the solution (p. 5, italics in original).

When the ambiguities in Figure 5.4 are properly resolved, Leijonhufvud represented the price adjustment process in a single negative feedback loop (shown as the lower part of Figure 5.5) linking p (price) and x (the gap between supply and demand).

The Walrasian feedback structure strives to hone in on the market-clearing price by mutual adjustments of price and supply and demand. Marshall dealt with the problem of how the economy determines the "right" rate of output. Again Leijonhufvud proposed an iterative feedback view:

At each successive step, a new "trial" quantity is taken as given. The reaction of consumers and producers to the value of this "independent" variable, we describe in terms of: (a) the demand-price—the maximum price at which consumers are willing to absorb this rate of output, and (b) the supply-price—the minimum price required to induce producers to continue this rate of output (p. 7).

Leijonhufvud then sketched a two-loop feedback mechanism similar in

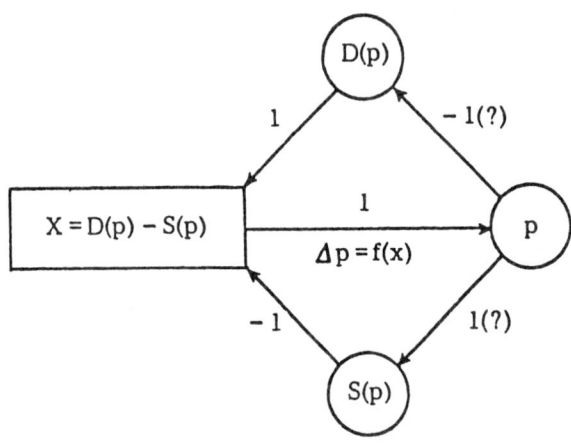

FIGURE 5.4: Leijonhufvud's representation of the feedback loop structure of the Walrasian adjustment mechanism for market price. D(p) and S(p) represent demand and supply as functions of price. Source: Leijonhufvud (1970, p. 6).

form to Figure 5.4 but focusing on changes in output. The resulting "Marshallian homeostat" seeks a rate of output that will assure equality of supply price and demand price.

> Just as the Walrasian homeostat based its regulation of price on information about quantities, the Marshallian regulates quantity on the basis of information about "prices"—or, more exactly, marginal valuations on both sides of the market (p. 8).

Leijonhufvud used this feedback view to discuss several special cases of economic interest, including industries exhibiting decreasing costs, "backward-bending" supply schedules and "downward-sloping supply-price" schedules (pp. 9–10).

However, an assumption made in the course of describing the Marshallian homeostat bothered him, and his concern leads to two interesting observations about Leijonhufvud's feedback view. It is a question of what information is available to producers:

> Producers' decisions on changing output are geared to the relationship between their supply-price and the *actual* price. That is the way we *must* formulate it. The assumption made earlier [in the Marshallian homeostat], however, was that actual price always equals the demand-price for the output-rate prevailing at the moment.
>
> How can this be? The assumption, of course, must be that the Walrasian homeostat is in operation and brings it about—and is operating "very fast" (very fast indeed) to bring the *observable*, actual price into line with the demand-price associated with the output of the moment.... Consequently, the Walrasian feedback mechanism is *contained within* the Marshallian (pp. 10–11, emphasis in original).

Thus one significant consequence of Leijonhufvud's concern about information available to producers is the conclusion that the Walrasian and Marshallian economic worlds are not separate or dichotomous. Leijonhufvud concluded they overlap. He captured them both in simplified form in the multi-loop structure shown in Figure 5.5.

Leijonhufvud observed that with minor modifications the multi-loop structure in Figure 5.5 could be used to analyze "stock-flow markets" and simple accumulation processes, and the significance of time lags in producing "oscillatory malfunction" (p. 12). But as it stands Leijonhufvud observed that it is still inadequate as a description of even the simple case of an isolated market. To improve on its assumption that price is unique, the model shown in Figure 5.5 needs a "Jevon's homeostat" included within the Walrasian: "the elimination of price-discrepancies through competition is in itself a communication process the nature of which ought to be explicitly represented" (p. 13). In addition, it still has the objectionable feature that "noone" sets price. Furthermore, it does not allow the rate of sales to differ from the rate of output (p. 12). Presumably, for the latter Leijonhufvud would add inventories.

But perhaps the most important observation we can derive from this development is the underlying insistence itself on basing economic theory on information available to economic actors. In a footnote Leijonhufvud asserted:

> Our theoretical conclusions in economics are sound only in so far as the economic choices asserted to be made are made on the basis of information that can, within the model itself, be shown to be available to the decisionmakers in question (footnote, p. 11).

That insistence figures prominently in the work of other feedback authors. It is particularly pronounced in the work of Simon and Forrester, but it is also a fundamental consequence of the stress in the cybernetics thread on the importance of information in control. It has not, however, been central to economic theory building. In fact, it tends to be seen as contradictory to some of the tenets of economics based upon optimization principles and rational expectations. For Leijonhufvud it apparently comes from the emphasis on the role of information in communication and control in the cybernetics thread. It is central to his most significant use of a feedback perspective—his perception of the "Keynesian revolution which did not come off" (1968, p. 397), to which we now turn.

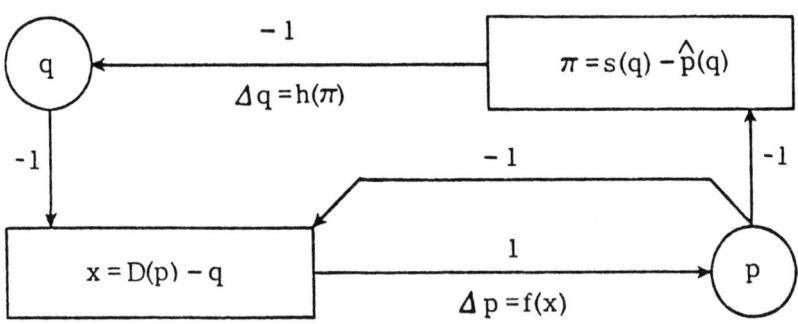

FIGURE 5.5: The Walrasian feedback mechanism coupled with the Marshallian. q = quantity produced; p = market-clearing price; s(q) = supply price; p(q) = demand price = actual price p; h, f = functions representing market adjustment rules. Source: Leijonhufvud (1970, p. 11).

Leijonhufvud's interpretation of the economics of Keynes

As the title implies, Leijonhufvud's book *On Keynesian Economics and the Economics of Keynes* (1968) was an attempt to reorient economists to what Leijonhufvud saw as the real meaning and significance of Keynes's *General Theory*. His view of Keynes is profoundly illuminated by ideas contained in our two feedback threads. He argues
- that some of the differences between the *General Theory* and the "Classics" are most meaningfully explained in feedback terms;
- that Keynes's theory must be interpreted dynamically;
- that a focus on the role of information, communication and control in economic dynamics is at the heart of Keynes's thinking.

Our previous discussion of Leijonhufvud's treatment of Walrasian and Marshallian stability illustrates the way he contrasts Keynes's views with the Classics. A central difference is that Keynes saw naturally occurring *positive* feedback loop processes at work, which are not emphasized in earlier economic theory. Discussing the Keynesian consumption function, for example, Leijonhufvud asserts:

The constraint on household income, emerging through the inability of traders to sell all they want at "the" prevailing market price, leads to a reduction in effective demand in product markets *with further ("multiplier") repercussions on aggregate income*. This deviation-amplifying feedback loop is characteristic of Keynesian quantity-adjustment models. The decline in income tends to proceed below the level to which it was brought by the initial shock. In the more common terminology (which is, however, sometimes misleading), the contraction is "cumulative." A "Classical" general equilibrium model, in contrast, presents the reassuring picture of an economic system equipped exclusively with deviation-counteracting feedback mechanisms (1968, pp. 56–57, emphasis in original).

Once prices and wages start falling, Leijonhufvud has Keynes saying, "the deviation-amplifying feedbacks characteristic of this system (take) hold," and "the deviation-counteracting price adjustments become less effective" (p. 57). Leijonhufvud's language here reflects the somewhat diverse traditions initiated by Myrdal (1944) (section 2.5) and Maruyama (1963) (section 4.4).

Similarly, Leijonhufvud contrasts in feedback terms Keynes's view of "involuntary unemployment" with the notion of "frictional" unemployment (involving a pool of people between jobs):

Keynes' involuntary unemployment is fundamentally a product of the cumulative process which he assumed the initial increase in unemployment would trigger. The assumed deviation-amplifying feedbacks involved in this process cannot be explained in terms of an isolated labor-market model—the entire money-using system must be considered (p. 81).

Citing other examples throughout the book, Leijonhufvud concludes that Keynes's theory did not dismiss "classical economic theory," but rather augmented it with what amounts to a feedback view. The main theme of the classical thinkers from Adam Smith onward, Leijonhufvud asserted, had been to show how an economy of many commodities and actors could be nicely self-regulating.

Keynes' objective was to show how such a system could malfunction in a way consistent with those observations of an economy in depression.... In Keynes' theory, then, the homeostatic mechanisms postulated by the Classics are neither "out of order" nor just missing (pp. 160–161).

They are intermingled with "deviation-amplifying" feedback loops that, in Keynes/Leijonhufvud's view, help to explain economic deviations as dramatic as a depression.

A fundamental implication of this interpretation of Keynes as a feedback thinker, is, in Leijonhufvud's view, that Keynesian economic theory cannot be approached from an equilibrium point of view. It is essential to approach his ideas dynamically:

A *purely static interpretation of his "unemployment equilibrium" is impossible*. His theory *must* be interpreted as dynamic in nature and, on today's usage, as dealing with unemployment *dis*equilibrium (p. 161, emphasis in original).

The "Keynes and the Classics" issues are better approached from a dynamic than a comparative static perspective (p. 389).

As we have seen, this emphasis on a dynamic view linked with the feedback perspective is most prevalent in the servomechanisms thread.

Coupled with a dynamic view, however, Leijonhufvud stressed the importance of the role of information and communication in the *General Theory*, an emphasis we have seen most centrally in the cybernetics thread. For example, to explain how an economy avoids the production of "false quantities," Walras had tried to show that production decisions are based upon knowledge of prices that would hold in equilibrium. To do that he had introduced an imaginary bargaining process involving "fictive tickets." Leijonhufvud explained that the fictive tickets device "ensures that Walras' system is suitably equipped with negative feedback mechanisms" so that "the tâtonnement process will eliminate excess demand 'errors.'" But to make it all work, Walras had to ignore the fact that production takes time (p. 73).

Keynes, however, did not ignore the time element and was led by it, in Leijonhufvud's view, to the importance of economic expectations:

The entire Chapter 5 of the *General Theory* is devoted to one simple point—that current production cannot be guided by what amounts to perfect knowledge, but must be based on expectations:

> The *actual realized* results of the production and sale of output will only be relevant to employment in so far as they cause a modification of subsequent expectations.... Meanwhile the entrepreneur ... has to form the best expectations he can ... ; *he has no choice but to be guided by these expectations, if he is to produce at all by processes which occupy time* (Keynes *General Theory*, pp. 46 & 47).
>
> The main point of this, however, was simply to deny producers the benefit of perfect information (Leijonhufvud 1968, p. 73, emphasis in original).

Throughout the *General Theory*, Leijonhufvud saw "the extraordinary extent" of Keynes's preoccupation with such "information problems" (p. 351).

In Leijonhufvud's view, Keynes explicitly linked this focus on information with a consideration of the importance of time in entrepreneurial decisions. For example, Keynes held that prices on long-lived assets are "much more likely to be 'grossly' mistaken" than prices on labor services and "liquid" commodities. The conclusion comes from thinking about the information used to set prices. Significant would be the "speed and reliability of the feedback mechanisms on which different markets can depend to reveal mistakes before prices and stock supplies have deviated very far from the respective warranted time-paths." He quoted Keynes as saying,

> It is of the nature of long-term expectations that they cannot be checked at short intervals in the light of realized results (Keynes, *General Theory*, Chapter 5, cited in Leijonhufvud 1968, p. 351).

An important insight about economic dynamics follows from this train of thought: the more liquid commodities and labor services would not be important sources of major economic ups and downs. Changes in inventories, for example, might amplify a depression, Leijonhufvud observed, but in Keynes's view they would not cause it (p. 351).[2]

Summarizing in a Postscript chapter his point of view about Keynes as an information feedback theorist, Leijonhufvud contrasted the Newtonian character of Walrasian economics with the cybernetic character of Keynes's view (pp. 389–397). The former appears to apply well to an economy in equilibrium but does not explain disequilibrium phenomena. Economic fluctuations, unemployment, or other undesirable phenomena become viewed as errors or unnatural events in the machinery—"slipping a cog" or "throwing a spanner in the works." In Keynes's view, as interpreted by Leijonhufvud, such disequilibrium phenomena are natural potential consequences of the way information moves through the system and influences economic behavior.

For our focus on the evolution of the feedback concept we can learn several things from Leijonhufvud's work. Most obviously, we have a well-documented argument that Keynes was a "pre-feedback thinker," in the tradition of a number of social scientists whose work we surveyed in section 2.5. Second, we see a natural blending of ideas from our two feedback threads: citations and language from both threads, a focus on dynamics explained by endogenous feedback structure from the servomechanisms thread, and a focus on communication and control from the cybernetics thread. It would be pointless to try to decide in which thread Leijonhufvud "belongs." Third, we see a very strong argument for interpreting economic phenomena in information feedback terms—literally, for replacing the prevailing paradigm of optimization and rational expectations with a feedback perspective that is a healthy blend of the traditions initiated by Phillips, Tustin, Goodwin, and Wiener.

We should expect such a blending of the threads, but it is important not to miss a selection process in it. Requisite variety is not in evidence, nor are the black boxes and networks that appear in some of the writings in the cybernetics thread. Neither, however, are the differential equations that appear in much of the early work in the servomechanisms thread. Since the work of Koopmans, Tinbergen, Samuelson, and Klein, difference equations had become the standard form for expressing dependence and interdependence in economics. The selection process thus appears to be explainable in terms of characteristics of the problems Leijonhufvud was trying to address and the tools and terminology with which he was previously familiar.

It is, I believe, an open question whether such an intellectual selection process reflects a "survival of the fittest." In some long run, the feedback ideas that persist are likely to be those with the most promise. But there seems to be no guarantee that in the short run less promising aspects of a new view will not be selected. The prevailing wisdom will strengthen certain aspects of a new view, those that are most easily assimilated, even if they will prove in the long run to be the less promising aspects. We must be careful about reading too much into what we observe people select from the feedback perspectives we are tracing.

The Feedback View of John Culbertson

Leijonhufvud's blend of feedback thought and economics gives the feeling of being potentially powerful and insightful. Culbertson (1968) starts with a similarly promising perspective, but his resulting use of the feedback concept is seriously flawed. Speculating on the sources of Culbertson's difficulties is an interesting exercise that may produce some insights for the evolution of the feedback perspective.

Written in the same year as Leijonhufvud's book on the economics of Keynes, Culbertson's *Macroeconomic Theory and Stabilization Policy* argues strongly for a feedback view of economic dynamics. Although he cited authors in the cybernetics thread (notably Ashby), his initial feedback-related comments tend to place him in the servomechanisms thread in the social sciences—not surprising given the heavy role economists played in the early work in that thread. He defined a feedback system in terms that identified it with interdependence and mutual causality:

It is useful to think of economic systems as belonging to a class of phenomena that we can call "interaction systems" or feedback systems. These are systems involving coherently structured processes of mutual interaction, or mutual influence, that suffice to produce a characteristic behavior over time (Culbertson 1968, p. 17).

This phrasing also links feedback significantly to dynamic behavior patterns, a link we have come to associate with the servomechanisms thread. Indeed, like Leijonhufvud and the early economists in the servomechanisms thread, Culbertson argued strongly for a dynamic view that emphasizes disequilibria and instability:

The market economies, far from being highly stable, have been a major dynamic force, transforming the earth in the past two centuries. The explosive course of economic growth has multiplied the population and dramatically changed the physical environment. The nonviability of stable systems in such an atmosphere of competition is illustrated in the economic sphere, where the stable traditional socioeconomic systems have proved unable to survive, a struggle now being underway to determine what class of unstable system is to replace them (pp. 38–39).

In marked contrast to Leijonhufvud, however, Culbertson claimed that Keynes had a static view based on the assumption that economies are stable. He hoped to improve on Keynesian economic models by taking a more endogenous view—another aspect of a feedback perspective whose explicit presence tends to characterize the servomechanisms thread. He argued for including expectations and policy variables within the "systematic feedbacks" influencing the behavior of the economy:

The dynamic theory must incorporate any systematic *dependence of expectations upon other variables of the model* in order to have a full statement of the feedback system. . . . The expectation-formation function of fully dynamic theory must express the dependence of expectations upon environmental changes such as actually occur, rather than upon *ceteris paribus* changes in one or another aspect of the environment such as form the stuff of static theory.

In short, the most thoroughly static theory may emphasize expectations. The crucial consideration is whether the expectations are themselves explained in terms of feedback or are given externally in a *ceteris paribus* world of comparative statics (pp. 41–42, emphasis in original).

Culbertson envisioned that his dynamic, endogenous, feedback-informed perspective would cure a number of biases he had observed in static macroeconomic theory. He expressed three of them directly in feedback terms. First, he argued, presuming stability leads to oversights and omissions:

Theory formulation in terms of the static framework plus the stability hypothesis leads to systematic omission of feedbacks from the model in order to make it stable. Feedbacks characteristically omitted from the Keynesian model are those involving investment, policy variables, expectation formation, and demand for credit. The stability thus forced upon the model a priori is the source of the incapacity for growth that is reflected in the stagnation thesis (p. 42).

Second, he objected to what he saw as a tendency to an exogenous point of view:

The variables the dependence of which upon feedback is thus omitted are specified as exogenous and thereby by convention become the central explanatory variables of the theory (p. 42).

Here he cited investment decisions and policy variables in the Keynesian view. Leijonhufvud might disagree, or he might say that such a "Keynesian view" is not the view of Keynes, but we can leave that argument to the economists.

Culbertson's third point about feedback in macroeconomic theory contrasted stability and adaptability and noted that an incorrect focus could result in misspecification of the system:

[A static view] may systematically lead to understatement of feedbacks and to representing the system as more stable—and less adaptable—than is really the case. Interpretation of economic behavior in terms of equilibrium response to given environments may lead to a systematically different specification of feedbacks than would interpretation in terms of immediate response to changes in environment. The latter is relevant to specification of feedbacks, but the corpus of existing economic theory relates to the former (pp. 42–43).

Finally, we should note that Culbertson's view also shares the modeling and policy analysis tendencies we have previously observed in the servomechanisms thread:

> In general, the effects of a policy action upon an interaction system are not direct and obvious, since its meaning and effects depend entirely upon the role that it plays within the specific interaction system. ... Justifiable diagnosis and prescription can only proceed in terms of an adequate representation of the interaction system in question, its characteristic behavior, and the manner in which this behavior will be altered by any given policy action. The road to understanding, then, must lie through achieving an ability to characterize, to theorize about, to understand interaction systems. The analytical problem is one of comparative dynamics. ... The relevant formal system for quantitative dynamic representation of interaction systems is the system of differential equations or, with our discontinuous treatment of time, difference equations (pp. 18, 19).

Notwithstanding his differences with Leijonhufvud about Keynes, Culbertson's basic endogenous view striving to derive dynamic behavior from feedback structure echoes a number of themes we have seen before. With this familiar-sounding feedback perspective, he proceeded to formulate a generic feedback model in terms of difference equations, which he then used as a framework for a detailed macroeconomic model developed later in the book. It is here that significant differences in Culbertson's feedback view begin to appear.

Culbertson's generic feedback model

The generic model consists of three or four terms, which Culbertson characterized by the "type" or "class" of feedback they represented. First there is a term for exogenous variables, defined by Culbertson as variables without "systematic feedback" from the system to the variable. Such exogenous variables could be of at least two types—"constant" or "trend governed." A trended variable could be represented by an equation of the form[3]

$$X_t = b^t X_0,$$

where t represents the time period, X_0 the initial value of X, and $b > 0$ the trend parameter. If $b > 1$, the trend is upward; if $b < 1$ the trend is downward. In the category of "constant" exogenous variables, Culbertson put what he called an "exogenous shift" variable, a_t, which could take on arbitrary new values in any time period.

The remaining variables in Culbertson's generic model are feedback variables. He saw three distinct types. First, he identified "ordinary" feedback, which he represented in difference equation form as

$$Y_t = cY_{t-1} \quad \text{or} \quad Y_t = cY_{t-2}$$

(and so on to higher order lags). The sign of c determines the polarity of the feedback loop:

$c > 0$ implies "positive ordinary feedback,"

$c < 0$ implies "negative ordinary feedback."

Positive ordinary feedback is familiar to economists, he noted, as the usual representation of the dependence of consumption C on income Y from one time period to the next:

$$Y_{t-1} \to C_t \to Y_t, \quad \text{so} \quad Y_{t-1} \to Y_t.$$

The relationship appears in this difference-equation form in Samuelson's multiplier/accelerator model (see section 2.2). For justification of labeling such "circularity of action" as feedback, Culbertson at this point footnoted Ashby (1956).

Putting an exogenous (non-trended) disturbance together with a trended variable and an "ordinary feedback" term, he exhibited the general form of ordinary feedback models as

$$Y_t = a_t + b_t + cY_{t-1}.$$

Then, surprisingly, Culbertson *rejected* "net ordinary negative feedback."

> When $c < 0$, involving net negative ordinary feedback in the system, the marginal effect of the disturbance involves departures in Y that are positive and negative in alternate periods. Evidently the set of experiences we wish to describe does not include anything like this, so we exclude negative values of c from our domain of reference. We are concerned only with economies with net positive feedback (p. 21).

In a footnote he mentioned the possibility of an overacting policy that immediately overcompensates for all disturbances, giving rise to "chatter" about a trend line. Even so he noted that values of c less than -1 would produce "explosive alternating behavior," which "seems even further from our domain of reference." Negative ordinary feedback does not make economic sense in Culbertson's model.

He softened the prohibition against negative ordinary feedback somewhat by saying that it applied to the "net feedback of the system as a whole." There could be particular mechanisms in the system that involve negative ordinary feedback, but apparently they could not be allowed to be dominant (my term, not Culbertson's). He did suggest one meaningful interpretation to "temporary negative feedback":

> This can reverse the direction of movement of the system, a resumed positive feedback then leading to continued movement of total demand in the new direction. Such temporary negative feedbacks may provide a framework for analysis of factors determining boundaries of self-feeding movements of the system (p. 22).

There are a number of serious puzzles here that we should resolve, but first let us complete Culbertson's view of feedback in economic systems.

Culbertson envisioned another type of feedback which he labeled "accelerator feedback." He defined it by the difference equation

$$Y_t = d(Y_{t-1} - Y_{t-2}), \quad d > 0.$$

The connection to Samuelson's multiplier/accelerator formulation is obvious. It is interesting to note that Culbertson did not describe accelerator feedback as a mix of his "ordinary" positive and negative feedbacks. It is presented as an entirely different class. He did note, however, that there had to be some constraints on the accelerator coefficient d. If $d > 1$, then the accelerator feedback is "explosive"; if $d < 0$ then again we have "one-period chatter that does not seem relevant to the class of system here under consideration" (p. 22).

Culbertson listed "feedback control mechanisms" as yet another, entirely distinct class of feedback:

> This is a feedback mechanism making the behavior of the system responsive to an externally set value through an error-controlled negative feedback or error-controlled servomechanism. That is, such a mechanism applies feedback in a way defined outside of the interaction-system, thereby acting as a regulator—planned or unplanned—of the performance of the system (p. 22).

Such feedback control mechanisms might be "deliberately constructed" policy mechanisms, or "natural" like those occurring in ecosystems. Presumably, the distinction between this class of feedback and those described previously is the specification of exogenously determined goals in those structures labeled "feedback control mechanisms."

Any of these classes of feedback could operate either "continuously" or "discontinuously." To Culbertson the distinction had nothing to do with difference versus differential equations: he captured both "continuous" and "discontinuous" feedback in terms of difference equations. He reserved the word "discontinuous" for a feedback structure that switches on and off, as if it has a "dead-zone" in which it is not turned on. Such discontinuous feedback operates, he suggested, to alter coefficients "reflecting definable shifts in the economy's system of dependencies." They would operate

> only when the economy is outside a defined corridor and moving away from or parallel to the corridor. That is, once the economy is turned around and moving back toward its corridor, although it still remains outside the corridor, the feedback-control mechanism does not operate to impede this movement (p. 24).

The presumption appears to be that such reversing feedbacks would be "ordinary negative feedback," which would reverse the direction of economic variables every period unless they switch off immediately after switching on.

With all the influences and types of feedback he described, Culbertson's generic economic model takes the following form:

$$Y_t = a_t + (g + g')Y_{t-1} + (d + d')(Y_{t-1} - Y_{t-2}).$$

Here a_t represents exogenous disturbances and/or trends. The coefficients g and d have fixed (positive) values. The coefficients g' and d' are usually zero but switch to negative for one period whenever Y_t falls outside the "corridor" described by the interval $[Y^*_{t-1}(1 - x), Y^*_{t-1}(1 + x)]$. Thus if x were 0.2, for example, the discontinuous negative feedback terms would switch on for one period if Y_t moves away from Y* by more than plus-or-minus 20 percent.[4]

Equipped with this framework, Culbertson proceeded to develop a macroeconomic model consisting of many terms, each of which can be identified as one of the types in this generic equation.[5] His goal was to explain economic fluctuations in terms of "shifting values of feedbacks":

> To explain the behavior of regular bounded fluctuations in terms of the model evidently calls for shifting values of one or more feedbacks to explain the reversal of self-feeding movements of the system—or some external factor operating to the same effect. Where this shifting feedback operates with some regularity—as is relevant to a condition of regular economic fluctuations—this can be represented by a feedback-control mechanism.... We have noted that shifting values of feedbacks affecting government expenditures or taxes might be associated with an effort of government to have an anticyclical fiscal policy.... [F]eedbacks governing investment and consumption might be associated with expectation-formation functions based upon past economic fluctuations, leading people to anticipate cyclical turning points and thereby contribute to causing them (pp. 24–25).

He gave some other examples of "shifting feedbacks" that could explain the "bounding of economic fluctuations," but we need not pursue any more details.

Difficulties in Culbertson's feedback view

It is initially astonishing that Culbertson banned "net ordinary negative feedback." We have seen that negative feedback explanations in the social sciences can be traced back hundreds of years. Furthermore, the polarity of loops emphasized and analyzed in the cybernetics thread was for twenty years almost exclusively negative. It is natural to ask why he went counter to these traditions, with which he was no doubt familiar. One possible answer is his conviction that past economic theory had overemphasized the stability of economic variables; he clearly intended to emphasize instabilities, and he undoubtedly knew that positive feedback loops are destabilizing influences. I am inclined to believe, however, that the real source of his difficulties (and there are more than just his departure from established feedback traditions, as we shall see momentarily) lies in his use of difference equations.

The trouble begins when Culbertson notes that a negative value of c in $Y_t = cY_{t-1}$ causes Y to produce extremely unrealistic behavior for an economic variable, oscillating abruptly—changing sign, in fact—in each successive time period. He reacts by banning negative values of c. Instead, he should have begun to doubt the appropriateness of his use of difference equations to capture causal-loop structure.

Observing that a negative value of c causes $Y_t = cY_{t-1}$ to switch signs every period led Culbertson to interpret negative feedback as a switching mechanism. From this perspective, oscillations look like the result of switches, threshold effects, and abrupt changes in active structure. Culbertson came to see economic variables bouncing back and forth between the extremes of a "corridor" of permissible values. Negative loops switch on momentarily to reverse the directions of variables, keeping them in their respective "corridors."

Some might argue that such a switching process actually takes place in an economy as it fluctuates. The extremes of Culbertson's "corridors" are, after all, something like the nonlinearities that limit movement in more continuous perspectives (the logistic equation of Verhulst or the Lotka-Volterra equations are perhaps reasonable analogues). Yet there is a fundamental flaw in Culbertson's interpretation of the role of negative loops in economic phenomena. If he were trying to explain an oscillation with, for example, the structure of a pendulum system, his bias toward seeing things in terms of momentary negative loop "switches" would mislead him completely.

The feedback structure of a pendulum or a mass on a spring captured in the usual differential equations consists of a single negative feedback loop, shown in Figure 5.6. (Present in this negative feedback explanation of the oscillatory motion is Hooke's law—that the restoring force is proportional to displacement from equilibrium—and Newton's second law—that force is the rate of change of momentum.) This continuous view forces the realization that there are important accumulations in a pendulum or mass-on-a-spring system. (Here we have singled out position and momentum.) The system oscillates not because of threshold effects or switches reversing its direction, but because of the peculiar structure of this pure negative loop with two principal accumulations (integrations). The pendulum or mass-on-a-spring is perhaps the clearest example, but we should note that very similar negative loop structures are at the heart of the economic and business analyses of Simon (1952), Phillips (1954), and Forrester (1961).[6]

We need not argue that Culbertson's economic theory is wrong to realize the important issue here. The real point is that his discrete, difference equation representation of feedback structure could totally mislead him about connections between system structure and dynamic behavior. Conceivably, an inappropriate differential equation or LISP representation could also be misleading. What we learn from Culbertson's use of the feedback concept is that

- the form of the representation of a feedback system can seriously influence conclusions about the connections between feedback structure and dynamic behavior.

Although the result in Culbertson's case may seem dramatic, the idea is not really new to us. A number of feedback scholars in both threads

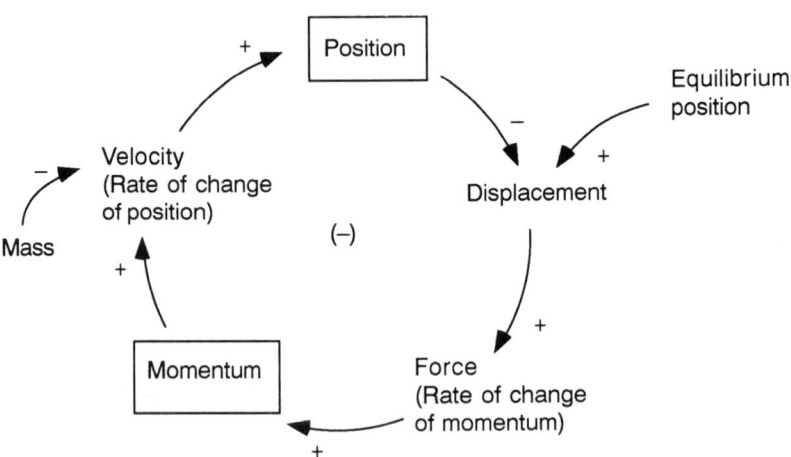

FIGURE 5.6: Feedback structure underlying the dynamics of an undamped pendulum.

have in fact argued for formal models on the grounds that even words can be an inappropriate representation, leading to false conclusions about the underlying causes of the behavior of complex systems.

At the risk of diverting attention from the importance of the choice of representations, I should note that I believe Culbertson is quite wrong in a number of particulars in his use of the feedback concept. Culbertson's choice of difference equations, largely forced by econometric tradition, convinced him that every negative loop is an oscillator. Few of our feedback thinkers would agree. The classical characterization of a negative loop is a goal-seeking structure, which may or may not oscillate, depending on circumstances that are frequently characterized in terms of the extent of accumulations or delays around the loop.[7]

Culbertson concluded that negative loops must consequently "switch on and off." Again, few of our feedback thinkers would agree. Endogenously evolving shifts in loop dominance, yes, but not with the abruptness or instantaneous character of a switch. As long ago as Tustin (1953) and Phillips (1954) and probably earlier, some economists have tried to argue that aggregate economic systems are more accurately represented as continuous systems. On this point I think Culbertson's particular use of the discrete difference equation view can be proved incorrect. If the period assumed between times t and t + 1 is made smaller, as one could certainly do, the length of time that any negative loop "switch" would be active would have to shorten. In the extreme, it would have to shorten to the point that "ordinary negative feedbacks" in Culbertson's model could be active for only a matter of minutes or seconds a year. I cannot see how that interpretation of the aggregate behavior of actors in an economy can be made meaningful.

Finally, Culbertson is simply wrong when he asserts that "shifting values of feedbacks" are necessary to explain "regular bounded fluctuations." A pendulum does not oscillate because of switches in its circular causal structure, and neither do the linear economic oscillating systems analyzed by Tustin, Phillips, and Simon. To be sure, there are nonlinear systems that trace their repeating behavior to shifts in loop dominance (e.g., the Lotka-Volterra equations and limit-cycle systems), but not all oscillators are of this sort. To assume that the turn-arounds in economic cycles come from "shifts" and "switches" leads the researcher to look for economic mechanisms that may well not be there. A pendulum does not cease moving in one direction and start back because of a shift in structure. Neither does any system characterized by a major negative loop containing essentially two significant accumulations. One would look in vain for an event that is the "cause of a turn-around." Perhaps even more likely, one would be sorely tempted to attribute spurious causal significance to events that merely correlate with observed peaks and valleys.

Culbertson's use of the feedback concept serves to alert us to the potentially great significance of the form of representation of system structure. The choice of representation sets up a number of implicit, usually unexamined *a priori* presumptions.[8] Further comments on the significance of the choice of the representation of feedback structure are contained in Chapter 6.

5.4 System Dynamics

In the thirty years since the publication of Forrester's *Industrial Dynamics* (section 3.3), the approach he outlined has been applied far beyond its original corporate context. A field of study has grown up, and the name has evolved to "system dynamics" to indicate its perceived generality. The field has academic and applied practitioners worldwide, degree-granting programs at a few major universities, newsletters and journals, an international society, and a large and growing body of literature.[9] Its book-length applications include studies of the dynamics of regional planning, research and development, urban stagnation and decay, commodity cycles, problems of growth in a finite world, economic development, economic fluctuations, community drug policy, human services delivery, energy lifecycles and transitions, dynamics and management of ecosystems, and various corporate policy studies.[10] There are at least ten texts in the field, not counting nonEnglish works or translations.[11] One of its publications, *The Limits to Growth* (Meadows et al. 1972), the controversial popular report of a global study sponsored by the Club of Rome, has been translated into 22 languages.

To do justice to the extent of the field of system dynamics is not possible here. Instead, I shall select particular aspects of published work in the field that contribute in significant ways to the evolution of the feedback concept in the social sciences. I will deal with these aspects under the headings of Principles of System Structure, Insights About Behavior of Complex Systems, Compensating Feedback, and Nonlinear Dynamics, each of which will be discussed in terms of system dynamics publications of the last twenty years.

Principles of System Structure

Over time Forrester honed his particular vision of the structure of a dynamic social system. By 1968 he had distilled it to the point that he could describe it in three pages (Forrester 1968a, 1969) and could summarize it in a list of "principles of systems" (Forrester 1968b). Naturally in these publications, dynamic system structure involves feedback, but that is only one of its characteristics. The theory came to be described and exemplified in a four-tiered hierarchy (Forrester 1969, p. 12; see also 1968a, p. 83 and 1968b, p. 4-17):

Closed boundary around the system
    Feedback loops as the basic structural elements within the boundary
        Level (state) variables representing accumulations within the feedback loops
        Rate (flow) variables representing activity within the feedback loops
            Goal |
            Observed condition | as components
            Detection of discrepancy | of a rate variable
            Action based on discrepancy |

It is interesting to note the place of these ideas in the evolution we have been tracing. The concept of the closed boundary signals the system dynamicist's endogenous point of view, which we saw frequently emphasized in the servomechanisms thread. It also serves indirectly to show Forrester's independence of von Bertalanffy and the general system theorists, whose work contributes to the cybernetics thread. A "closed system" in general systems theory is a system that experiences no interchange of material, energy, or information with its environment—a corked bottle at constant external conditions, for example. In contrast, Forrester's concept represents a system that is not "materially closed," but rather "causally closed"—the closed boundary separates the dynamically significant inner workings of the system from the dynamically insignificant external environment:

Formulating a model of a system should start from the question "Where is the boundary, that encompasses the smallest number of components, within which the dynamic behavior under study is generated?" . . . Thinking in terms of an isolated system forces one to construct, within the boundary of his model, the relationships which *create* the kinds of behavior that are of interest (1968b, p. 4-2, emphasis in original).

Thus Forrester thinks of a "closed system" in *causal* terms, as a set of interacting components that in and of themselves cause, or generate, the dynamic behavior they exhibit. He summarized this point of view as his first principle of system structure, the principle of the closed boundary:

In concept a feedback system is a closed system. Its dynamic behavior arises within its internal structure. Any interaction which is essential to the behavior mode being investigated must be included inside the system boundary (pp. 4-1,2).

But it is a misleading summary, which has caused confusion (see, e.g., Eden and Harris 1975, pp. 130–131). The two views of closed systems—materially closed and causally closed—are related but are significantly different. No serious system dynamics model is closed in the general system theory sense. Every one exchanges material with its environment—the little clouds representing sources and sinks in Forrester-like flow diagrams represent stocks of material outside the system boundary. Because of such exchanges, Forrester's "closed boundary" systems are, in von Bertalanffy's terms, "open systems." This terminological confusion is one more bit of evidence that these two authors and their ideas of system closure are not in the same feedback thread.[12]

Even more significant, however, is the *importance* Forrester places on the concept of the closed causal boundary. It appears at the top of his hierarchy of system structure. The endogenous point of view has top billing. There is a good reason: although he does not say so, it is possible to argue for a feedback perspective solely on the basis of the assumption of a closed causal boundary. Without causal loops, all variables must trace the source of their variation ultimately outside the system. We would be forced into an exogenous view of the causes of system behavior. Assuming instead that the causes of all significant behavior in the system are contained within some closed boundary forces causal influences to feed back upon themselves, forming causal loops. In Forrester's extreme endogenous point of view, all significant dynamic variation comes from the interactions of variables internal to the system. Exogenous explanations of system behavior are simply not of interest. Feedback loops enable the endogenous point of view and give it structure.

In his teaching text (1968b), Forrester also elevated his notions of levels and rates in feedback loops to principles of system structure:

(1) A feedback loop consists of two distinctly different types of variables—the levels (states) and the rates (actions). Except for constants, these two are sufficient to represent a feedback loop. Both are necessary.
(2) Levels integrate (or accumulate) the results of action in a system. . . .
(3) Levels are changed only by the rates. . . .
(4) Levels and rates are not distinguished by units of measure. . . . The identification must recognize the difference between a variable created by integration and one that is a policy statement in the system.
(5) Rates [are] not instantaneously measurable. . . . No rate can, in principle, control another rate without an intervening level variable.
(6) Rates depend only on levels and constants. . . .

(7) Level variables and rate variables [in a feedback loop] must alternate....
(8) Levels completely describe the system condition (Forrester 1968b, pp. 4-6–4-12).

The source of these principles about levels and rates is clear: Forrester is describing in nonmathematical terms the engineer's and the mathematician's notion of a state-determined system. Integrations and rates of change are presented as necessary components of feedback loops. They appear not as mathematical entities, however, but as formalizations of processes of accumulation and change in real systems. These principles of system structure appear in a teaching text, not a philosophical work, so we should not expect alternative structural views, but nonetheless it is significant that they are presented as the way feedback-loop structure must be captured. Difference equations that fail to distinguish accumulations and rates of change are not, according to these principles, an appropriate representation of a feedback system. Neither, presumably, would be diagrams or verbal descriptions of feedback loops that ignore their rate/level substructure.

It is noteworthy here that a form of representation of a feedback loop has been raised to the level of a principle of feedback structure. To be sure, we have noted before that Forrester by choice adopted a continuous point of view (see section 3.3). The choice of representation appears to be a consequence of that point of view. Nonetheless, as we have seen, these are choices that are not shared by all social scientists making use of the feedback concept. We shall return to the question of the significance of system representation in Chapter 6.

Insights About Behavior of Complex Systems

The expressed goal of the system dynamics approach is understanding how a system's feedback structure gives rise to its dynamic behavior. Increasingly, Forrester came to emphasize what he perceived to be generic insights, derived from specific modeling studies but, because of their general structure and behavior, apparently applicable to a wide range of settings.

The first published statement of such generic insights appeared in *Urban Dynamics* (1969) in a chapter on "Notes on Complex Systems." The urban dynamics modeling study grew out of an extended series of discussions at MIT between Forrester, former mayor of Boston John Collins, and a number of others with practical experience in urban affairs. Forrester contributed the approach, while the others supplied the knowledge of "the pressures, motivations, relationships, reactions, and historical incidents needed to shape the theory and structure of the specific social system" (1969, p. ix). The result of the collaboration was a system dynamics model that reproduced the pattern of growth, stagnation, and decline characteristic of many large urban centers.

Experiments with the model suggested that some favorite policies for reviving the cities are either neutral or actually detrimental. Increased low-cost housing, for example, was found to exacerbate the decline of the central city. A government-sponsored job program for the underemployed turned out in the long run to be, at best, ineffective. A job-training program providing marketable skills for 19,000 underemployed people each year resulted in a net increase in upward mobility of only 11,000 per year. External financial aid to the city, which initially generated increased tax revenue per capita with no increase in city tax rates, showed modest improvements, but surprisingly eventually forced up the taxes needed from internal sources within the city (pp. 51–70).

Policies found to help the city return to economic health include the reasonable notions of incentives for new business construction and more rapid removal of declining business structures, as well as the surprising and controversial ideas of the discouragement of worker-housing construction and more rapid demolition of slum housing (pp. 71–106). One might expect that these combined policies work because they banish the poor from the city. In fact, however, the opposite is true: in the long run the revival policy increases the net migration of underemployed into the city (p. 103). The policy works because it restores the city's ability to be a "socio-economic converter." It results in greater upward mobility of the underemployed into skilled workers and skilled workers into managerial and professional roles.

Forrester concluded that such counterintuitive results are to be expected from complex systems. Since he defined a "complex system" to be "a high-order, multiple-loop, nonlinear feedback structure," it is evident that feedback loops are perceived to be a major source of puzzling behavior and policy difficulties.

> Intuition and judgment, generated by a lifetime of experience with the simple systems that surround one's every action, create a network of expectations and perceptions that could hardly be better designed to mislead the unwary when he moves into the realm of complex systems. One's life and mental processes have been conditioned almost exclusively by experience with first-order, negative feedback loops (1969, p. 109).

But complex, multi-loop systems capable of shifting loop dominance behave differently. Cause and effect in such systems, Forrester asserts, are not closely related in time and space (p. 110). Worse, there are many "coincident symptoms" that look like causes but are merely correlational.

> In a situation where coincident symptoms appear to be causes, a person acts to dispel the symptoms. But the underlying causes remain. The treatment is either ineffective or detrimental. With a high degree of confidence we can say that the intuitive solutions to the problems of complex social systems will be wrong most of the time. Here lies much of the explanation for the problems of faltering companies, disappointments in developing nations, foreign-exchange crises, and troubles of urban areas (p. 110).

Forrester thus concludes that our difficulties with understanding and managing complex systems stem fundamentally from their multi-loop nature. The reader may recall that Merton made a very similar suggestion many years before in his paper on "The Unanticipated Consequences of Purposive Social Action" (Merton 1936; see section 2.5). Forrester (1971a) repeats and extends this line of thinking on the "counterintuitive behavior" of social systems.

In addition to their counterintuitive tendencies, Forrester identified six other properties of complex systems that had emerged from his simulation studies up to and including particularly the urban work.

- Complex systems are remarkably insensitive to changes in many system parameters (1969, p. 110).

The claim is an observation about the nonlinear, high-order, multi-loop simulation models he and his colleagues and students at MIT had investigated over the previous ten years. It appeared to be a property of such systems that changes of as much as 50 percent or more in most parameters often had little effect on the patterns of behavior exhibited by the model. Forrester interpreted the result in real terms, noting that the behaviors of diverse firms, different economies, and distinct cities had fundamental similarities in spite of their obvious "parameter" differences.

- Complex systems counteract and compensate for externally applied corrective efforts. (pp. 109, 111).

The problem, Forrester asserted, is that corrective programs "shift the system balance so that the corresponding natural processes encounter more resistance and reduce the load they were carrying." His example was the job-training program mentioned above, which was only 60 percent effective because of a decline in naturally occurring upward mobility. He concluded that "Probably no active, externally imposed program is superior to a system modification that changes internal incentives and leaves the burden of system improvement to internal processes" (p. 111). A nice conclusion for an endogenous point of view.

- Complex systems resist most policy changes (p. 110).

The reason is a combination of parameter insensitivity and systemic compensation. Simulating policy changes involves either changing parameters or altering model structure. Interpreted in real terms, the two previous characteristics of complex systems imply that most policy changes will not produce the results expected.

- Complex systems contain influential pressure points, often in unexpected places, from which forces will radiate to alter system balance (p. 109).

Occasionally, complex systems are sensitive to certain parameter changes, and sometimes compensating feedback does not foil a policy intervention and may even aid it. But in Forrester's modeling experience, such successful policies are often hard to find and "must be discovered through a careful examination of system dynamics."

- Complex systems often react to a policy change in the long run in a way opposite to how they react in the short run (pp. 109, 112).

Forrester noted that in the urban model the short-run effects of the underemployed training program, the low-cost housing program, and the slum-housing demolition program were all directly opposite to their long-term effects. "Worse-before-better" makes beneficial policies hard to implement and maintain to the point where they bear fruit. "Better-before-worse" makes policies that are detrimental in the long run hard to abandon.

- Complex social systems tend toward a condition of poor performance (p. 112).

A consequence of all of the preceding properties, the drift to low performance results primarily from misunderstanding what will improve things, doing things that are beneficial in the short run, and redoubling our efforts when results begin to worsen. Forrester concluded, "Again the complex system is cunning in its ability to mislead."

These principles of complex systems are now part of the lore of system dynamics. They appear again in Meadows (1982), where they are accompanied by causal-loop diagrams exposing the generic feedback structure underlying each one. To Forrester's list, Meadows adds addiction, official addiction (shifting the burden to the intervener), and high-leverage policies pushed in the wrong direction.

To system dynamicists, such generic insights relating feedback structure, dynamic behavior, and policy analysis are ample reason to press the feedback point of view that spawned them. The insights acquire part of their force from the fact that they are indeed empirical observations about the behavior of complex systems—although we must remember that the complex systems for which they are unquestionable empirical results are nonlinear, multi-loop, simulation models structured around levels and rates. The fit between the models and the structure and behavior of real systems may be perceived to be very close, but transferring insights from models to reality still requires leaps of logic and faith. Skeptics may wish to argue that these insights may pertain only to the simulation models.

An example of the derivation of an insight relating feedback structure and dynamic behavior may help to show how much logic and how much faith are really involved. Shortly before embarking on his urban work, Forrester published a brief article on corporate growth that has become something of a system dynamics classic. "Market Growth as Influenced by Capital Investment" (Forrester 1968a) presented the elements of the systems dynamics approach in the context of a modeling study of a firm experiencing erratic and stagnant growth in an essentially unlimited market. The assumption of an unlimited market is of course not realistic, but it makes one of Forrester's propositions abundantly clear: the firm's own internal policies are responsible for its erratic and stagnant behavior. Although a real firm might tend to blame such troubles on market weaknesses, there are none in the model. Thus to the extent that the model can be viewed as a reasonable abstraction of some aspects of the structure and behavior of real firms, it gives substance to Forrester's endogenous point of view.

One insight attributed to analysis of the structure and behavior of the market growth model concerns potentially pernicious effects of adaptable or "sliding" goals. Figure 5.7 shows two simulations of the model. In Figure 5.7a, growth in salesmen, orders booked, and production capacity, although erratic, is eventually evident. The results in figure 5.7b, however, are much less promising. The firm is losing production capacity, the sales force peaks and declines toward the end of the run, and orders booked are on a distinct downward trend.

The only difference between these two runs is a slight change in the firm's policy for ordering production equipment. This firm is assumed to base its capacity acquisition plans on its recent average delivery delay, the time it takes to process and ship an order. In both simulations, orders for production capacity are progressively increased whenever the observed delivery delay increases beyond a target delay set by management. In (a) the target or acceptable delivery delay is held fixed; in (b) the assumption is made that the target delivery delay is a long-term average of past performance. In (b) the goal slips, and with it goes the firm's potential for growth.

The feedback structure responsible for this disappointing behavior from sliding goals has been abstracted by Meadows (1982) (see Figure 5.8). The significant "state of the system" in the market growth situation is the perceived delivery delay. The negative loop in the figure is a

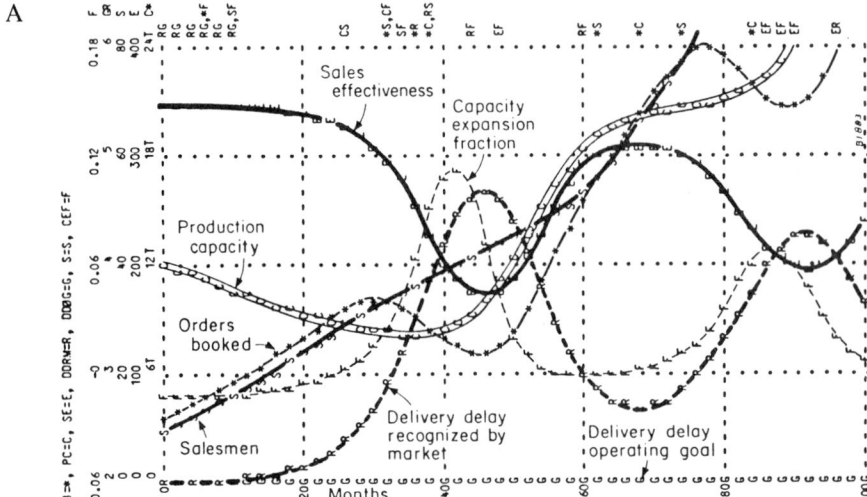

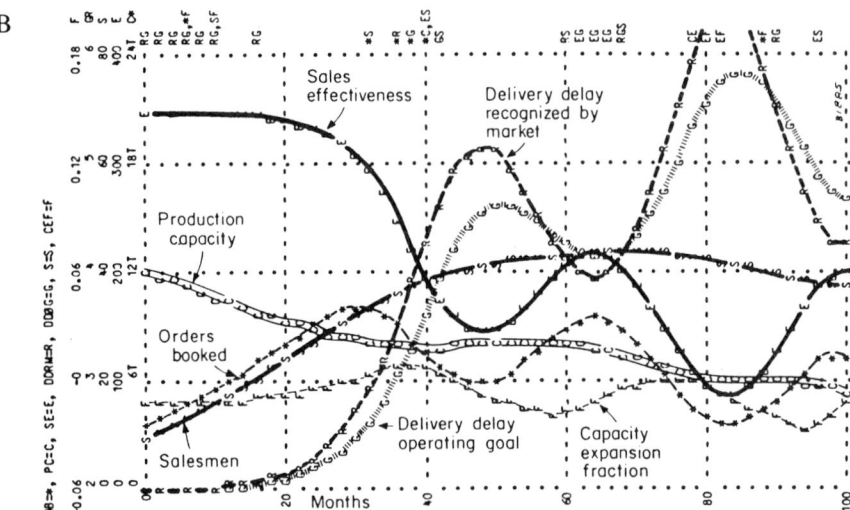

FIGURE 5.7: Behavior of the market growth model under two different capacity expansion policies: (a) fixed goal for delivery delay; (b) delivery delay target based on past performance. Source: Forrester (1968a, pp. 99, 100).

goal-seeking process which increases orders for production capacity to try to hold the perceived delivery delay at or near the desired target value. But if the desired target value is itself dependent upon the past performance of the system, a positive loop is created around the negative loop, which weakens the system's control capabilities. In Forrester's original analysis:

> The goal structure of the organization is now floating. It simply strives to achieve its historical accomplishments. For the more subtle goals in an organization, this striving to equal the past, and conversely being satisfied if one equals the past, is a strong influence.
> The result of changing the goal structure from a fixed goal to a goal set by tradition is shown here in [Figure 5.7b] . . . After delivery delay rises, the operating goals rises after a time delay. . . . This means that delivery delay does not produce the degree of concern that it did when the goal was fixed and low. As a consequence, expansion does not seem so justified or so important. In [Figure 5.7b] the goal structure continues to collapse. Delivery delay continues to rise, the traditional goal rises after it, the discrepancy is never great enough to produce active expansion of capacity, and there is a constant erosion of capacity (Forrester 1968a, p. 223).

The derivation of the structure/behavior insight is thus a combination of simulation experiments, model analysis, and thought about the real system.

Meadows (1982, pp. 104–105) applied this generic insight to inflation, air quality standards, and trash on city streets. She argued that in each case there are natural human tendencies to set standards based upon past performance, with the result that "performance is very likely

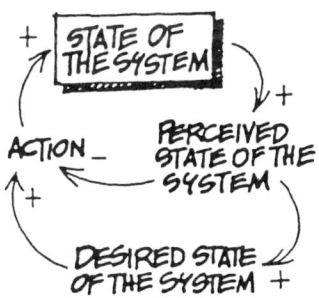

FIGURE 5.8: Generic feedback structure underlying a drift to low performance caused by sliding goals. Source: Meadows (1982, p. 105).

to drift downhill." Personally, I find the insight persuasive and have applied it with students to the interesting question of the source and dynamics of a student's goals in a course—and of the teacher's goals for the student. Presumably these applications, and the original insight relating feedback structure and dynamic behavior in the formal market growth model, would have to be judged like all insights, on the extent to which they open lines of thinking for people. Insights do require leaps. Formal validation of a quantitative model would probably not add much to a skeptic's belief in the "sliding goals" principle.

Compensating Feedback

Perhaps the most significant feedback theme repeatedly encountered in the system dynamics literature is the notion of compensating feedback. The phrase refers to a tendency observed in system dynamics models for a parameter or structural change to call forth natural adjustments of other factors within the model that counteract the direction of the imposed change. A feedback system has the capacity to compensate for imposed changes, pushing itself back toward its original condition. Ashby referred to the phenomenon indirectly in his notion of "ultrastability." In the system dynamics literature, however, compensating feedback is seen in less theoretical terms as a natural characteristic of complex systems, with dramatic policy implications.

A simple but graphic example of compensating feedback appeared in Forrester's study of the "global problématique"—rising population, industrialization, and pollution in the face of declining natural resources and the ability of areas of the globe to feed themselves (Forrester 1971b). A policy of ZPG (zero population growth) achieved by reductions in global birth rates was simulated as if it were instituted world-wide in 1970. It halted population growth only for a decade or so. Growth resumed. One reason for the return to growth is a compensating negative feedback loop in the model that links population, per capita standard of living, and birth and death rates. A stable population, combined with rising global industrialization, creates a rising per capita standard of living that feeds back to reduce infant mortality and increase average life spans. Family sizes that stabilize population in 1970 would result in a growing population in a later, more affluent globe. The step down in the crude birth rate in 1970 is thus compensated for by a feedback loop that acts to raise the net birth rate and restore population growth. Forrester observed that the phenomenon raises serious questions about the effectiveness of birth control alone as a means of controlling population. The feedback loop that compensates here is not a mysterious mechanism, but it is frequently overlooked by those of us who advocate ZPG as a prime cure for global difficulties. We learn from the simulation that what we really intend is a global tendency to maintain very low or zero net growth in spite of natural feedback tendencies in the system to defeat that aim. We also should learn the importance of including a population endogenously in any analysis of policies intended to improve its lot.

In a system dynamics study of heroin use in a community (Levin, Hirsch, and Roberts 1975), a large number of potentially compensating feedback effects for heroin- and crime-control policies were noted. Several of them work through the price system. Stepped-up police activity that succeeds in reducing the supply of heroin to the community ought to reduce heroin-related crime, but instead, the authors suggested, it increases it, in the short to medium term at least. The reason is obvious: the price of heroin rises, and addicts who support their habits with criminal activity have to resort to a higher frequency of revenue-raising crimes. More subtly, if it becomes too difficult to import heroin into the community, suppliers may switch to substitutes such as methadone which can be produced within the community and which are even more difficult to control. Both of these are potentially compensating effects for policies aimed at controlling crime by reducing the heroin supply.

Analyzing the opposite policy produced more evidence for the importance of the notion of compensating feedback. Suppose the community tries to reduce heroin-related crime by providing addicts with

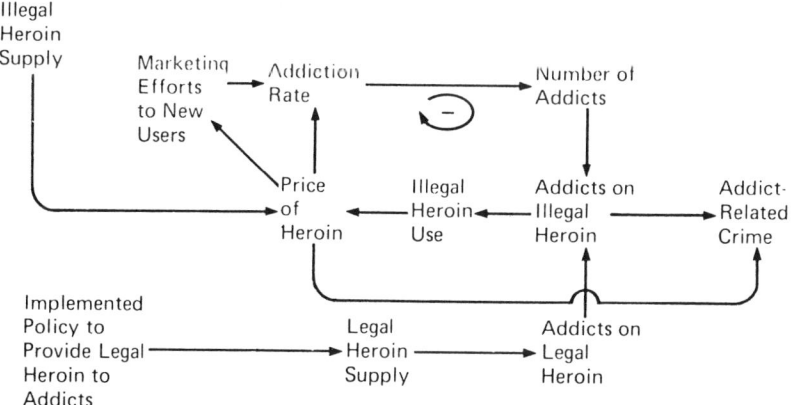

FIGURE 5.9: Compensating feedback loop active in a policy to provide legal heroin to addicts. Source: Levin, Hirsch, and Roberts (1975, p. 11).

legal heroin. The authors of the study sketched the feedback structure shown in Figure 5.9 and argued that the increased supply would cause prices to fall. The rate of addiction would go up, they argued, for two compounding reasons. Each long-term addict who had built up a considerable tolerance and need for the drug and who converts to the legal supply would free up enough heroin to support the habits of four or five new users. Furthermore, price declines would mean that pushers would more aggressively recruit new buyers. The authors concluded:

> Thus, while legal heroin might reduce addict-related crime, it would probably significantly increase the total number of addicts. Note that [Figure 5.9] indicates the presence of another negative feedback loop, one that leads to the undesired policy consequence. The increase of addicts on legal heroin quickly reduces the number of addicts on illegal supply; this diminishes illegal heroin use, decreases heroin price, and induces an increased rate of new addictions. The increased flow of new addicts adds to the total addict pool, increasing those on illegal heroin. The loop is initiated by a force aimed at decreasing illegal heroin addiction, but ends on an up-beat of increased illegal addiction (Levin, Hirsch, and Roberts 1975, pp. 11–12).

The mechanism of these compensating effects for the policy intervention is thus seen to be a negative feedback loop. Within the field of system dynamics, the search for compensating feedback loops, and the interpretation of policy interventions in terms of them, is now a fixture of the methodology.

Nonlinear Dynamics

The notion of nonlinearity and the importance of representing nonlinearities accurately has been fundamental to the perspective of the thinkers in the servomechanisms thread. As seen in section 2.2, the mathematical concept has an intuitive and useful interpretation in feedback loop terms. It is nonlinearity that shifts the dominance among feedback loops in a system. More precisely, nonlinearities in mathematical models can endogenously change over time the causal significance of variables in a dynamic system. Nonlinearities can change the structures that are active or influential. Thus nonlinearities are viewed by system dynamics practitioners as vital determinants of the interesting or problematic behavior of a dynamic social system.

Nonlinearity is a critically important idea, emphasized repeatedly by feedback authors we have investigated, but the situation is now known to be much more dramatic than early writers could have imagined. Discovered by a diverse group of mathematicians and physicists in the

1960s and 1970s, and only now becoming widely known, is the potential of nonlinear mathematical models to exhibit apparently *random* behavior. Even some very simple nonlinear systems have the potential to show dynamic patterns that are extraordinarily complex. The phenomenon is known as deterministic chaos, and it is the focus of a great deal of excitement and research today. (For a nontechnical overview see Gleick 1987.)

Two developments in this burgeoning field are significant for the evolution of the feedback concept in the social sciences and the two conceptual threads we have been following.

First, as a consequence of our earlier discussions (e.g., sections 2.2 and 3.3) we have the inescapable conclusion that the nonlinear models that exhibit deterministic randomness and chaotic behavior are feedback systems. Figure 5.10, for example, shows a representation of the loop structure of the famous Lorenz model:

$$\frac{dx}{dt} = \sigma y - \sigma x,$$

$$\frac{dy}{dt} = x(r - z) - y,$$

$$\frac{dz}{dt} = xy - bz.$$

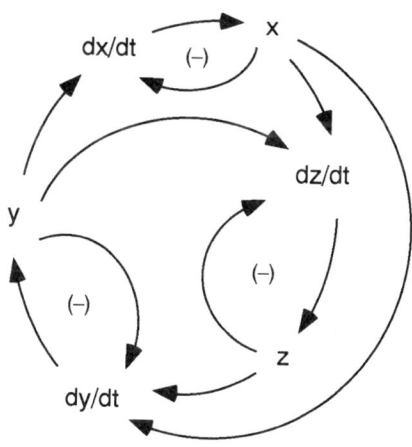

FIGURE 5.10: Feedback loop structure of the Lorenz model.

The model was originally designed to be a highly simplified picture of certain weather-related variables in an atmosphere heated from below: x is related to convection, while y and z represent horizontal and vertical temperature variations (Lorenz 1963). As shown in Figure 5.10, the Lorenz model contains six loops: three little loops involving x, y, and z individually, and three major loops involving x and y, y and z, and x, y, and z. Because x and y can take on both positive and negative values, the polarities of the major loops can shift back and forth. For particular values of the parameters, this stock-and-flow feedback structure exhibits deterministic chaos and extreme sensitivity to initial conditions ($\sigma = 10$, $r = 28$, $b = 8/3$ is one such parameter set.)

The closed-loop feedback character of such chaotic systems has been noted by some (see, e.g., *System Dynamics Review*, Special Issue on Chaos 4(1–2) 1988). But just as it was in Maxwell's time, the loop structure of chaotic models is not central to the mathematical analysis of such systems. Yet since nonlinear models can show chaos, and nonlinear models can endogenously change dominant structure, it is tantalizing to consider whether the complex behavior patterns arise as consequences of shifting loop dominance. It has been shown, for example, that a classic kind of bifurcation in a simple continuous nonlinear system occurs because of a shift in dominance between positive and negative loops at an equilibrium point (Richardson 1984). The evolution of chaos in a model of migrating populations (Mosekilde et al. 1985; Mosekilde, Aracil, and Allen 1988; Reiner, Munz, and Weidlich 1988) has been intuitively analyzed in terms of the feedback loop structure of the system (Richardson and Sterman 1988). There exists the possibility that the feedback concept and loop dominance may provide intuitive insight into the causes of complex nonlinear dynamics.

Furthermore, we now face the likelihood that the enormous range of feedback systems social scientists have observed includes structures that can endogenously generate unpredictable behavior. Recent work by John Sterman, for example, has found that the decision rules people actually use in simple dynamic games involving nonlinearities and circular causality can generate deterministic chaos (Sterman 1987, 1988, 1989). We have a new *source* of randomness. Tustin's seminal question (section 3.3), whether the "closed sequences of dependence" in an economy account for business cycles, is now being extended: Can nonlinear feedback structures in economic interactions be responsible for any of the *unpredictability* we observe in economic cycles, without invoking exogenous random disturbances? Given the ubiquity of feedback thinking, and all sorts of puzzling social system behavior, similar questions will be raised in many corners of the social sciences. Answering them will require the creation of increasingly sophisticated nonlinear models of puzzling social and policy systems.

A second development emerging from the literature on complex nonlinear dynamics adds further support to the separation of feedback threads we have been tracing. Ilya Prigogine and his colleagues in the so-called "Brussels school," with others, have cast a new light on the phenomenon of *self-organization* in complex systems (Prigogine and Stengers 1984; Yates 1987).

The ideas began in thermodynamics and physical systems. A mixture of two different gases in a container with a hot side and a cold side is seen to separate, with one of the gases concentrating on the hot wall and the other on the cold wall (Nicolis and Prigogine 1977, p. 3). A liquid contained between two cylinders is observed to form and reform layers of various sorts as one of the cylinders spins faster and faster (Abraham 1987). As a parameter smoothly changes in a simple nonlinear model, the behavior of its variables moves in sudden shifts to oscillations of increasingly long period, eventually leading to an infinitely long period of random, chaotic behavior (Feigenbaum 1980/1983). Or if the parameter moves in the other direction, the model's behavior is seen to move from patternless, unpredictable ups and downs to stable oscillations of definite period and amplitude. Patterned behavior appears to self-organize out of chaos.

The reader will recall that there was considerable interest in the cybernetics thread in the phenomena of self-reference and self-organization (see sections 3.2 and 4.3; also Yovitz and Cameron 1960, von Foerster and Zopf 1962, and Varela 1975, 1979). Now we have renewed interest in the notion of self-organization, but among the literature and practitioners in the servomechanisms thread. The two feedback threads in the social sciences would appear to converge on the issue of self-organization.

Yet there are dramatic differences, recognized somewhat by the authors themselves. Yates, for example, in his preface to a volume of papers on *Self-Organizing Systems: The Emergence of Order* that grew out of an invited conference held in Yugoslavia in 1979, writes:

> Many natural systems become structured by their own internal processes; these are the self-organizing systems, and the emergence or order within them is a complex phenomenon that intrigues scientists from all disciplines. Unfortunately, complexity is ill-defined. Global explanatory constructs, such as cybernetics or general systems theory, which were intended to cope with complexity, produced instead a grandiosity that has now, mercifully, run its course and died (Yates 1987, p. xi).

In deep contrast to the earlier work in the cybernetics thread, the writers in this volume looked at self-organization dynamically and employed formal mathematical models to represent and explain complex self-organizing behavior.

Put most succinctly, of those who focused on self-organization, writers in the cybernetics thread focused on self-organization of *structure*, while writers in the servomechanisms thread focused on self-organization of *behavior*. In the cybernetics thread, self-organization was discussed verbally. The metaphor for change was the *rewriting* of system structure—exogenous or endogenous changes in the *rules* governing the behavior of a social system. In the writings of people in the Brussels school and the authors in the Yates volume, self-organization emerges out of nonlinear mathematical models of complex systems. The model itself is not rewritten and does not change over time, but endogenously shifts its dominant or active structure because of nonlinearities it encounters when operating far from equilibrium. The behavior-over-time generated by the model changes character, moving among regions of stability and instability, organized patterns and disorganized chaos.

Just as the word *feedback* takes on different meanings in the two threads of feedback thinking in the social sciences, so does the phrase *self-organization*. And the differences again center on the verbal versus mathematical models and a focus on discrete events versus dynamic patterns over time.

Summary

The system dynamicist sees four overriding benefits from looking at societal problems from a feedback perspective. First, feedback loops—circular causal loops of positive and negative polarity composed of accumulations and their rates of change—are seen as the structure underlying dynamic behavior. Second, feedback loops are seen to be responsible for the counterintuitive behavior and policy resistance observed in real social systems. Consequently, the system dynamicist hopes that by understanding their roles in creating such behavior, people can improve over time their ability to create workable and desirable social policy. Third, feedback loops and the notion of loop dominance provide an intuitive and accessible description of mathematical models created to study the behavior of complex social systems, even those capable of extremely complex nonlinear dynamics and deterministic chaos. Fourth, and perhaps most important, feedback loops enable the creation of self-contained theory. In the system dynamicist's endogenous point of view, patterns of dynamic behavior are understood as consequences of the internal (feedback) structure of causally closed systems. Such feedback-based endogenous theory is seen to be much stronger, much more worthy of analysis, than theory that relies on external influences to explain behavior.

5.5 Summary and Directions

In section 3.3 we characterized the beginnings of the servomechanisms thread in the social sciences as follows:
- The patterns of behavior of a dynamic system were traced to its feedback structure.
- Formal dynamic models were employed.
- The dynamic behavior of a feedback system was considered to be difficult to discern without the aid of formal mathematical models.
- Feedback loops were seen as an intrinsic part of the real system, not merely as possible mechanisms of external system control.
- Positive loops were present in the analyses, along with negative loops.
- Well-intentioned policies were seen to have the potential to create or exacerbate the problem behavior they were intended to cure.
- Nonlinearities were perceived to be a persistent characteristic of real socioeconomic feedback systems. Consequently, they were considered to be a necessary characteristic of reliable formal models of such systems.
- The work tended to be directed toward policy analysis.

In addition there were two characteristics that both feedback threads in the social sciences shared:
- Feedback was seen as the mechanism of systems that adapt over time.
- The concept of feedback was employed to conceptualize system structure.

All of these characteristics have continued in the servomechanisms thread. However, with the exception of the spread of the field of system dynamics itself, the influence of this thread on feedback thinking in the social sciences appears to be slight compared to the influence of the cybernetics thread.

Econometrics and the Servomechanisms Thread

In the 1940s and 1950s Gunnar Myrdal outlined a program for the social sciences involving the quantitative modeling of "interlocking, circular and cumulative changes" and "the causal inter-relations within the system as it moves under the influence of outside pushes and pulls and the momentum of its own internal processes" (Myrdal 1944, p. 1068; 1957, pp. 27, 30; see section 2.5). Essentially, there are only two expressions of that approach to the analysis of dynamic phenomena in social systems today: system dynamics and econometrics. One traces its origins to statistical traditions in the social sciences, while the other traces to engineering servomechanisms. The two approaches share a loop view of circular causal interactions and interdependence in social systems. Both approaches can be said to be feedback perspectives on social system dynamics. Both have by now come to be seen as somewhat related quantitative, computer-oriented approaches to socioeconomic behavior. But there are elements of the two approaches that reflect their different origins.

One important distinction between the two approaches stems from a technical difference in basic model structure. Econometric models are formulated almost entirely of linear difference equations, while system dynamics simulation models are composed of nonlinear differential (or integral) equations. An econometric model thus expresses a discrete, period-to-period or quarter-to-quarter view of economic or social change, while a system dynamics model tries to capture a continuous perception of accumulations and change. In addition, the linear nature of econometric models means that they cannot exhibit changes in loop dominance such as we observed, for example, in the logistics equation structure or the Lotka-Volterra predator-prey system. Both of these characteristics of econometric models are consequences, essentially, of the form of equations required to fit whole models to data using formal statistical procedures.

A repeated theme, however, in both of the feedback threads that we have been tracing is that social systems exhibit changes in feedback structure over time—cause maps are redrawn, new information changes the state of things, or nonlinearities shift loop dominance, altering active system feedback structure. In this sense, the econometric tradition is somewhat outside both of the feedback threads that we have been tracing. Econometric models cannot alter structure endogenously, either by nonlinearities or by self-referential rewrites of equations, so they must rely on exogenous influences to capture structural change. I personally prefer a more endogenous view, but my purpose here is not to criticize but to observe there is a difference. A characteristic of the servomechanisms thread from its earliest beginnings in the work of Goodwin and Tustin is an endogenous point of view. The focus is inward, and the attempt is to derive system dynamics and even structural change from internal circular causal processes. The econometric tradition is a mixed endogenous-and-exogenous view.

Directions

In spite of these differences with the servomechanisms thread, some aspects of econometrics are seen to be part of the legacy of Tustin and Phillips, particularly the use of optimal and adaptive control techniques with econometric models. Furthermore, Simon's conceptualization of feedback structure in organizations (section 5.2) links his feedback view to econometrics. Given these perceived connections, but acknowledging that little of econometrics actually traces to engineering servomechanisms, there appear to be four main directions that the servomechanisms thread now points:

- system dynamics—computer simulation of continuous, nonlinear feedback systems, emphasizing an endogenous point of view
- econometrics—statistical modeling and estimation of socioeconomic systems for prediction and policy analysis
- control theory in economics—application of optimal and adaptive control techniques from engineering to econometric models
- causal-loop diagramming—striving to derive behavioral implications intuitively from causal-loop diagrams of circular causal feedback systems.

The last entry in this list, causal-loop diagramming, matches a similar tendency growing out of the cybernetics thread. Many such efforts exist. It would be hard to determine direct connections to, or separations from, either feedback thread, unless there is telltale evidence in the form of references to "stability and instability" or "patterns of behavior."[13]

Notes

1. Aoki (1976), Aoki and Marzollo (1979), Baum and Howrey (1981), Chow (1970), Fair (1974, 1978), Kendrick (1976), Pitchford and Turnovsky (1977). See also bibliographies in Cochrane and Graham (1976) and Athans and Kendrick (1974), and also publications of the Society for Economic Dynamics and Control.

2. In a much later independent development, Forrester and the MIT System Dynamics Group have come to very similar conclusions about mechanisms underlying economic long waves (Forrester 1976, 1977, 1979; Sterman 1984).

3. Culbertson used Greek letters for some coefficients and time-dependent variables. For ease of reading I have substituted English letters throughout.

4. Culbertson gave complicated equations for g' and d' which I am omitting in favor of a verbal description of what they mean. Both of these switch coefficients are related to Y_{t-1} or Y_{t-2}, so additional feedback loops are present.

5. He derived the model in two equations, which he then combined into the following:

$$Y_t = C_0 + I_0 + G_0 - (c_Y + i_Y)T_0 + f'h(M_0 - M_0D)$$
$$+ [g_Y + (c_Y + i_Y)(1 - t_Y) + f'h(s_Y - m_Y)]Y_{t-1}$$
$$+ [g_A + c_A + i_A - t_A(c_Y + i_Y) + f'(s_Y - 1)](Y_{t-1} - Y_{t-2}),$$

where

$$f' = \frac{c_R + i_R}{hm_R}; \quad c_R, i_R, m_R < 0; \quad f' > 0.$$

The interested reader can consult Culbertson (1968, p. 176) for the details. Here we merely note that the model is in the form of the generic model described in the text.

6. See section 3.3. For a discussion linking the mass-on-a-spring structure to employment, inventories and backlogs, see Richardson (1981, pp. 338–343).

7. For the deepest investigation that I know of of the nature of the feedback structure underlying oscillations, see Graham (1977).

8. See Meadows (1980).

9. For extensive bibliographies see Lebel (1981); Legasto, Forrester, and Lyneis (1980). For current publications see issues of *Dynamica*, the *System Dynamics Newsletter* (MIT), and the *System Dynamics Review*.

10. Hamilton et al. (1963); Roberts (1964); Forrester (1969); Meadows (1970); Forrester (1971); Meadows et al. (1972); Meadows and Meadows (1974); Meadows et al. (1974); N. Forrester (1972); Mass (1975); Levin, Hirsch, and Roberts (1975); Levin et al. (1976); Naill (1977); Choucri (1981); Boyce (1980); Gutierrez and Fey (1980). For others see Legasto, Forrester, and Lyneis (1980), pp. 259–261.

11. Forrester (1961, 1968b); Goodman (1974); Alfeld and Graham (1976); Coyle (1976); Coyle and Sharp (1976); Richardson and Pugh (1981); Roberts et al. (1983); Richmond, Peterson, and Vescuso (1987); Richmond, Vescuso, and Peterson (1987); Hanneman (1988).

12. These differences in what it means for a system to be "closed" suggest that Forrester's ideas progressed independently of the general systems movement. However, it should be noted that von Bertalanffy also referred to feedback systems as closed systems (von Bertalanffy 1951). I argued in section 3.2 that that determination conflicts with von Bertalanffy's own characterization, but the fact remains that superficially Forrester and von Bertalanffy (erroneously?) agree that feedback systems are "closed." To conclude that Forrester knew of von Bertalanffy's definitions of closed and open systems (and is therefore connected to him by the scholarly communications network), we would have to conclude that both Forrester and von Bertalanffy became confused in applying the concepts to feedback systems. For one of them to be confused is highly unlikely; the probability that both are confused on the subject is nil. I conclude that Forrester's development of the closed boundary idea in system dynamics is essentially independent of the general systems movement.

13. The decision-support work of Colin Eden and his colleagues in England is an interesting example (Eden and Harris 1975; Eden, Jones, and Sims 1979, 1983). In his more recent work, causal-loop diagrams or influence diagrams are used to bring out and improve the "cognitive maps" of people faced with a decision. In that usage the diagrams have a very transitory character, much as

in Weick's usage. Yet Eden's early work (Eden and Harris 1975) references authors in both feedback threads and makes extensive use of system dynamics analyses in which feedback diagrams aim to capture real, persistent, underlying causal structure. Eden's own use of the feedback concept seems to have evolved toward a flexible use of influence diagrams, mixing events and persistent pressures, as suits the cognitive maps of his decision support clients.

Chapter 6
Issues and Implications

> "I meant 'there's a nice knock-down argument for you!'"
> "But 'glory' doesn't mean 'a nice knock-down argument,'" Alice objected.
> "When *I* use a word," Humpty Dumpty said, in a rather scornful tone, "it means just what I chose it to mean—neither more nor less."
> "The question is," said Alice, "whether you *can* make words mean so many different things."
>
> Lewis Carroll, *Through the Looking Glass*

6.1 Introduction

In the preceding chapters we have investigated a particular pattern of thought that permeates an astonishing variety of investigations and accomplishments in scholarly activity over the last two hundred years. In this book alone we have learned that feedback thinking has been applied to

- control
- stability and instability
- economic self-regulation
- structure of mathematical models
- dynamics of epidemics
- predator-prey interactions
- population dynamics
- arms races
- economic cycles
- economic dynamics
- kinesthesia and proprioception
- homeostasis
- purposeful behavior
- evolution
- dialectic processes
- historicist and functionalist explanations
- corporate management
- industrial dynamics
- theory-building in psychology
- stimulus and response
- better-before-worse behavior
- morphostasis and morphogenesis
- speculation
- panics in crowds
- inflation
- persistence of a political system
- vicious circles
- race relations
- bandwagon effects
- suicidal and self-fulfilling prophecies
- bank failures
- unanticipated consequences of social action
- self-hood and autonomy
- dynamics of cultural contact
- organization studies
- depression
- urban dynamics
- heroin and crime
- global dynamics

knowledge of results
policy analysis
worse-before-better behavior

research and development
addiction
policy resistance

These represent only some of those I know about, and my knowledge represents only a fraction of the actual applications of the loop concept underlying feedback and circular causality. Certainly then, the first implication of this investigation is that the concepts of positive and negative feedback are powerful unifying notions that illuminate the structure of arguments, explanations, and causal views in virtually every corner of the social sciences. Furthermore, the feedback concept, in the generality with which it is defined in section 1.2, offers the potential of creating a visual, nonquantitative bridge between mental and formal models in the natural and social sciences. The inescapable conclusion is that the loop concept underlying feedback and circular causality is one of the most profound and most penetrating fundamentals in all social science.

However, it is not a unified loop perspective. Amid the diversity of these applications is a diversity of opinions about what makes the loop concept powerful in social science. In our two feedback threads, which now show signs of unraveling into several more strands, there is further diversity about the characteristics of a feedback perspective that is perceived to hold the greatest promise for the social and policy sciences. How can these diversities be resolved? What can be learned from the different interpretations and uses of the feedback concept we have uncovered?

Some might argue that the diversity is healthy. Species diversity is generally good for ecosystems, and that probably holds true for ecosystems of ideas as well as organisms. However, it appears to me that the diversity of loop views we have encountered has weakened the general perception of the power of any one of them. In spite of the enthusiasm of the writers we have surveyed, the feedback concept has not become as central to the education of citizens and social scientists as this investigation would suggest it should be.[1] We could speculate on the reasons: educational change happens deliberately slowly, unidirectional causal views are hard to break, and so on. I suggest, however, at least part of the fault lies in our failure to perceive that there are different loop perspectives hidden in the word feedback. If all loop views are seen to be essentially the same, then the weaknesses perceived in any one of them would be attributed to them all. The accumulated perceptions of general weaknesses could unfairly inhibit the perceived power of any particular feedback approach. So, while acknowledging healthy diversity, I believe it is important to try to pull out of it some lessons for future promising directions for the feedback concept in the social sciences.

At this point, however, there is no doubt confusion about the essential components of the feedback perspectives we have encountered. One way to cut through that confusion is to imagine a no-holds-barred discussion between proponents of different feedback perspectives. Themes that crop up in such an argument can crystallize for us the fundamental issues that separate the different threads.

The next section, then, presents two arguments for why the feedback concept has not reached its potential in the social sciences. For clarity they will be phrased without scholarly tact or reservation. The first part presents the argument of a devotee of the servomechanisms thread; the second presents a corresponding or counterargument by a scholar in the cybernetics thread. Both arguments presume two premises: that the loop concept underlying feedback and circular causality is of fundamental importance in the social sciences, and that as such it should play a central role in the education of future citizens, scholars, and policy analysts. Both arguments also assume the reader's familiarity with the descriptions and analyses contained in the preceding chapters—the data, as it were, supporting these differing interpretations of feedback thought.

It is important to emphasize that the purpose of these arguments is not controversy or debate. Instead, they are a method of clarifying the foundations of the different feedback perspectives. Section 6.3 will draw out from these two arguments themes or issues of significance for the future evolution of feedback thinking in the social sciences. Section 6.4 will reflect on these issues and conclude this investigation.

6.2 Two Arguments

The partisan presentations in this section are designed to identify most clearly the significant issues and implications of two separate feedback threads in the social sciences. Each paragraph in each of the following arguments should be prefaced with phrases like "the system dynamicist believes..." or "the TOTE theorist takes the view that..." or "the typical feedback thinker in the cybernetics thread thinks...". But in the interests of brevity, I have frequently left out such qualifiers. Read each section as an opinionated statement written from the perspective of a devotee of the appropriate feedback thread.

The points of view are written as arguments pointing out weaknesses in the other thread's development of the feedback concept. For emphasis, each partisan speaker also occasionally contrasts those perceived weaknesses with corresponding strengths perceived in his or her own thread. Again, however, the purpose is not controversy but clarity about the current state of the evolution of feedback thinking in the social sciences.

Weaknesses in the Feedback Perspective of the Cybernetics Thread, as Viewed from the Servomechanisms Thread

A scholar attuned to the development of the feedback concept in the servomechanisms thread sees the following weaknesses in the feedback points of view expressed in the cybernetics thread in the social sciences.

(1) Focus on stability and homeostasis

The cybernetics thread has emphasized the concepts of stability and homeostasis, rather than dynamics. The question "Is the system stable?" has only two answers: "yes" and "no." The "no" answer holds whenever one can at best say "sometimes." With only two possible answers to the question, scholars in the cybernetics thread have tended to consider only two extremes: Is the system dominated by deviation-amplifying processes or by deviation-counteracting processes? The question easily becomes ill-posed: Is the system itself deviation-amplifying or deviation-counteracting? In a complex multi-loop system capable of shifting loop dominance, the question has no unique answer. It is meaningless to say that an urban system "is deviation-amplifying." At some times in its development, certain aspects of the behavior of an urban area are dominated by positive feedback loop processes. At other times, or for other aspects of behavior, negative loop processes dominate. Furthermore, it is possible to construct simple, realistic systems composed of nothing but negative loops that are very unstable, showing exploding oscillations. Dominance by negative loops is not sufficient to guarantee stability. Moreover, there is considerable debate over what "stability" means.[2]

In contrast, the emphasis in the servomechanisms thread is on patterns of dynamic behavior. The question "How does this feedback system behave?" has a great variety of possible answers, and it cannot be answered reliably without mathematics or simulation except in trivial circumstances. But if stability is the question people have been habituated to answer, and if they think that it is possible to determine stability without determining actual patterns of dynamic behavior, then they will try to do so. They will bypass the task of deriving dynamic behavior from feedback structure, and they will miss what others discover who make the effort to derive behavior from structure. It is likely that they will come to conclude that the question of stability and the

322 Issues and Implications

conclusions they draw from it are not very useful in analyzing societal problems or social systems.

It appears that some in the cybernetics thread believe that the most one can conclude from the feedback concept is stability or instability—that what is observed in social systems is more determined by randomness or unexplainables than by structure, or that complexity prevents reliable analysis. If the most we can say is whether a system is stable or not, then there is yet another reason not to try to derive dynamics from feedback structure.

The focus on stability and homeostasis in the cybernetics thread thus tends to discourage investigations of the connections between feedback structure and patterns of dynamic behavior. Tustin's question, for example (section 3.3)—whether "closed sequences of dependence" can account for business cycles—would never be asked in the cybernetics thread. Social scientists in that thread of feedback thinking will miss one of the central premises or conclusions of the servomechanisms thread—that patterns of dynamic behavior are consequences of feedback structure.

(2) Focus on discrete events

Discreteness gives information the character of moves in a game—no information, then a burst of information, then no information, and so on. "Messages" are seen to have a discrete nature. Feedback in such a discrete view becomes "knowledge of results," the result of a Test in a TOTE unit, a response or perhaps a stimulus, a message, or, in general, knowledge of an "event." But discrete events are seen to have a high level of uncertainty associated with them. They have a heavy "noise component," a low "signal-to-noise ratio." That fact leads social scientists focusing on discrete events in feedback loops to emphasize the random component in the behavior of social systems.

Discreteness appears in the cybernetics thread in a number of ways. McCulloch and Pitts (section 3.2) started one discrete view with the binary off/on devices they envisioned as the building block of neural nets that so influenced Beer and Deutsch (sections 4.2 and 4.5). That line of thinking results in a view of the "state" of a system as one particular vector of values of all the variables of interest in the system. A string of n zeros and ones has 2^n possible "states." A difference between the actual "state" of such a system and a desired "state" shows up as a mismatch in one or more corresponding locations in the strings. It is this view of the state of a system that leads to Ashby's concept of variety and the Law of Requisite Variety (section 3.2). A discrete view is essential for Ashby's concepts: the "variety" of any continuous variable is infinite. Furthermore, the emphasis is forced outward, on the incoming signals of high variety that must be dealt with.

It is no accident that the concepts of positive and negative feedback in these discrete views are seen to be less significant than the content of specific messages, the character of a particular event, or the variety in system inputs and the ability of the system to generate matching variety. To scholars in the servomechanisms thread, the perception of feedback structure is useful to explain and predict patterns of behavior. They would not attribute to it any special competence to predict an event to explain why a given event occurred at a given time. Loops of positive and negative polarity may connect with dynamic behavior in the system dynamics sense, but the connection in discrete settings is far less clearly defined.

System dynamicists and others in the servomechanisms thread deliberately "move one step back from events": they smooth discrete events into patterns of continuous behavior, and they blend discrete decisions into continuous policies and pressures. They do so because they believe that it is policy structure and dynamic behavior that are connected through the feedback concept, not individual events and decisions. They view separate events and decisions as riding on the surface of an underlying tide of policy, pressures, and dynamic pattern. Without such a step backward from discrete events and decisions, they would say, the feedback concept has, if any, only a brief significance.

Some social scientists may find that step unacceptable. To many, events and decisions are the things to be explained and predicted. If one wants to understand discrete events and decisions, the continuous view of dynamic patterns and policies is unacceptable, and the power of the feedback concept that goes with that continuous view will not be perceived.

We have also seen a discrete view, in the form of difference equations, used by people who are sociologically at least in the servomechanisms thread (section 5.3). The discreteness of that representation also appears to weaken the feedback perspective. It contributed to Ashby's conclusion that the feedback concept is "inadequate in itself" for "understanding the general principles of dynamic systems (section 3.2). It led Culbertson (section 5.3) to interpret negative feedback as a "switch" and to view oscillations as self-reinforcing processes moving between "floors" and "ceilings." There are such limit cycles, but there are also oscillators that do not have that structure. Scholars in the servomechanisms thread argue that a continuous representation, in terms of differential or integral equations or digital computer simulations of such continuous processes, captures more accurately the continuous ebb and flow of quantities, perceptions, and pressures that underlie events and decisions in aggregate social systems.

It is frequently argued that data, particularly economic data, are not available continuously, so formal representations should be in terms of discrete difference equations. The discrete points in time they reflect are all that is observed. A feedback thinker in the servomechanisms thread arguing for a continuous point of view would note that the observed data can be seen as a sample from an essentially continuous stream. A continuous model could be sampled, just as continuous reality is sampled. The sampling idea preserves the continuous perspective in which feedback and circular causality have their greatest potential.

Presumably it is possible to represent feedback systems in terms of difference equations and learn something of significance. However, such a representation makes it easy for social scientists to miss a feedback principle that emerges naturally from a continuous representation, namely, the principle that in every feedback loop there must be at least one accumulation (integration) (section 5.3). The failure to acknowledge that principle is behind both Ashby's and Culbertson's difficulties with difference equations. Powers made use of it obliquely when he criticized a discrete stimulus-response interpretation of the feedback loop section (section 4.7). The scholar in the servomechanisms thread concludes that discrete representations, while not necessarily "wrong," fail to facilitate reliable conclusions from complex circular causal structures.

(3) The "brain" model

The cybernetics feedback thread in the social sciences is not synonymous with cybernetics, but nonetheless some of the associations in the formal multi-discipline of cybernetics carried over with the feedback concept. One of these associations was the brain as a model for societal structure and behavior. The neural nets of McCulloch and Pitts (section 3.2), and their ties to the structure of digital computers, led many social scientists in the cybernetics thread to focus on communication networks rather than on the deviation-amplifying and deviation-counteracting loops in circular causal processes. That focus drew attention away from the concept of circular causality in social science. The "brain model" is not a causal model. It led to views of social systems emphasizing circular patterns of communication in information networks and de-emphasizing or ignoring circular causality.

The network model is perceived to be very suggestive, but it does not rely for its power on the perception of signed causal loops. When loops are perceived in the networks employed by Deutsch, Beer and others, they are seen as looping channels of communication, not as positive and negative circular causal processes. Channels of communication are seen as only the start of an understanding of societal phenomena. The real significance to the social scientist in the cybernetics thread is contained in the content of messages communicated around the loop. To those in the servomechanisms thread, however, the feedback concept acquires its power from associating a loop of influences with a concept of loop polarity. The notion of loop polarity goes beyond the mere identification of a circular channel of communication. Polarity adds a causal view to the feedback concept. The brain model, by emphasizing communication networks and de-emphasizing causality, weakened the use of the concept of circular causality in the social sciences.

Another aspect of the brain model, discussed under (2), is its use of binary, logical, off/on devices, the neurons of brains and the binary bits of digital computers. The brain model thus helped to reinforce a tendency to employ a discrete, event-oriented view with the feedback concept.

Perhaps the most significant legacy of the brain model in the social sciences is the point of view about complexity that it reinforced. The brain is so complex that understanding how its structure produces its observed behavior is perceived to be almost hopeless. Using the brain as a model for organizations, corporations, the economy, and other complex systems led naturally to the point of view that such complex, probabilistic systems cannot be understood. The best that might be done, as Beer so clearly asserted (section 4.2), is to use the concept of feedback (and other cybernetic ideas) to manage or control such systems with or without understanding. The brain model thus helped to give prominence to the black box point of view. To a scholar in the servomechanisms thread, emphasizing the use of feedback to understand complexity, that black box point of view is a serious weakness in the cybernetics thread.

(4) The black box

A number of authors in the cybernetics thread in the social sciences assumed that causal understanding, or even circular causal understanding, of complex systems is impossible. Beer, for example, argued for some time that the only way to model a complex system is to manipulate inputs, observe outputs, and construct a binary white box that matches the observed patterns of inputs and outputs to the black box system (section 4.2). To control a complex system, one builds such a white box, designs a controller for it, and uses that controller for the real black box system.

The white box is something like a formal simulation model used as laboratory tool. We have seen many in both feedback threads who advocate such formal models. But the black box approach denies the possibility of constructing a model that structurally or causally looks like the real system. A modern black box approach, which stems partly from Wiener's *Extrapolation, Interpolation and Smoothing of Stationary Time Series* (1949), is ARIMA time series analysis.[3] In this powerful technique, statistical models composed of autoregressive and moving average terms essentially "write themselves" to conform to given time-series data. In the process they make no attempt to represent actual system structure. An ARIMA model may be a white box, in the sense that its structure is explicit and knowable, but it is not a causal model. It is not composed of interacting positive and negative circular causal processes intended to match perceived characteristics of social reality. Like all black box models, ARIMA models are models of behavior, not of structure.

The black box concept led to two tendencies that weakened the feedback perspective in the social sciences. First, it combined with other tendencies in the cybernetics thread to suggest that the appropriate use of feedback in social science was in add-on control structures. From the point of view of the servomechanisms thread, that it a very restricted view of the power of the feedback concept . It misses, by choice, a perspective of the internal circular causal processes of the system trying to be controlled. Furthermore, the attempt to control, even using feedback, without understanding internal feedback structure may be doomed to failure. If social systems are policy resistant and counterintuitive because of their circular causal structure, then policy interventions made without considering that internal structure will likely fail.

Philosophically, the black box concept contributed to a self-fulfilling prophecy. If one assumes that it is impossible to use the concept of feedback and circular causality to understand the structure and behavior of a real system, then one will not even try. The conclusion that one cannot understand will be reinforced by the attempts people make to manage social systems in the absence of understanding. To a scholar in the servomechanisms thread, the black box concept weakened the potential of the feedback concept in the social sciences because it promised to be a substitute for understanding.

(5) Perception of feedback as an analogy

Some social scientists view the use of the feedback concept in social science as "borrowing an idea from engineering." In the early literature in the cybernetics thread, feedback and the communications network were indeed thought of as analogies. Other social scientists in both feedback threads see feedback and circular causality as facts of life of living systems. Perception of feedback as an analogy represents a weak intellectual commitment to the idea as an explanatory tool. Weak commitment, as Kuhn (1962/1970) suggests, means rapid abandonment in the face of minor anomalies. The perception of feedback as an analogy rather than as a fact of life in living systems weakened efforts to understand the role of feedback and circular causality in social systems.

(6) Overemphasis on randomness

In the cybernetics thread there is a definite tendency to favor probabilistic models over deterministic ones. Beer, for example, argued that exceedingly complex systems are obviously stochastic and that approaching them as if they were deterministic is pointless. He pointed out that different economists will give different causal theories of the same economic behavior, and he concluded that causal understanding of exceedingly complex systems is impossible. Feedback thinkers with that point of view will tend not to formulate causal models, mental or formal. They will tend not to take the steps necessary to see links between feedback structure and dynamic behavior in complex systems. They will instead emphasize the apparent randomness or unpredictability of social systems.

Yet causal models are what we mean by "understanding." The concept of "randomness" can be viewed as a substitute for causal understanding. We may be able to discern patterns in randomness (probability distributions) and make substantive use of them (statistic inference), and we may be able to use noise to stimulate behavior modes latent in a system's structure, but randomness is inherently not a "causal" concept. An overemphasis on randomness guarantees an underemphasis on circular causality and a growing perception of the inadequacy of the feedback concept.

There is a long tradition in the sciences to use the notion of randomness to "explain" observed deviations from theory. The word "explain" is used, but it is clear that randomness in this sense is invoked as an "explanatory principle" that merely puts a label on a lack of fundamental understanding. We all use explanatory principles to some extent. The question here that is relevant to the evolution of the feedback concept is: how should the concepts of randomness, feedback and circular causality be integrated to best serve the social and policy sciences?

System dynamicists choose to use randomness to stimulate some aspects of behavior modes that emerge from their circular causal theories. Others, who place greater emphasis on the role of randomness, make stochastic processes rather than feedback structure the object of study. They learn a great deal about randomness and the way real systems filter random signals, but they will not come to "understand" how the feedback structure of the system accomplishes that filtering.

Scholars in the servomechanisms thread prefer to emphasize causality over randomness. They believe that anyone who intervenes in a system to have a certain effect must have a causal model of the system which implies that the intervention will cause the desired effect. Policy modeling, mental or formal, thus has to be causal. To say the desired effect will result with probability p is still predominantly a causal, not a stochastic, view. Formulators of circular causal feedback models do not claim that reality is deterministic. Rather, they merely show their preference for trying to "understand." Policy analysts and implementors of policy always have that preference.

(7) Input/output point of view

In the Macy meetings (section 3.2), feedback came to be defined as "the alteration of input by output." The loop notion is there, of course, as input is processed to become output, and some information about output is fed back to alter input. But the input/output wording tends to slight the causal loop aspect of the feedback and to emphasize the importance of influences outside the loop. The input/output point of view weakens the feedback perspective because it draws attention away from endogenously determined behavior and toward the anticipation of exogenous inputs. The emphasis on inputs and outputs in the cybernetics thread has inhibited the development of an endogenous point of view in social science founded on the feedback concept.

Weaknesses in the Servomechanisms Thread, as Viewed from the Perspective of the Cybernetics Thread

A social scientist in the cybernetics thread is likely to see the following weaknesses in the feedback point of view expressed in servomechanisms thread.[4]

(1) Deterministic point of view

The behavior of real systems has an obvious stochastic component. Feedback thinkers in the servomechanisms thread build models of social reality that ignore or down play the natural variability of social systems. Random processes in nature mean that management through deterministic understanding is not possible.

Furthermore, some in the servomechanisms thread do not make use of the advances in statistics that make estimation of models and prediction of time series possible. Societal feedback control mechanisms can be designed to handle various forms of natural variability, but the deterministic view in the servomechanisms thread has held scholars in that thread back from this promising use of feedback control principles.

(2) Use of formal models of fixed structure based on engineering methods

Many authors in the servomechanisms thread interpret their views of feedback structure in formal models composed essentially of differential or integral equations. System dynamics DYNAMO models (section 3.3), as discrete digital computer simulations of continuous systems, are in this class as well. Such representations are borrowed from the physical sciences and are inappropriate for the social sciences, which deal with systems of much greater complexity and variability and much more ambiguous structure.

Furthermore, social systems are living systems. Unlike physical systems they change over time. No model of fixed structure can represent a social system over time. The perceived inability of continuous simulation models to change their own structures is one of the reasons a number of thinkers in both feedback threads have moved in the direction of artificial intelligence and LISP-like simulations. A formal model built in these methodologies can rewrite its own structure and evolve over time, as does the reality of any living system. Models built in the servomechanisms thread can represent growth in quantities but they fail to capture the growth in structure and complexity that accompanies growth in size in living systems.

(3) Closed systems

Social systems are inherently open systems, influenced by inputs from outside. To scholars in the cybernetics thread, the endogenous point of view that characterizes the perspective of the servomechanisms thread denies this essential property of social systems. The open-system property is fundamental to Wiener's insight that information is "negative entropy": like inflows of energy, inflows of information into a system can enable it locally to organize and overcome the natural processes of dissolution that threaten all systems (see section 3.2). In a more applied, policy-oriented context, the open-system property is also fundamental to Ashby's observation that a regulator that fails to try to predict inputs will never be perfect (section 3.2). The endogenous view therefore is guaranteed to result in management control policies that are less powerful than ones that take into account the existence and nature of external influences.

The failure to recognize the fundamental open-system character of social systems has prevented scholars in the servomechanisms thread from understanding of the significance of variety in social systems. As Ashby defined it, variety is the number of discernibly different states or conditions a system can exhibit (see section 3.2). To many scholars in the cybernetics thread, it is input variety that creates dynamic consequences that the system tries to control. They believe that the concept of negative feedback as a control mechanism only reaches its potential in living systems when it is combined with Ashby's Law of Requisite Variety. Properly applied, Ashby's law shows what must be controlled and what resources must be assembled to achieve control. Part of the failure of the servomechanisms thread to understand the significance of the concept of variety is its tendency to continue to represent system structure in terms more appropriate to engineering systems than to living systems.

(4) Limited concept of Information

In the servomechanisms thread, feedback is limited to positive or negative polarity. The information content of the feedback is essentially ignored. But what is important in social systems is the information contained in communications. The significant unit of feedback is the message and its meaning. The TOTE formulation of a feedback loop is thus far richer than the simple polarity loop of circular causality, because it can represent and transmit meaning beyond deviation-amplification and deviation-counteraction. Maruyama (section 4.4) showed that the meaning incorporated in a small set of rules could account for the growth and morphogenesis of a very complex structure, but more must be involved than simple determinations of positive and negative feedback. To scholars in the cybernetics thread, meaning is too important to be reduced to two options.

The limited concept of information in the servomechanisms thread also ignores the advances in communication theory. Shannon's Tenth Theorem, for example, which is Ashby's Law in a more advanced form, shows that a system's regulatory or selection capability is limited by its information-handling capacity (Shannon and Weaver 1949; Ashby 1962). The theorem places a bound on our aspirations for regulating complex systems. Proceeding in ignorance of the bound, policy analysts in the servomechanisms thread could expend a great deal of effort in fruitless causes trying to achieve degrees of regulation that are patently impossible given the information channel capacities inherent in the system.

(5) Focus on patterns of behavior, not on decisions and events

Many feedback thinkers, particularly those in the servomechanisms thread, are forced by the representations they use to focus on patterns of behavior. The resolution of their analytical lens is not sufficient to reach to the detail of day-to-day decisions and the events they shape and are shaped by. Yet events and decisions are the important stuff of management, government, problem-solving, interpersonal relations, group dynamics, corporations, history—the entire sweep of the social sciences. All the most significant questions deal with the phenomena surrounding individual events and decisions.

An emphasis on patterns of behavior has another rather obvious limitation: it restricts the kinds of problems addressed to those that have discernible patterns. In all social science issues there is undoubtedly some form of movement over time, but patterns are frequently not present. Even a common phenomenon, such as revolution, looks very different in different countries around the globe. Analysis has to be grounded in the specifics of individual situations to be insightful and useful.

(6) Failure or inability to address issues of self-reference

Feedback in the servomechanisms thread is employed to address dynamic questions. However, the most significant concepts to emerge from considering the processes of information feedback are those of self-reference, including reflection, self-organization, self-transformation, and autopoesis (self-generation). How do social systems reflect on their own structure and behavior? How do they reorganize themselves? How does organization form out of chaos? Answers to these questions are critical for understanding and predicting how individuals, groups, corporations, governments, societies, and even the globe will evolve over time. It is significant that the global models of Forrester

(1971b) and Meadows et al. (1972) did not deal with these issues. To the scholar in the cybernetics thread, the concept of feedback limited to loops of positive and negative polarity is less appropriate for addressing questions of self reference than loop concepts associated with the feedback of messages and meaning. The formal models judged more appropriate to these investigations are those, such as artificial intelligence models in LISP, that can process, act on, and even create linguistic structures.

Comment

The preceding arguments could conceivably generate heated debate. However, we shall not pursue that debate. The importance of the arguments in this section is not whether one is a true perception and the other false. Instead, the importance for us of the points raised in these two arguments lies in the consistent world views they reveal. Section 6.3 exposes those world views and discusses their implications for the future evolution of feedback thinking in the social sciences.

6.3 Themes and Issues

The partisan arguments in the preceding section, together with careful consideration of the development of the two feedback threads in Chapters 3, 4, and 5, reveal two self-consistent feedback points of view, which may be deep enough to deserve the term worldviews. They provide different perspectives on a number of issues that have been present, implicitly and explicitly, in the developments through Chapter 5 and that became more obvious in the arguments in section 6.2. Eleven of these issues or themes are listed on the left of Table 6.1. In the two columns at the right are brief descriptors of points of view expressed by various social scientists in the cybernetics and servomechanisms threads. This section is devoted to an explanation of the ideas contained in this table. The significance of the points of view contained in these eleven issues is the subject of the concluding section 6.4.

The terms in Table 6.1 should be somewhat familiar. The reader ought to be able to attach to these tendencies people, books, and articles we have investigated. Neither column describes any one particular individual or work, however. It might be best to view the descriptors in Table 6.1 as "ideal types" or essences of tendencies we have observed in the diversity of applications of feedback and circular causality in the social sciences. Let us turn to what these issues mean and how they have arisen in the evolution of the feedback concept in the social sciences.

Explanations of the Issues in Table 6.1

Point of view

Any view of social reality can be placed on a continuum that ranges from an extreme exogenous point of view to an extreme endogenous point of view. At its most extreme, an exogenous point of view involves no feedback or circular causality. Behavior in such an open-loop view

TABLE 6.1 Issues and their interpretations in the two feedback threads in the social sciences

Issue	*Cybernetics thread*	*Servomechanisms thread*
Point of view	Exogenous/endogenous	Endogenous
Attitude toward complexity	Complex systems not understandable in causal terms	Feedback is the key to understanding complex systems
Explanation function of feedback	Stability, structural change	Patterns of dynamic behavior
Units of Description	Event, decision	Pattern of behavior, policy structure
Representation	Linguistic, discrete	Quantitative, continuous
Units of Analysis	Communications net, message	Causal loop, link and loop polarity
Causal view	Cause-effects as stimuli & responses of cause-effects	Behavior as a consequence of feedback structure
Structural change	Rewrites	Nonlinearities
Role of computing machines	Source of analogies, system monitor, controlling brain, simulator	Simulator, deducing machine, laboratory tool
Management problems	Variety, stochastic variation	Counterintuitive behavior of feedback systems
	Anticipating and compensating for disturbances	Understanding internally generated problems
Source of dynamics	Filtered randomness	Feedback structure

is seen to be caused by exogenous factors. All dynamics and change in such a view must trace to influences outside the system under consideration. A single regression equation in the form $Y = b_0 + b_1X_1 + b_2X_2 + \ldots + b_pX_p + e$ is an example of a model at the extreme open-loop end of the exogenous/endogenous continuum. Most macroeconomic models are somewhere near the middle of this continuum, deriving some of their dynamics from exogenous variables modeled as functions of time and some from resulting interactions of variables in feedback loops. Any of Forrester's published models is an example of an extreme endogenous point of view, in which all dynamic behavior traces to circular causal interactions of variables embedded in feedback loops.

In the servomechanisms thread we observed considerable emphasis on an endogenous point of view. Feedback thinkers in this thread made explicit efforts to construct endogenous theories of social and economic behavior based upon feedback and circular causality. The importance of feedback loops in the analyses of authors in the cybernetics thread shows that they, too, take a somewhat endogenous point of view. In contrast, however, the emphasis in the cybernetics thread on random disturbances and variety in system inputs shows that the point of view in that thread is a mixture of exogenous and endogenous perspectives. The point of view of a scholar in the cybernetics thread is more toward the middle of the endogenous/exogenous continuum than toward the endogenous extreme.

The issue here labeled "point of view" thus concerns the extent to which one strives to derive behavior endogenously from the internal nature or structure of a system and the complementary extent to which one attributes behavior to external influences.

I believe that the concept of the endogenous point of view is one of the most important ideas in the evolution of feedback thinking in the social sciences. As we have noted (see sections 3.3 and 5.4), an extreme endogenous point of view *forces* a feedback perspective. The idea of circular causality that is the focus of this study can thus be viewed as a consequence of this more fundamental idea, the endogenous perspective. The real issue that underlies feedback and nonfeedback points of view is the question of whether to take an endogenous or an exogenous view of social system behavior. As we have noted, the feedback concept enables the endogenous point of view and gives it structure.

Attitude toward complexity

We noted a dramatic difference between Beer's approach to the management of complex systems and Forrester's (see sections 3.3, 4.2, and 5.4). Following Ashby, Beer asserted that exceedingly complex systems inherently "cannot be described in a precise and detailed fashion. . . . The methods we should use to handle exceedingly complex systems are those of input-manipulation, output-classification; they are not those of 'cause-and-effect' analysis" (Beer 1959, p. 52). The feedback concept can be used, along with other cybernetic notions, to control or manage complex systems, but the systems themselves cannot be understood in causal terms, circular or otherwise. In contrast, what Beer asserted cannot be known in complex systems authors in the servomechanisms thread attempt to capture explicitly in more or less deterministic equations. While both Beer and Forrester would say they emphasize information feedback, one ignores the system's internal circular causal structure that is the primary focus of the other. The reason for the difference is, quite literally, what each considers knowable.

The distinction in attitudes toward complexity is most evident in the work of Beer and Forrester, but it may also be discerned in the work of other authors in the cybernetics and servomechanisms threads. The servomechanisms thread maintains the tendency to analyze the internal circular causal feedback structure of complex systems. In the cybernetics thread, circular causality is largely replaced by the concept of the communication network and the feedback of discrete messages. It is reasonable to presume that individual events, decisions, and messages are perceived to be less predictable in strictly causal terms, and the most one can hope for in exceedingly complex systems is the identification of channels of communication. Where circular causality does appear in the cybernetics thread, as in the work of Weick (section 4.6), feedback structure is perceived to be very changeable, again a reasonable presumption if complexity is seen as a barrier to causal understanding.

The result of the difference in attitude toward complexity is a marked difference in the ways in which the feedback concept is employed in the two threads, as the following categories illustrate.

Explanatory function

We have observed differences between the two feedback threads in what the feedback concept has been employed to explain. In the servomechanisms thread, the concept is used to explain patterns of dynamic behavior in social systems. In the cybernetics thread, the feedback concept is used to account for the persistence and stability of conditions in a social system and for the structural changes that appear to be involved in learning, evolution, and social reorganization. These differences in the explanatory function of the feedback concept go hand-in-hand with the next five themes listed in Table 6.1.

Units of description

The concepts and terms used to describe and analyze societal phenomena differ in the two feedback threads. The units or "atoms" used to describe behavior in the cybernetics thread have tended to be the event and the decision. We have seen various emphases on the "message" (Deutsch), "tests and operations" (Miller, Galanter, and Pribram), "stimuli and responses" (Easton and others), and feedback as "knowledge of results" (Annett and others), each of which can be classified under one or the other of the generic terms "event" and "decision." In contrast, the unit of behavioral description emphasized by social scientists in the servomechanisms thread is the "pattern of behavior." Thus the system dynamicist works with "behavior modes," which are expressed in graphs of quantities varying over time.

In this regard, Forrester's distinction between "policy" and "decision" is significant (Forrester 1961, p. 97; see section 3.3). A decision is a discrete event. A policy is a persistent framework, a set of decision rules, within which arises an ongoing stream of decisions. Forrester described this distinction in terms of the conceptual "distance" of the viewer. Observed from a sufficient distance, a decision stream is seen not as a sequence of discrete decisions but as a pattern of behavior that reflects a policy structure. Thus the units in which systems are described in the servomechanisms thread are policies, not decisions, and patterns of behavior, not events, and those patterns are usually expressed in terms of graphs over time. These more aggregate units or atoms of system description are consistent with the usually more distant perspective of the servomechanisms thread.

I believe that the distinctions between events, patterns of behavior, decisions, and policy structure are among the most important observations in this investigation of the evolution of feedback thinking. The units of system description employed in the different feedback threads can be seen to be the foundation underlying many of their other characteristics. I will reserve further comments for section 6.4.

Representation

The units in which social systems are described prescribe to a large extent the form of representation employed in mental and formal models of those systems. A focus on events and decisions is consistent with, and probably forces, a discrete, linguistic representation. Words are very handy, if not essential, for describing events and decisions. In contrast, a focus on patterns of behavior is rather inextricably linked with a more continuous, quantitative representation. Dynamic patterns of behavior are naturally captured in graphs over time.

Quantitative representations can be either discrete or continuous, as, for example, difference or differential equations. We have observed that the choice of representation can make a profound difference in the conclusions one draws about the meaning and significance of feedback loops (see section 5.3). Though the conclusion stemmed from observations about difference versus differential equations, it holds as well for the more significant distinction between linguistic and quantitative representations.

Units of analysis

In particular, the choice of a linguistic or quantitative representation is tied to the units of feedback analysis one employs. We have noted that in the cybernetics thread the feedback loop is frequently interpreted as a communications circuit. Many such circuits form a "communications net," the central concept employed by Deutsch and others in the cybernetics thread. Feedback analyses based upon that view and a linguistic representation come naturally to emphasize the message as a basic unit of analysis. The event-oriented view settles on the message loop. Schön (1980) has used the term "back talk" for the phenomenon, a neat and deliberate revision of "feedback."

We have observed that this message loop perspective tends to emphasize the structure of communications networks and the content of messages and to de-emphasize, strictly speaking, the notion of loop polarities. (Further comments on this observation are contained in section 6.4). In contrast, the more continuous quantitative representations employed in the servomechanisms thread attempt to capture not communications networks but circular causal structures, and the basic unit of system analysis is the polarity of a feedback loop, not the message. Events and messages are below the level of aggregation of the circular causal view of the servomechanisms thread. In its quantitative representations, events and messages appear abstracted as increases and decreases in variables that are intended to reflect the persistent pressures and patterns of change such information would produce.

Causal view

A consequence of these distinctions is that there is actually a subtle difference in the causal views emphasized in the two feedback perspectives. The discrete, event-oriented, language-based perspective of the cybernetics thread tends to emphasize the causal role of individual messages and moves in social systems. Events are the behavior of interest, and they are caused by other events, or more precisely, by information about other events. The acknowledged closed-loop nature of social systems leads to the realization that any cause is an effect and any effect is a cause. In the cybernetics thread we have "cause-effects," in the form of stimuli and responses, individual messages, and other discrete events in closed-loop communication networks.

However, individual cause-effects simply do not appear in the more continuous, pattern-oriented perspective of the servomechanisms thread. If the focus is on dynamic patterns of behavior, the causal mechanism is seen to be the loop structure of the system. The "cause" of an arms race is viewed not as a given event or even a given sequence of events, but as a feedback structure dominated by self-reinforcing positive loops, within which events take place. The causal view in this thread is summarized in the assertion that "patterns of dynamic behavior are consequences of feedback structure."

The two perspectives thus differ in what they emphasize as "causes" of societal phenomena: the cybernetics thread emphasizes discrete messages and events in circular communication networks; the servomechanisms thread emphasizes causal links and loops in persistent circular causal feedback structures. These different emphases are natural consequences of the different units of description and analysis and the different modes of representation in the two threads.

Structural change

Another consequence of the choice of representation of feedback systems is the way structural changes are handled. The issue is very important to scholars in both feedback threads. Social systems are perceived to change, learn, and evolve over time. Without some means of changing structure, the feedback loop perspective, however dynamic its emphasis, is seen to be too static or rigid to reflect reality.

Scholars in the cybernetics thread have tended to capture such structural changes linguistically and sometimes diagrammatically, by "rewriting" or "redrawing" system structure. The changes are sometimes invoked exogenously, as when Weick shows us how to "discredit a cause map," and sometimes endogenously. Some artificial intelligence computer programs, for example, have the ability to actually rewrite parts of their own computer code. Seeing that society "invents" new structure as it evolves, some see in this capability a centrally important development for theory-building in the social sciences.

The issue of structural change is no less important to scholars in the servomechanisms thread, but in their quantitative representations of system structure such phenomena are captured in nonlinearities. Nonlinear models have the property that they can shift loop dominance and endogenously change the structure of feedback loops that are active over any given period of simulated time. Indeed, from a feedback loop perspective, this ability to shift loop dominance is the fundamental reason for advocating nonlinear models of social system behavior.

It is true that a nonlinear model can adapt and change over time, and the significance of its ability to do so deserves to be more widely understood. But such adaptations are built into nonlinear models from the start. A system dynamics model written in DYNAMO or STELLA (Richmond, Peterson, and Vescuso 1987) invents no new structure as it runs, while presumably an artificial intelligence model written in LISP can actually change as it runs the program that determines how it runs. However, by endogenously changing its active structure over time, a nonlinear model can expose the behavioral implications of any structural change envisioned by the modeler. Such advances in understanding the relationships between feedback structure and dynamic behavior are the goal of scholars in the servomechanisms thread.

Role of computing machines

Both feedback threads have attached special significance to computers in the social sciences. In the servomechanisms thread, computers have been urged as tools for system simulation, for playing out over simulated time the implications of assumptions built into formal mathematical models. Computer simulation takes the place of closed-form solutions of mathematical models, enabling the investigation of models that approach the circular causal complexities and nonlinearities that scholars in this thread perceive in real social systems. Authors in the cybernetics thread have also urged the use of computer simulation, emphasizing the potential of LISP simulations of symbolic and linguistic processes for the investigation of human psychological and problem-solving processes. As simulation tools, computers are seen in both threads as more or less infallible deducers of assumptions that people build into computer models.

But computing machines also have had other roles in the cybernetics thread. They have functioned as a source of analogies for the structure of human systems. We have the information-processing model of human cognition and the notion of a social system as a communication net containing binary off/on devices modeled on the binary logic of computers and neural nets. They have also been advocated as monitors and brains of real social systems, as in the work of Beer. Neither of these uses has been significant in the servomechanisms thread.

Management problems

The management and control problems perceived or emphasized in the two feedback threads differ as well. Authors in the cybernetics thread came to focus on the variety in events and messages that impinge on a system from without and are generated from processes within. Control of important variables in a system in the face of more or less random disturbances is a central concern. Feedback appears as the mechanism of control. A similar interest in management and control appears in the writings in the servomechanisms thread, but the emphasis is less on dealing with variety and disturbances and more on understanding the way interconnected sets of positive and negative feedback loops can combine to exacerbate or even create the problems perceived in social systems. Feedback loops may be mechanisms for achieving control, but they are also perceived in this thread as a major source of policy problems and counterintuitive behavior in social systems.

Source of dynamics

The final entry in Table 6.1 draws a distinction between the two threads in what is emphasized as the source of dynamics and change in social systems. The event-oriented perspective of scholars in the cybernetics thread, with its emphases on messages and decisions, leads naturally to a view of social system dynamics that gives prominence to the role of randomness. Behavior in this view can be thought of as "filtered randomness," probabilistic variation shaped by the structure of the social system. The pattern-oriented perspective of authors in the servomechanisms thread emphasizes the more deterministic role of circular causal feedback structure. Behavior in this view is the consequence of feedback structure, perhaps under the influence of random disturbances that stimulate behavior modes latent in that structure.

The reader should observe that the distinctions between perceived sources of dynamics in the two threads are almost spurious. As described, both characterizations of the sources of system dynamics are essentially the same. The difference is only one of emphasis: is the focus more on the stochastic variation and variety that is filtered by the system, or is the focus more on the circular causal feedback structure that accomplishes that filtering? The cybernetics thread tends to emphasize the stochastic variation and the filtering; the servomechanisms thread tends to emphasize the internal structure of the filter. Yet in spite of the apparent closeness of the two perspectives on the sources of system dynamics, their differences have serious implications. Depending on the point of view taken, a social scientist will perceive different problems, develop and utilize different conceptual and technical tools, and formulate different kinds of solutions, as we have seen throughout this investigation.

Significance of These Themes

Let me emphasize again that my purpose in drawing out these distinctions in feedback views is not controversy but clarity. The points of view crystallized above are neither right nor wrong. They are self-consistent ways of looking at societal phenomena from perspectives that emphasize loops of information feedback and circular causality. Various scholars would perhaps argue strongly for the promise of one view over the other, but the fact that there are proponents of both views suggests that deciding permanently in favor of one view may be impossible. How can their differences be justified? Can they be resolved? Should they be resolved? How can we learn from the contrasts drawn here the most promising directions for the application of the feedback concept in the social sciences? These difficult questions are the subject of the next section, which concludes this investigation.

6.4 Reflections and Conclusions

We have reached the point of asking how the differences we have observed can be reconciled. What implications can be perceive in the preceding analyses for the future use of the feedback concept in the social and policy sciences?

Reflections on Continuity, Distance, and Units of Description and Analysis

The central issue or theme that serves to distinguish the various feedback perspectives we have observed is, I believe, the category I have termed "units of description." Phenomena are described in the cybernetics thread in terms of events, decisions, and messages. In the servomechanisms thread, social systems are described in terms of dynamic patterns of behavior and persistent policy structure. Whether one's focus is on events or dynamic patterns can be viewed as a primary factor that determines the other characteristics of a scholar's feedback perspective. Selecting an event-orientation leads naturally to discrete, linguistic representations; a focus on patterns is naturally consistent with more continuous, quantitative representations. The event-orientation is also more consistent with the interpretation of a feedback loop as a communications network in which individual messages assume major importance; an emphasis on dynamic patterns is more consistent with the concept of feedback as the persistent circular causal structure underlying events and messages. We have the message loop and the causal loop, and we can view the choice between them as a consequence of the units in which a scholar chooses to describe phenomena in social systems. The primacy of the choice of "unit of description" is debatable, of course, but it has a logical consistency about it that makes it appealing as a summary distinguishing feature of the two feedback threads.

The choice of the unit of description and analysis may be a consequence of an even more fundamental choice: the "distance of perspective" of a social scientist making use of the feedback concept. The idea of "conceptual distance" has appeared explicitly in the work of some of the feedback thinkers in this investigation. Forrester (1961), for example, emphasized that his feedback perspective stemmed from viewing the system under investigation from "a very particular distance," not so close as to be concerned with the action of a single individual, but not so far away as to be ignorant of the internal pressures in the system (see section 3.3). Easton (section 4.5) also addressed the issue:

> By standing so far off from a system that we do not see the multiple component units at the points of inputs and outputs, we shall greatly simplify the processes, it is true. It is as though we were initially reconciling ourselves to using a telescope rather than a microscope because we are not yet sufficiently confident of the units and processes that we want to lay open to detailed analysis. But this is the very task of qualitative model construction, initially (Easton 1965b, p. 377).

Thus the distance of a feedback perspective is a metaphorical concept roughly parallel or synonymous to the "resolution" of a social scientist's descriptive and analytical lenses.

The concept of the distance of a perspective is attractive as a distinguishing feature of our two feedback threads because of its importance in the servomechanisms thread. To see positive and negative loops in the persistent causal structure underlying events and decisions, it seems essential to distance oneself from everyday details. But not too far. A particular distance is required, one that blurs events into patterns of behavior and perceives policy structure in the flow of decisions. Unfortunately, there does not seem to be a consistent distance of perspective characterizing the feedback views in the cybernetics thread. A focus on events and decisions suggests a close perspective, while the descriptions of communication nets we have seen are at a very distant, broad-brush level of detail. I conclude that the concept of conceptual distance is useful for isolating an essential characteristic of one of our two feedback threads, but not the other.

A third possible way of summarizing the distinctions we have observed between the two feedback threads focuses on continuity and discreteness. The message loop perspective characterizing much of the work in the cybernetics thread emphasizes discrete events and messages. The causal loop perspective of the servomechanisms thread strives to capture more persistent pressures underlying the pattern of events. Thus there is a theme of discreteness versus continuity underlying the two feedback threads we have traced.

It is important to realize how thoroughly the distinction between discrete and continuous perspectives has permeated the slice of intellectual history we have surveyed. It looks like one of the issues that could really kindle heavy debates among our various feedback thinkers. The reader will recall that Dewey (1896) objected that the reflex arc concept in psychology "breaks continuity and leaves us nothing but a series of jerks." He argued for a more continuous view in which stimulus and response do not follow discretely one after the other, but are "contemporaneous phases of one and the same coordination." Tustin and Phillips (section 3.3) explicitly argued for a continuous view of the multiplier-accelerator structure, rather than a discrete difference equation representation, because of the continuity that aggregation brings to the behavior of economic variables. Yet discreteness is essential for Ashby's concept of variety and the Law of Requisite Variety because variety is a count of possible states or conditions a system or its environment can exhibit.

The distinction between discrete and continuous perspectives also arose explicitly in Miller, Galanter, and Pribram's discussion of analog and digital psycho-physical processes (section 4.3). They expressed the point of view that psycho-motor processes would be under the control of continuous feedback circuits, while higher-level cognitive processes would have a more discrete, language-like character. In contrast, deliberately emphasizing neural current instead of neural impulses, Powers (section 4.7) took exception to a discrete view of psychological processes and tried to capture all of them in continuous terms. Indeed, discreteness and continuity are at the heart of the differences between the TOTE unit and Powers's use of the feedback concept. Furthermore, like Tustin and Phillips, Powers argued for a continuous view to match the level of aggregation of neural signs to the behavior he wished to analyze.

Easton's brief discussion of "triggers" and "causes" of behavior in political systems is also connected to the distinction between discreteness and continuity. An event that triggers a revolution is a discrete happening. The "causes," Easton suggested, would be "underlying smoldering discontent," a clearly more continuous pressure that serves to magnify the triggering event into a major historical phenomenon (section 4.5). Even Weick (section 4.6) flirted with the distinction between discrete and continuous when he stopped to observe that he had "substituted the word 'variable' for the word 'event' to signify that an event can take on a variety of values" (Weick 1979, p. 77). By now it is superfluous to also point out that Forrester explicitly advocated a continuous point of view of the dynamics of social systems.

The distinction between a discrete view and a continuous view can be phrased in various ways. We have discussed it in terms of units of system description and analysis and the choice of representations, and we have pictured it as a consequence of the distance of perspective one takes in viewing social phenomena. However it is depicted, it should be seen as a choice available to social scientists. One can choose to focus on events and decisions or patterns and policy structure; one can choose the distance from which to analyze social behavior; one can choose to see the feedback concept embodied in message loops or in causal loops.

It is tempting to say that how one chooses is a matter of worldview or fundamental perspective or methodological priors. Partly, the choice of perspective is indeed a matter of personal philosophy. Those who prefer to take the more continuous perspective, for example, can even appeal to Tolstoy's *War and Peace*.[1]

The progress of humanity, arising from an innumerable multitude of individual wills, is continuous in its motion.

The discovery of the laws of this motion is the aim of history. But in order to arrive at the laws of continuous motion due to the sum of all these individual wills, the human mind assumes arbitrary, disconnected units. The first proceeding of the historian is taking an arbitrary series of continuous events to examine it apart from others, while in reality there is not, and cannot be, a beginning to any event, but one event flows without break in continuity from another (Tolstoy 1869, p. 766).

And further, on causation—

The combination of causes of phenomena is beyond the grasp of the human intellect. But the impulse to seek causes is innate in the soul of man. And the human intellect, with no inkling of the immense variety and complexity of circumstances conditioning a phenomenon, any one of which may be separately conceived of as the cause of it, snatches at the first and most easily understood approximation, and says here is the cause. In historical events, where the actions of men form the subject of observation, the most primitive conception of a cause was the will of the gods, succeeded later on by the will of those men who stand in the historical foreground—the heroes of history. But one had but to look below the surface of any historical event, to look, that is, into the movement of the whole mass of men taking part in that event, to be convinced that the will of the hero of history, so far from controlling the actions of the multitude, is continually controlled by them. . . . Causes of historical events—there are not and cannot be, save the one cause of all causes. But there are laws controlling these events; laws partly unknown, partly accessible to us. The discovery of these laws is only possible when we entirely give up looking for a cause in the will of one man (Tolstoy 1869, p. 918).

Those who prefer the more distant, continuous perspective would find in these remarks an eloquent statement of the philosophical position that justifies their approach to societal phenomena. The issue is a classic one. Those who favor the more discrete perspective emphasizing individual decisions and events would perhaps quote Thomas Carlyle, who insisted that "The history of the world is but the biography of great men" (Carlyle 1840).

Distance, Continuity, and the Nature of a Loop Perspective

My reason for dwelling on these issues—the event- or behavior-orientation, the conceptual distance, and the discreteness or continuity of a feedback perspective—is that I believe they explain an observation that touches the heart of this investigation. We have noted that as the feedback concept evolved there was a tendency of some feedback thinkers in the social sciences to down play the concept of loop polarity. We have numerous examples, but witness, for emphasis, Herbert Simon's evolution of the loop concept towards "means/end" analysis (section 5.2). The notion is central to his path-breaking work in the study and simulation of human problem-solving. Weizenbaum (1976, p. 249) noted that Simon described it in terms of an "incongruity giving rise to action to reduce the incongruity by some means," a phrasing that sounds classically like a negative feedback loop. Weizenbaum saw fundamental connections between Simon's view of human problem-solving and Forrester's perspective on social systems (and critiqued both for being simplistic), but we should note an important distinction. The word "feedback" and the concept of positive and negative loops of circular causality do not, to my knowledge, appear anywhere in Simon's problem-solving publications after Simon (1957). The loop concept in some form may be there, but in the analysis of logical patterns and problem solving its application leads not to a focus on the connections between feedback structure, loop polarities, and the behavior of a system, but rather to a focus on specific sequences of language-like problem-solving moves.

The point is that from a close perspective in which events and decisions are the units of description and analysis, the concept of loop polarity is apparently not seen to be particularly useful. That conclusion is both an observation and an inference: we have seen the tendency to down play loop polarities in event-oriented perspectives, and upon reflection the tendency in that context makes sense. Even if I can perceive a positive loop process work in a back-and-forth argument between my children, it is extremely hard to use a feedback perspective based on loop polarities to tell how the events in that argument will proceed or how it will end.

I conclude that the distinction between discrete events and continuous behavior, which stems partly from a choice of conceptual distance, is critically important for determining how the loop concept will be employed in the social sciences and what will be learned from its application. From an event-oriented perspective, concepts such as variety, reflection, and self-transformation will appear to hold more promise for the analysis of social systems than will the concept of an interconnected complex of causal feedback loops of positive and negative polarity. Only from a more distant perspective in which events and decisions are deliberately blurred into patterns of behavior and policy structure will the notion that "behavior is a consequence of feedback structure" arise and be perceived to yield powerful insights.

Purpose and Problem Focus

We have noted that we have a choice among perspectives to take. At the risk of calling up the imaginable Tolstoy/Carlyle debate, I would suggest, however, that the choice is not usually free. It should depend upon the purpose and problem focus of the analysis. It seems reasonable that the focus of a study of problem-solving behavior must be on language-like, discrete moves—thought events and actual observable problem-solving stimuli and responses. However, such a focus would be inappropriate, I believe, in a study of the dynamics of an economy, for example, for the same reasons that Dewey, Tustin, Phillips, Forrester, Powers, and others offered. The behavior of aggregations of individuals does not have the discrete, abrupt movement that one associates with the steps in an individual's solution to a problem or with the moves in a chess game. Aggregates, like population or GNP or even glucose in an individual's blood stream, look continuous over time.

I would suggest, therefore, that the most distant feedback loop perspective, emphasizing persistent causal structure, loop polarities, and dynamic behavior patterns, is most appropriate for the study of aggregate systems, in which behavior can be productively considered to be essentially continuous. In that more distant view, patterns of policy resistance, compensating feedback, and counterintuitive behavior can be seen as consequences of the multi-loop structure of social systems.

Correspondingly, I would suggest that the closed-loop view that I have previously labeled the message loop perspective is more appropriate for the study of the detailed behavior of individuals or groups in which behavior can be productively seen as a sequence of discrete moves. In this view, we have dialectic processes, conversational planning, and the implications of reflection and self-transformation.

The aggregate systems, individuals, and groups in these suggested principles need not refer to people. Presumably, the same ideas hold true for loop perspectives on the behavior of cellular systems or individual cells, for example. The important distinction is whether the behavior of interest is more productively viewed as a sequence of discrete moves, which is best described and analyzed in logical or linguistic terms, or as a pattern of dynamic behavior over time, which is best described graphically.

In the end, the choice of loop perspectives ought to come down to what one hopes to see. The social scientist taking the more distant perspective that emphasizes feedback structure underlying patterns of dynamic behavior can hope to see the problems and complexities arising from compensating feedback, policy resistance from feedback effects, and the ramifying effects of multi-loop nonlinear systems. He or she pays for such insights, however, by forgoing a view of individual events and decisions. The social scientist taking a closer loop perspective that emphasizes individual events, decisions, and actions can hope to see and understand phenomena associated with dialectical processes, conversational planning, reflection, self-organization, self-transformation, and perhaps even the societal analogs of linguistic paradox. The social scientist taking this view pays for these insights into events and decisions by forgoing the insights about patterns of behavior and policy structure that a more distant view can reveal.

Notes

1. There have been some attempts to give a central role to the concept of feedback in social science texts; see in particular, Kuhn (1975), Buckley (1967), and, although it has an engineering emphasis, ESCP (1971). Some of the feedback works investigated or mentioned in this study also qualify as texts, e.g., Forrester (1961), Klir and Valach (1967), and Weick (1969/1979). But such

texts are swamped into insignificance by the number of social and policy science works that make no mention of feedback or circular causality. The concept does not play the important role that the authors reviewed in this study would assign to it.

2. N. Forrester (1982) identified five different stability criteria used in economic analyses. Furthermore, he showed that the results of implementing a given business cycle stabilization policy in a dynamic feedback model can be simultaneously stabilizing under one criterion and destabilizing under another.

3. An "AR" or autoregressive model is a statistical model of the form $z_t = a_1 z_{t-1} + a_2 z_{t-2} + \ldots + a_p z_p + e_t$ in which z at time t is regressed on previous values of z itself. An "MA" or moving average model is a statistical model of the form $z_t = e_t + b_1 e_{t-1} + b_2 e_{t-2} + \ldots + b_q e_{t-q}$ in which the e_i are white noise random processes (random error terms that propagate through time). An "ARMA" model is a mixture of the two types. An "ARIMA" model involves differencing ($x_t = z_t - z_{t-1}$) as well as autoregressive and moving average terms.

4. The following discussion draws somewhat on points of view contained in Ashby (1962), Richards (1980), and Alker (1981), as well as on ideas contained throughout the preceding investigation of the cybernetics thread.

5. Quotations from Tolstoy have been handed out to beginning system dynamics students for a number of years to orient them to the more distant, continuous perspective that system dynamicists take. The quotations were originally brought to Forrester's attention by his daughter Judith.

Bibliography

Abraham, Ralph (1987). Dynamics, a Visual Introduction. In Yates (1987).
Adolph, Edward F. (1968). *Origins of Physiological Regulations* (New York: Academic Press).
Aftalion, Albert (1909). La réalité des surproductions générales. *Revue d'économie politique* 209–210; cited in Mayr (1971).
――― (1927). The Theory of Economic Cycles Based on the Capitalist Technique of Production. *Review of Economic Statistics* 165; cited in Goodwin (1951a).
Alfeld, Louis E. and Alan K. Graham (1976). *Introduction to Urban Dynamics* (Cambridge, MA: MIT Press).
Alker, Hayward R. (1966). Causal Inference and Political Analysis. In *Mathematical Applications in Political Science*, vol. 2 (Dallas: Arnold Foundation, SMU Press).
――― (1981). From Political Cybernetics to Global Modeling. In R. Merritt and B.M. Russell, eds., *From National Development to Global Community* (New York: Allen and Unwin), pp. 353–378.
Allen, R.G.D. (1955). The Engineer's Approach to Economic Models. *Economica* 22: 158–168.
――― (1956). *Mathematical Economics* (London: Macmillan).
Annett, J. (1969). *Feedback and Human Behavior* (Baltimore: Penguin Books).
Apostle, H.G. (1966), ed. and trans. *Aristotle's Metaphysics* (Bloomington: Indiana University Press).
Aoki, Masanao (1976). *Optimal Control and System Theory in Dynamic Economic Analysis* (New York: North Holland).
Aoki, Masanao and Axel Leijonhufvud (1976). Cybernetics and Macroeconomics: A Comment. *Economic Inquiry* 14: 251–258.
Aoki, Masanao and Angelo Marzollo (1979). *New Trends in Dynamic System Theory and Economics* (New York: Academic Press).
Ashby, W. Ross (1945). The Effect of Controls on Stability. *Nature* 155: 242–243.
――― (1946a). The Behavioral Properties of Systems in Equilibrium. *American Journal of Psychology* 59: 682–686.
――― (1946b). The Physical Origin of Adaptation by Trial and Error. *Journal of General Psychology* 32: 319–323.
――― (1947a). Automatic Development of Equilibrium in Self-Organizing Systems. *Psychometrika* 12: 135–140.

——— (1947b). Interrelations between Stabilities of Parts within a Whole Dynamic System. *Journal of Comparative and Physiological Psychology* 40: 1–7.
——— (1952). *Design for a Brain* (New York: John Wiley and Sons).
——— (1954). The Application of Cybernetics to Psychiatry. *Journal of Mental Science* 100: 114–124.
——— (1956). *Introduction to Cybernetics* (New York: John Wiley and Son).
——— (1962). Principles of the Self-Organizing System. In von Foerster and Zopf (1962), pp. 255–278. Reprinted in Buckley (1968), pp. 108–118.
Ashton, R.H. (1976). Deviation-Amplifying Feedback and Unintended Consequences of Management Accounting Systems. *Accounting, Organizations and Society* 1: 289–300.
Athans, Michael and D. Kendrick (1974). Control Theory and Economics: A Survey, Forecast and Speculations. *IEEE Transactions on Automatic Control* 19,5: 518–523.
Balzac, H. (1850). *Lettres à l'Étrangere*, vol. 2; cited in Trésor de la Langue Française (Paris: Institut National de la Recherche Scientifique, 1976).
Bateson, Gregory (1935). Culture Contact and Schismogenesis. *Man* 35: 178–183. Reprinted in Bateson (1972).
——— (1936/1958). *Naven* (Cambridge: Cambridge University Press, 1936); 2nd edition 1958 (Stanford, CA: Stanford University Press).
——— (1949). Bali: The Value System of a Steady State. In *Social Structure: Studies Presented to A.R. Radcliffe-Brown*, Meyer Fortes, ed. (Oxford: Clarendon Press). Reprinted in Bateson (1972).
——— (1972). *Steps to an Ecology of Mind* (New York: Ballantine Books).
Bateson, Gregory, D.D. Jackson, J. Haley, and J.H. Weakland (1956). Toward a Theory of Schizophrenia, *Behavioral Science* 1: 251–264.
Baum, Christopher and E. Phillip Howrey (1981). An Examination of Postwar U.S. Stabilization Policy: Monetary and Fiscal Policy in an Accelerationist World. Presented at the Third Economics and Control Conference, The Society for Economic Dynamics and Control, Lyngby, Denmark, June 1981.
Baumol, William J. (1970). *Economic Dynamics*, 3rd ed. (London: Macmillan).
Beer, Stafford (1955). The Productivity Index in Active Service. *Applied Statistics* 4,1: 1–14.
——— (1956). Operational Research and Cybernetics. *Proceedings of the First International Conference on Cybernetics*, Namur.
——— (1959/1967). *Cybernetics and Management* (London: English Universities Press).
——— (1966). *Decision and Control: The Meaning of Operational Research and Management Cybernetics* (London: John Wiley and Sons).
——— (1968). *Management Science: The Business Use of Operations Research* (Garden City, NY: Doubleday).
——— (1972/1981). *Brain of the Firm: The Managerial Cybernetics of Organization* (London: Allen Lane, The Penguin Press, 1972). 2nd edition (Chichester: John Wiley and Sons, 1981).
——— (1975). *Platform for Change* (London: John Wiley and Sons).
——— (1979). *The Heart of Enterprise: The Managerial Cybernetics of Enterprise* (Chichester: John Wiley and Sons).
Bell, Charles (1826). On the nervous circle which connects the voluntary muscles with the brain. *Proceedings of the Royal Society* 2: 266; cited in A Note by the Editors, Heinz von Foerster, Margaret Mead, Hans L. Teuber, *Cybernetics: Transactions of the Eighth Conference* (New York: Josiah Macy, Jr. Foundation, 1951), p. xvi.

Bellman, Richard and Robert Kalaba (1964), eds. *Mathematical Trends in Control Theory* (New York: Dover Publications).
Bernard, Claude (1878–79). *Leçons sur les phénomènes de la vie communes aux animaux et aux végétaux* (Paris: J.B. Bailliere).
Bode, H.W. (1960). Feedback—the History of an Idea. Presented at the Symposium on Active Networks and Feedback Systems, Polytechnic Institute of Brooklyn, April 19–21, 1960. Reprinted in Bellman and Kalaba (1964).
Boulding, Kenneth (1951). A Conceptual Framework for Social Science. *Papers of the Michigan Academy of Science, Arts and Letters* 37. Reprinted in Boulding (1968).
——— (1953). *The Organizational Revolution: A Study in the Ethics of Economic Organization* (New York: Harper).
——— (1956). General Systems Theory—The Skeleton of Science. *Management Science* 2,3: 197–208. Reprinted in Boulding (1968).
——— (1968). *Beyond Economics* (Ann Arbor: The University of Michigan Press).
——— (1971). The Dodo Didn't Make It: Survival and Betterment. *Bulletin of the Atomic Scientists* 27,2: 19–22.
Boyce, Stephen G. (1980). Management of Forest for Optimal Benefits (DYNAST-OB). U.S. Department of Agriculture Forest Service, Research paper SE-204 (Asheville, NC: Southeast Forestry Experimental Station).
Braun, M. (1978). *Differential Equations and Their Applications*, 2nd ed. (New York: Springer-Verlag).
Brown, Gordon S. and Donald P. Campbell (1948). *Principles of Servomechanisms* (New York: John Wiley).
Buckley, Walter (1967). *Sociology and Modern Systems Theory* (Englewood Cliffs, NJ: Prentice-Hall).
——— (1968), ed. *Modern Systems Research for the Behavioral Scientist* (Chicago: Aldine Publishing Company).
Bunge, M. (1963). *Causality: The Place of the Causal Principle in Modern Science* (New York: Meridian Books, World Publishing Co.).
Campbell, Donald T. (1970). Natural Selection as an Epistemological Model. In R. Naroll and R. Cohen, eds., *A Handbook of Method in Cultural Anthropology* (Garden City, NY: Natural History Press).
——— (1972). On the Genetics of Altruism and the Counter-hedonic Components in Human Culture. *Journal of Social Issues* 28,3: 21–37.
——— (1974). Evolutionary Epistemology. In P.A. Schlipp, ed., *The Philosophy of Karl R. Popper*, vol. 14-I (LaSalle,IL: Open Court).
Cannon, Walter B. (1926). *American Journal of Medical Science* 171.
——— (1929). Organization of Physiological Homeostasis. *Physiological Reviews* 9: 399–431.
——— (1932). *The Wisdom of the Body* (New York: W.W. Norton).
Carlyle, Thomas (1840). *On Heroes, Hero-Worship and the Heroic in History*. (London: The Carlyle's House Memorial Trust, 1903).
Chomsky, A. Noam (1956). Three Models for the Description of Language. *IRE Transactions on Information Theory*, IT-2,3: 113–124.
——— (1957). *Syntactic Structures* (The Hague: Mouton).
——— (1959). Review of B.F. Skinner's *Verbal Behavior*. *Language* 35: 26–58.
Choucri, Nazli (1981). *International Energy Futures: Petroleum, Prices, Power, and Payments* (Cambridge, MA: MIT Press).
Chow, G.C. (1970). Optimal Stochastic Control of Linear Economic Systems. *Journal of Money, Credit and Banking* 2: 291–302.

Cochrane, James L. and John A. Graham (1976). Cybernetics and Macroeconomics. *Economic Inquiry* 14: 241–250.
Cooper, W.W. (1951). A Proposal for Extending the Theory of the Firm. *Quarterly Journal of Economics* 65: 87–109.
Coyle, R. Geoff (1976). *Management System Dynamics* (New York: John Wiley).
Coyle, R. Geoff and John A. Sharp (1976). *System Dynamics—Problems, Cases, and Research* (New York: John Wiley).
Craik, Kenneth J.W. (1943/1952). *The Nature of Explanation* (Cambridge: Cambridge University Press).
Culbertson, John M. (1968). *Macroeconomic Theory and Stabilization Policy* (New York: McGraw-Hill, 1968).
Cyert, R.M. and J.G. March (1963). *A Behavioral Theory of the Firm* (Englewood Cliffs, NJ: Prentice-Hall).
Dennis, Wayne (1948), ed. *Readings in the History of Psychology* (New York: Appleton-Century-Crofts).
Descartes, René (1972). *Traité de l'Homme*. French text with translation and commentary by Thomas Steele Hall (Cambridge, MA: Harvard University Press).
Deutsch, Karl W. (1948). Toward a Cybernetic Model of Man and Society, from Some Notes on Research on the Role of Models in the Natural and Social Sciences. *Synthese* 7: 506–533. Reprinted in Buckley (1968), pp. 387–400.
——— (1953, 1966). *Nationalism and Social Communication* (Cambridge, MA: MIT Press).
——— (1963). *The Nerves of Government* (London: The Free Press of Glencoe, Collier-Macmillan).
Dewey, John (1896). The Reflex Arc Concept in Psychology. *Psychological Review* 3: 357–370. Reprinted in Dennis (1948), pp. 355–365.
Easton, David (1957). An Approach to the Analysis of Political Systems. *World Politics* 9: 383–400.
——— (1965a). *A Framework for Political Analysis* (Englewood Cliffs, NJ: Prentice-Hall).
——— (1965b). *A Systems Analysis of Political Life* (New York: John Wiley).
Eden, Colin, and John Harris (1975). *Management Decision and Decision Analysis* (New York: John Wiley).
Eden, Colin, Sue Jones, and David Sims (1979). *Thinking in Organizations* (London: Macmillan).
——— (1983). *Messing About in Problems* (Oxford: Pergamon Press).
Elster, Jon (1975). *Leibniz et la formation de l'esprit capitaliste* (Paris: Aubier Montaigne).
Emery, F.E. (1969), ed. *Systems Thinking* (Baltimore: Penguin Books).
English, H.B. and Ava C. English (1958). *A Comprehensive Dictionary of Psychological and Psycho-Analytical Terms* (New York: Longmans).
Enke, S. (1951). Equilibrium Among Spatially Separated Markets: Solution by Electric Analogue. *Econometrica* 19: 40–47.
ESCP (1971). *The Man-Made World*, Engineering Sciences Curriculum Project. (New York: McGraw-Hill).
Fair, R.C. (1974). On the Solution of Optimal Control Problems as Maximization Problems. *Annals of Economic and Social Management* 3: 135–154.
——— (1978). The Use of Optimal Control Techniques to Measure Economic Performance. *International Economic Review* 19: 289–309.
Feigenbaum, E. and J. Feldman (1963), eds. *Computers and Thought* (New York: McGraw-Hill).

Feigenbaum, M.J. (1980/1983). Universal Behavior in Nonlinear Systems. *Los Alamos Science* 1:4. Reprinted in *Physica D* 7 (1983): 16–39.
Festinger, Leon (1957). *A Theory of Cognitive Dissonance* (Evanston, IL: Row, Peterson).
Forrester, Jay W. (1958). Industrial Dynamics: A Major Breakthrough for Decision Makers. *Harvard Business Review* 36,4: 37–66.
——— (1961). *Industrial Dynamics* (Cambridge, MA: MIT Press).
——— (1968a). Market Growth as Influenced by Capital Investment. *Industrial Management Review* (now the *Sloan Management Review*) 9,2: 83–105. Reprinted in Forrester (1975) and Roberts (1978).
——— (1968b). *Principles of Systems* (Cambridge, MA: MIT Press).
——— (1969). *Urban Dynamics* (Cambridge, MA: MIT Press).
——— (1971a). Counterintuitive Behavior of Social Systems. *Technology Review* 73 (Jan. 1971).
——— (1971b, 1973). *World Dynamics* (Cambridge, MA: MIT Press).
——— (1975). *Collected Papers of Jay W. Forrester* (Cambridge, MA: MIT Press).
——— (1976). Business Structure, Economic Cycles, and National Policy, *Futures* (June 1976): 195–214.
——— (1977). Growth Cycles. *De Economist* 4.
——— (1979). An Alternative Approach to Economic Policy: Macrobehavior from Microstructure. In Nake Kamrany and Richard Day, eds., *Economic Issues of the Eighties* (Baltimore: The Johns Hopkins University Press).
Forrester, Nathan B. (1972). *The Life Cycle of Economic Development* (Cambridge, MA: MIT Press).
——— (1982). *A Dynamic Synthesis of Basic Macroeconomic Theory: Implications for Stabilization Policy Analysis*, Ph.D. dissertation, Massachusetts Institute of Technology, Cambridge, MA.
Frank, L.K. et al. (1948). Conference on Teleological Mechanisms. *Annals of the New York Academy of Sciences* 50,4: 187–278.
Frisch, Ragnar (1933). Propagation Problems and Impulse Problems in Dynamic Economics. In *Economic Essays in Honor of Gustav Cassel* (London: Allen and Unwin). Reprinted in R.A. Gordon and L.R. Klein, eds., *Readings in Business Cycles* (Homewood, IL: Richard D. Irwin, Inc., 1965).
Garfield, Eugene (1964). *The Use of Citation Data in Writing the History of Science* (Philadelphia: Institute of Scientific Information).
Gerard, R. (1968). The Neurophysiology of Purposive Behavior. In von Foerster *et al.* (1968).
Gibbs, C.B. (1970). Servo-Control Systems in Organisms and the Transfer of Skill. In D. Legge, ed., *Skills* (Baltimore: Penguin Books).
Glasser, William (1965). *Reality Therapy, A New Approach to Psychiatry* (New York: Harper and Row).
——— (1981). *Stations of the Mind: New Directions for Reality Therapy* (New York: Harper and Row).
Gleick, James (1987). *Chaos: Making a New Science* (New York: Viking Penguin).
Gödel, Kurt (1931). Über formal unentscheidbare Sätze der *Principia Mathematica* und verwandter Systeme (On Formally Undecidable Propositions of *Principia Mathematica* and Related Systems). See Nagel and Newman (1956).
Goodman, Michael R. (1974). *Study Notes in System Dynamics* (Cambridge, MA: MIT Press).
Goodwin, Richard M. (1951a). Econometrics in Business-Cycle Analysis. In Alvin H. Hansen, ed., *Business Cycles and National Income* (New York: W.W. Norton).

——— (1951b). The Nonlinear Accelerator and the Persistence of Business Cycles. *Econometrica* 19: 1–17.

——— (1951c). Iteration, Automatic Computers, and Economic Dynamics, *Metroeconomica* (April 1951).

Graham, Alan K. (1977). *Principles on the Relationship between Structure and Behavior of Dynamic Systems*, Ph.D. dissertation, Massachusetts Institute of Technology, Cambridge.

Gutierrez, L.T. and W.R. Fey (1980). *Ecosystem Succession: A General Hypothesis and a Test Model of a Grassland* (Cambridge, MA: MIT Press).

Hamilton, H.R., S.E. Goldstone, J.W. Milliman, A.L. Pugh III, E.B. Roberts, and A. Zellner (1963). *System Simulation for Regional Analysis: An Application to River Basin Planning* (Cambridge, MA: MIT Press).

Hanneman, Robert (1988). *Modeling Social Systems: A System Dynamics Approach* (Beverly Hills, CA: Sage Publications).

Harrod, R.F. (1939). An Essay in Dynamic Theory. *Economic Journal* (March 1939).

Hebb, D.D. (1964). *A Textbook of Psychology* (Philadelphia: Saunders).

Henderson, L.J. (1913). *The Fitness of the Environment* (New York: The Macmillan Co.).

——— (1928). *Blood: A Study in General Physiology* (New Haven: Yale University Press).

——— (1935). Physician and Patient as Social System. *New England Journal of Medicine* 212: 819.

Hofstadter, Douglas R. (1979). *Gödel, Escher, Bach* (New York: Basic Books).

Holt, E.B. (1931). *Animal Drive and the Learning Process* (New York: Holt); cited in Miller, Galanter, and Pribram (1960), 44.

Homans, George (1950). *The Human Group* (New York: Harcourt, Brace and Co.).

Hume, David (1739). An Essay on the Human Understanding. In L.A. Selby-Bigge, ed., *Enquiries Concerning Human Understanding and Concerning the Principles of Morals*, by David Hume. 3rd ed. (Oxford: Clarendon Press, 1975).

——— (1752). On the Balance of Trade. In T.H. Green and T.H. Grose, eds., *Essays Moral, Political, and Literary* (London, 1898).

Hurwitz, A. (1895). On the Conditions Under Which an Equation has Only Roots with Negative Real Parts. *Mathematische Annalen* 46: 273–284. Reprinted in Bellman and Kalaba (1964).

Hutchinson, G. Evelyn (1948). Circular Causal Systems in Ecology; Conference on Teleological Mechanisms. *Annals of the New York Academy of Sciences* 50,4: 221.

James, H.M., N.B. Nichols, and R.S. Phillips (1947). *Theory of Servomechanisms* (New York: McGraw-Hill).

Jones, Richard W. (1973). *Principles of Biological Regulation: An Introduction to Feedback Systems* (New York: Academic Press).

Kahn, R.F. (1931). The Relation of Home Investment to Unemployment. *Economic Journal* (June 1931); cited in Keynes (1936).

Kaldor, N. (1940). A Model of the Trade Cycle. *Economic Journal* 50: 78–92.

Kalecki, M. (1935). A Macrodynamic Theory of Business Cycles. *Econometrica* 3: 327–344.

Kelly, C.R. (1968). *Manual and Automatic Control* (New York: Wiley).

Kendrick, David (1976). Applications of Control Theory to Macroeconomics. *Annals of Economic and Social Measurement* 5,2: 171–190.

Kermack, W.O. and A.G. McKendrick (1927). Contributions to the Mathematical Theory of Epidemics. *Proceedings of the Royal Statistical Society* 115: 700–721; discussed in Braun (1978), 428–434.
Keynes, John Maynard (1936). *The General Theory of Employment, Interest, and Money* (New York: Harcourt, Brace and World).
Klein, Lawrence R. (1950). *Economic Fluctuations in the U.S., 1921–1941* (New York: Wiley).
Klir, Jiri and Miroslav Valach (1967). *Cybernetic Modeling*. English translation by Pavel Dolan, edited by W.A. Ainsworth (London: Iliffe Books, Ltd.; Princeton, NJ: D. Van Nostrand).
Koopmans, T.C. (1936). *Linear Regression Analysis of Economic Time Series* (Haarlem).
Kormondy, Edward J. (1969). *Concepts of Ecology* (Englewood Cliffs, NJ: Prentice-Hall).
Kramer, Ernest (1968). Man's Behavior Patterns. In Milsum (1968).
Kuhn, Alfred (1963). *The Study of Society: A Unified Approach* (Homewood, IL: R.D. Irwin).
——— (1975). *Unified Social Science* (Homewood, IL: Dorsey Press).
Kuhn, Thomas (1962/1970). *The Structure of Scientific Revolutions* (Chicago: The University of Chicago Press).
Lakatos, Imre and Alan Musgrave (1970), eds. *Criticism and the Growth of Knowledge. Proceedings of the International Colloquium in the Philosophy of Science*, London, 1965 (Cambridge, UK: Cambridge University Press).
Landau, Martin (1972). *Political Theory and Political Science: Studies in the Methodology of Political Inquiry* (New York: Macmillan).
——— (1973). On the Concept of a Self-Correcting Organization. *Public Administration Quarterly* 1: 533–542.
Lebel, Jean D. (1981). System Dynamics. *Dynamica* 7,1: 7–31.
Legasto, A.A., J.W. Forrester, and J.M. Lyneis (1980), eds. *System Dynamics, TIMS Studies in Management Sciences* 14 (Amsterdam: North Holland).
Leibenstein, H. (1950). Bandwagon, Snob, and Veblen Effects in the Theory of Consumers' Demand. *Quarterly Journal of Economics* 64: 183–207.
Leibniz, Gottfried Wilhelm (1714). The Monadology. Reprinted in Philip P. Wiener, ed., *Leibniz: Selections* (New York: Charles Scribner's Sons, 1951).
Leijonhufvud, Axel (1968). *On Keynesian Economics and the Economics of Keynes: A Study in Monetary Theory* (New York: Oxford University Press).
——— (1970). Notes on the Theory of Markets. *Intermountain Economic Review* 1,1: 1–13.
Leontief, Wassily (1982). Letter, *Science* 217: 106–107.
Lerner, Daniel (1965). *Cause and Effect* (New York: The Free Press).
Lessnoff, M. (1974). *The Structure of Social Science* (London: Allen and Unwin).
Levin, Gilbert, G.B. Hirsch, and E.B. Roberts (1975). *The Persistent Poppy: A Computer Aided Search for Heroin Policy* (Cambridge, MA: Ballinger).
Levin, Gilbert, E.B. Roberts, G.B. Hirsch, D.S. Kligler, N. Roberts, and J.F. Wilder (1976). *The Dynamics of Human Service Delivery* (Cambridge, MA: Ballinger).
Levins, Richard (1974). The Qualitative Analysis of Partially Specified Systems. *Annals of the New York Academy of Sciences* 231: 123–138.
Lewin, Kurt (1947). Feedback Problems of Social Diagnosis and Action, Part II-B of Frontiers in Group Dynamics. *Human Relations* 1: 147–153. Reprinted in Buckley (1968), pp. 441–444.
——— (1951). *Field Theory in Social Science* (New York: Harper and Brothers).

Lorenz, E.N. (1963). Deterministic Nonperiodic Flow. *Journal of the Atmospheric Sciences* 20: 130–141.
Lotka, Alfred J. (1925). *Elements of Physical Biology*. Reprinted as *Elements of Mathematical Biology* (New York: Dover, 1956).
Lyapunov, A.M. (1892). *On the General Problem of Stability of Motion* (Kharkov [Russian]); translation by Abramovici and Shimshoni and printed as *Stability of Motion* (Princeton, NJ: Princeton University Press, 1966).
Lyneis, James (1980). *Corporate Planning and Policy Design, A System Dynamics Approach* (Cambridge, MA: MIT Press).
MacIver, Robert M. (1942/1964). *Social Causation* (New York: Ginn and Co.).
MacMillan, R.H. (1951). *An Introduction to the Theory of Control in Mechanical Engineering* (Cambridge, UK: Cambridge University Press).
Malthus, Thomas R. (1798). *First Essay on Population*. Reprinted (London: Macmillan, 1926, 1966).
March, J.G. and H.A. Simon (1958). *Organizations* (New York: John Wiley).
Maruyama, Magoroh (1963). The Second Cybernetics: Deviation-Amplifying Mutual Causal Processes. *American Scientist* 51: 164–179. Reprinted in Buckley (1968), pp. 304–316.
——— (1974). Paradigms and Communication. *Technological Forecasting and Social Change* 6: 3–32.
Mass, Nathaniel J. (1975). *Economic Cycles: An Analysis of Underlying Causes* (Cambridge, MA: MIT Press).
Maxwell, James Clerk (1868). On Governors. *Proceedings of the Royal Society of London* 16: 270–283. Reprinted in Bellman and Kalaba (1964).
Mayr, Otto (1970). *The Origins of Feedback Control* (Cambridge, MA: MIT Press).
——— (1971). Adam Smith and the Concept of the Feedback System: Economic Thought and Technology in 18th-Century Britain. *Technology and Culture* 12,1: 1–22.
——— (1986). *Authority, Liberty, and Automatic Machinery in Early Modern Europe* (Baltimore: Johns Hopkins University Press).
McCorduck, Pamela (1979). *Machines Who Think* (New York: W.H. Freeman).
McCulloch, Warren S. (1965). *Embodiments of Mind* (Cambridge, MA: MIT Press).
McCulloch, Warren S. and Walter H. Pitts (1943). A Logical Calculus of the Ideas Immanent in Nervous Activity. *Bulletin of Mathematical Biophysics* 5. Reprinted in McCulloch (1965).
Meadows, Dennis L. (1970). *Dynamics of Commodity Production Cycles* (Cambridge, MA: MIT Press).
Meadows, Dennis L., W.W. Behrens III, D.H. Meadows, R.F. Naill, J. Randers, and E. Zahn (1974). *Dynamics of Growth in a Finite World* (Cambridge, MA: MIT Press).
Meadows, Dennis L. and Donella H. Meadows (1974). *Toward Global Equilibrium: Collected Papers* (Cambridge, MA: MIT Press).
Meadows, Donella H. (1980). The Unavoidable A Priori: In Jørgen Randers, ed., *Elements of the System Dynamics Method* (Cambridge, MA: MIT Press).
——— (1982). Whole Earth Models and Systems. *Coevolution Quarterly* (Summer 1982): 98–108.
Meadows, Donella H., Dennis L. Meadows, J. Randers, and W.W. Behrens III (1972). *The Limits to Growth* (New York: University Books, a Potomac Associates Book).

Meehan, Eugene J. (1968). *Explanation in Social Science: A System Paradigm* (Homewood, IL: Dorsey Press).
Merton, Robert King (1936). The Unanticipated Consequences of Purposive Social Action. *American Sociological Review* (1936): 894–904.
——— (1940). Bureaucratic Structure and Personality. *Social Forces* 18: 560–568.
——— (1948). The Self-Fulfilling Prophecy. *Antioch Review* (Summer 1948): 193–210. Reprinted in R.K. Merton, *Social Theory and Social Structure* (New York: Free Press, 1948, 1957, 1968).
Mill, John Stuart (1846). *System of Logic* (New York: Harper).
——— (1848). *Principles of Political Economy, with Some of their Applications to Social Philosophy*. From the 5th London edition (New York: D. Appleton and Co., 1883).
Miller, George A., Eugene Galanter and Karl H. Pribram (1960). *Plans and the Structure of Behavior* (New York: Henry Holt).
Miller, James G. (1955). Toward a General Theory for the Behavioral Sciences. *American Psychologist* 10: 513–531.
——— (1971). Living Systems: The Group. *Behavioral Science* 16: 302–398.
——— (1978). *Living Systems* (New York: McGraw-Hill).
Milsum, John H. (1968), ed. *Positive Feedback: A General Systems Approach to Positive/Negative Feedback and Mutual Causality* (Oxford: Pergamon Press).
Minorsky, Nicholas (1922). Directional Stability of Automatically Steered Bodies. *Journal of the American Society of Naval Engineers* 34 (May 1922).
Morecroft, John D.W. (1985). Rationality in the Analysis of Behavioral Simulation Models. *Management Science* 31,7: 900–916.
Morehouse, N.F., R.H. Strotz and S.J. Horwitz (1950). An Electro-Analog Method for Investigating Problems in Econometric Dynamics: Inventory Oscillations. *Econometrica* 18: 313–328.
Mosekilde, Erik, Javier Aracil, and Peter M. Allen (1988). Instabilities and Chaos in Nonlinear Dynamic Systems. *System Dynamics Review* 4(1–2): 14–55.
Mosekilde, Erik, S. Rasmussen, H. Jørgensen, F. Jaller, and C. Jensen (1985). Chaotic Behavior in a Simple Model of Urban Migration. *Proceedings of the 1985 International System Dynamics Conference*, Keystone, Colorado.
Myrdal, Gunnar (1939). *Monetary Equilibrium* (London: W. Hodge and Co., Ltd.).
——— (1944). *An American Dilemma: The Negro Problem and Modern Democracy* (New York: Harper). Reprinted 1962.
——— (1957). *Rich Lands and Poor* (New York: Harper). Also published as *Economic Theory and Underdevelopment* (London: Duckworth).
Nagel, Ernest and James R. Newman (1956). Gödel's Proof. In James R. Newman, ed., *The World of Mathematics*, vol. 3 (New York: Simon and Schuster), pp. 1668–1695.
Naill, Roger (1977). *Managing the Energy Transition* (Cambridge, MA: Ballinger).
Needham, Joseph (1954). *Science and Civilization in China* (Cambridge).
Newell, Allan, J.C. Shaw, and Herbert A. Simon (1958). Elements of a Theory of Human Problem Solving. *Psychological Review* 65,3: 151–166.
——— (1959). Report on a General Problem-solving Program. *Proceedings of the International Conference on Information Processing*, Paris.

Newell, Allan and Herbert A. Simon (1956). The Logic Theory Machine: A Complex Information Processing System. *IRE Transactions on Information Theory* (Sept. 1956).
——— (1963). GPS, A Program that Simulates Human Thought. In Feigenbaum and Feldman (1963).
——— (1972). *Human Problem Solving* (Englewood Cliffs, NJ: Prentice-Hall).
Nicolis, G. and I. Prigogine (1977). *Self-Organization in Nonequilibrium Systems: From Dissipative Structures to Order Through Fluctuations* (New York: John Wiley).
Nyquist, H. (1932). Regeneration Theory. *Bell System Technical Journal* 11: 126–147.
OED (1933). *Oxford English Dictionary* (London: Oxford University Press).
——— (1972). *A Supplement to the Oxford English Dictionary* (London: Oxford University Press).
Ogata, Katsuhiko (1970). *Modern Control Engineering* (Englewood Cliffs, NJ: Prentice-Hall).
Pask, Gordon (1960). The Natural History of Networks. In Yovits and Cameron (1960).
Pearl, R. (1922). *The Biology of Death*, cited in Lotka (1925), p. 66.
Pearl, R. and L.J. Reed (1920). In *Proceedings of the National Academy of Science* 6: 275; cited in Lotka (1925), p. 66.
——— (1921). *Scientific Monthly* 194; cited in Lotka (1925), p. 66.
Phillips, A.W. (1950). Mechanical Models in Economic Dynamics. *Econometrica* 17: 283–305.
——— (1954). Stabilization Policy in a Closed Economy. *Economic Journal* (June 1954): 290–305.
——— (1957). Stabilization Policy and the Time-Forms of Lagged Responses. *Economic Journal* (June 1957): 265–277.
Pindyck, Robert S. and D.L. Rubinfeld (1976). *Econometric Models and Economic Forecasts* (New York: McGraw-Hill).
Pitchford, J.D. and S.J. Turnovsky (1977). *Applications of Control Theory to Economic Analysis* (New York: Elsevier).
Platt, John (1966). *The Step to Man* (New York: John Wiley).
Porter, A. (1950). *Introduction to Servomechanisms* (London: Methuen; New York: Wiley).
Powers, William T. (1973). *Behavior: the Control of Perception* (Chicago: Aldine).
——— (1978). Quantitative Analysis of Purposive Systems: Some Spadework at the Foundations of Scientific Psychology. *Psychology Review* 85,5: 417–435.
Powers, W.T., R.K. Clark, and R.I. McFarland (1960). A General Feedback Theory of Human Behavior. *Perceptual and Motor Skills Monograph*. Part 1: 11 (1, serial no. 7); Part 2: 11 (3, serial no. 7).
Price, D.J. (1965). Networks of Scientific Papers. *Science* 149: 510–515.
Prigogine, Ilya and Isabelle Stengers (1984). *Order Out of Chaos* (New York: Bantam Books).
Radcliffe-Brown, A.R. (1922). *The Andaman Islanders. A Study in Social Anthropology* (Cambridge, UK: Cambridge University Press).
Rapoport, Anatol (1968). A Philosophical View. In Milsum (1968).
Reiner, Rolf, Martin Munz, and Wolfgang Weidlich (1988). Migratory Dynamics of Interacting Subpopulations: Regular and Chaotic Behavior. *System Dynamics Review* 4(1–2): 179–199.
Reusch, J. and Gregory W. Bateson (1951). *Communication: The Social Matrix of Society* (New York: Norton).

Richards, Lawrence (1980). Cybernetics and Management Science. *Omega* 8(1): 71–80.

Richardson, George P. (1984). Loop Polarity, Loop Dominance, and the Concept of Dominant Polarity. *Proceedings of the 1984 International System Dynamics Conference*, Oslo, Norway.

——— (1986). Dominant structure. *System Dynamics Review* 2,1: 68–75.

Richardson, George P. and Alexander L. Pugh III (1981). *Introduction to System Dynamics Modeling with DYNAMO* (Cambridge, MA: MIT Press).

Richardson, George P. and John D. Sterman (1988). A Note on Migratory Dynamics. *System Dynamics Review* 4(1–2): 200–207.

Richardson, Lewis F. (1919/1935). Mathematical Psychology of War. *Nature* 135: 830; 136: 1025.

——— (1938). The Arms Race of 1909–13. *Nature* 142: 792.

——— (1939). Generalized Foreign Politics. *British Journal of Psychology*, Monograph Supplement, No. 23.

——— (1947/1960). *Arms and Insecurity* (London: Stevens and Sons).

Richmond, Barry, Steven Peterson, and Peter Vescuso (1987). *An Academic User's Guide to STELLA™* (Lyme, NH: High Performance Systems).

Richmond, Barry, Peter Vescuso, and Steven Peterson (1987). *STELLA™ for Business* (Lyme, NH: High Performance Systems).

Roberts, E.B. (1964). *The Dynamics of Research and Development* (New York: Harper and Row).

——— (1978), ed. *Managerial Applications of System Dynamics* (Cambridge, MA: MIT Press).

Roberts, Nancy, D.F. Andersen, M. Garret, R. Deal, and W. Shaffer (1983). *Introduction to Computer Simulation: The System Dynamics Approach* (Reading, MA: Addison-Wesley).

Rosenblueth, Arturo, Norbert Wiener, and Julian Bigelow (1943). Behavior, Purpose, and Teleology. *Philosophy of Science* 10: 18–24. Reprinted in Buckley (1968).

Routh, E.J. (1877). *A Treatise on the Stability of a Given State of Motion* (London: Macmillan).

Russell, Bertrand (1913). On the Notion of Cause. *Proceedings of the Aristotelian Society* 13: 1–26. Reprinted in Bertrand Russell, *Mysticism and Logic* (Garden City, NY: Doubleday Anchor, 1957).

Samuelson, Paul A. (1939). Interactions Between the Multiplier Analysis and the Principle of Acceleration. *Review of Economic Statistics* 21: 75–78.

Schön, Donald A. (1980). Conversational Planning. Essay in honor of Sir Geoffrey Vickers, available from the author, MIT, Cambridge, MA. 02139.

Shannon, Claude E. (1948). A Mathematical Theory of Communication. *Bell System Technical Journal* 27: 379–423, 623–656.

Shannon, Claude E. and Warren Weaver (1949). *The Mathematical Theory of Communication* (Urbana: University of Illinois Press).

Sherrington, Charles (1906). Integrative Action of the Nervous System. Quoted in Miller, Galanter, and Pribram (1960), pp. 23–24.

Shibutani, Tamotsu (1968). A Cybernetic Approach to Motivation. In Buckley (1968).

Simon, Herbert A. (1947/1957). *Administrative Behavior: A Study of Decision-making Processes in Administrative Organization* (New York: Macmillan, 1947). 2nd ed. (New York: Free Press, 1957).

——— (1952). On the Application of Servomechanism Theory in the Study of Production Control. *Econometrica* 20,2. Reprinted in Roberts (1978).

——— (1954). Some Strategic Considerations in the Construction of Social Science Models. In P. Lazarsfeld, ed., *Mathematical Thinking in the Social Sciences* (Glencoe, IL: Free Press).
——— (1957). *Models of Man* (New York: John Wiley).
——— (1962). The Architecture of Complexity. *Proceedings of the American Philosophical Society* 106,6: 467–482.
——— (1978). Rational Decision Making in Business Organizations. *American Economic Review* 69,4: 493–513.
Slack, Charles W. (1955). Feedback Theory and the Reflex Arc Concept. *Psychological Review* 62: 263–267. Reprinted in Buckley (1968), pp. 317–320.
Sluckin, W. (1954). *Minds and Machines* (London: Penguin).
Smith, Adam (1776). *An Inquiry into the Nature and Causes of the Wealth of Nations*, ed. Edwin Cannan (New York).
Sommerhoff, Gert (1950). *Analytical Biology* (Oxford: Oxford University Press).
Starkey, B.J. (1955). *Laplace Transformations for Electrical Engineers* (New York: Philosophical Library).
Sterman, John D. (1984). A Behavioral Model of the Economic Long Wave. *Journal of Economic Behavior and Organizations* 5 (1984).
——— (1987). Testing Behavioral Simulation Models by Direct Experiment. *Management Science* 33,12: 1572–1592.
——— (1988). Deterministic Chaos in Models of Human Behavior. *System Dynamics Review* 4(1–2): 148–178.
——— (1989). Misperceptions of Feedback in Dynamic Decision Making. *Organizational Behavior and Human Decision Processes* 43 (June).
Stinchcombe, Arthur L. (1968). *Constructing Social Theories* (New York: Harcourt, Brace and World).
Strotz, R.J., J.F. Calvert and N.F. Morehouse (1951). Analog Computing Techniques Applied to Economics. *A.I.E.E. Transactions* 70: 557–563.
Strotz, R.J., J.C. McAnulty and J.B. Naines (1953). Goodwin's Nonlinear Theory of the Business Cycle: An Electro-Analog Solution. *Econometrica* 21: 390–411.
System Dynamics Review (1988). Special Issue on Chaos. 4(1–2) (1988).
Takahashi, Yasundo, Michael J. Rabins, and David M. Auslander (1970). *Control and Dynamic Systems* (Reading, MA: Addison-Wesley).
Thomas, W.I. and D.S. Thomas (1928). *The Child in America: Behavior Problems and Progress* (New York: Knopf).
Tinbergen, Jan (1937). *An Econometric Approach to Business-Cycle Problems* (Paris).
——— (1939). *Statistical Testing of Business Cycle Theories: I—A Method and Its Application to Investment Activity; II—Business Cycles in the United States of America* (Geneva: League of Nations Economic Intelligence Service). Reprinted (New York: Agathon Press, 1968).
Tolman, Edward (1939). Prediction of Vicarious Trial and Error by Means of the Schematic Sowbug. *Psychological Review* 46: 318–336; cited in Miller, Galanter, and Pribram (1960), p. 44.
Tolstoy, Leo (1869). *War and Peace*. Translated by Constance Garnett (New York: Modern Library).
Troland, L.T. (1928). *The Fundamentals of Human Motivation* (New York: Van Nostrand); cited in Miller, Galanter, and Pribram (1960), p. 44.
Tustin, Arnold (1953). *The Mechanism of Economic Systems* (Cambridge, MA: Harvard University Press).

Varela, Francisco J. (1975). A Calculus for Self-Reference. *International Journal of General Systems* 2: 1–24.
────── (1979). *Principles of Biological Autonomy* (New York: Elsevier-North Holland).
Venn, John (1888). *Logic of Chance* (London); cited in Merton (1936).
Verhulst, P.F. (1838). Notice sur la loi que la population suit dans son accroisement; cited in Kormondy (1969), pp. 69–71.
Vickers, Geoffrey (1959a). *The Undirected Society* (Toronto: University of Toronto Press).
────── (1959b). Is Adaptability Enough? *Behavioral Science* 4: 223.
────── (1967). *Towards a Sociology of Management* (New York: Basic Books).
Volterra, Vito (1931). Leçons sur la théorie mathématique de la lutte pour la vie; discussed in Braun (1978), pp. 413–420.
von Bertalanffy, Ludwig (1945). Zu einer allgemeinen Systemlehre. *Deutsche Zeitschrift für Phylosophie* 18, 3/4.
────── (1950a). The Theory of Open Systems in Physics and Biology. *Science* 3: 23–29. Reprinted in Emery (1969), pp. 70–85.
────── (1950b). An Outline of General System Theory. *British Journal of the Philosophy of Science* 1: 139–164. Condensed and reprinted as Chapter 3 of von Bertalanffy (1968).
────── (1951). General System Theory: A New Approach to Unity of Science. *Human Biology* 23. 1: Problems of General System Theory, 302–312; 6: Toward a Physical Theory of Organic Teleology, Feedback, and Dynamics, 346–361.
────── (1955). General System Theory. *Main Currents in Modern Thought* 11,4: 75–83. Reprinted as Chapter 2 of von Bertalanffy (1968).
────── (1968). *General System Theory, Foundations, Development, Applications* (New York: George Braziller).
von Foerster, Heinz (1950–1953), ed. *Cybernetics: Circular Causal and Feedback Mechanisms in Biological and Social Sciences.* Transactions of the Cybernetics Conferences, vols. 6–10 (New York: Josiah Macy, Jr. Foundation).
────── (1967). Time and Memory. *Annals of the New York Academy of Sciences* 138: 866–873.
von Foerster, Heinz, and George W. Zopf, Jr. (1962), eds. *Principles of Self-Organization* (New York: Symposium Publications Division, Pergamon Press).
von Foerster, Heinz et al. (1968). *Purposive Systems* (New York: Spartan Books).
von Neumann, John (1966). *Theory of Self-Reproducing Automata* (Urbana, IL: University of Illinois Press).
von Wright, G.H. (1971). *Explanation and Understanding* (Ithaca, NY: Cornell University Press).
Weick, Karl (1969/1979). *The Social Psychology of Organizing* (New York: Addison-Wesley).
Weizenbaum, Joseph (1976). *Computer Power and Human Reason* (San Francisco: W.H. Freeman).
Wender, P.H. (1968). Vicious and Virtuous Circles: The Role of Deviation-Amplifying Feedback in the Origin and Perpetuation of Behavior. *Psychiatry* 31: 309–324.
Wiener, Norbert (1948). *Cybernetics: or Control and Communication in the Animal and the Machine* (Cambridge, MA: MIT Press).

——— (1949). *Extrapolation, Interpolation, and Smoothing of Stationary Time Series* (New York: John Wiley).
——— (1950/1954). *The Human Use of Human Beings: Cybernetics and Society* (New York: Houghton Mifflin, 1950; Garden City, NY: Doubleday Anchor, 1954).
Yates, F. Eugene (1987), ed. *Self-Organizing Systems: The Emergence of Order* (New York: Plenum Press).
Yovitz, M. and S. Cameron (1960), eds. *Self-Organizing Systems* (New York: Pergamon Press, 1960).

Index

Action feedback, 244
Addiction, 257
Administrative Behavior (Simon), 145
Advertising, 151
Aftalion, Albert, 70, 131
Alcohol, 47, 257
Allen, R. G. D., 130, 140–43, 150, 153, 159, 277
Allende Gossens, Salvador, 185–86
American Dilemma (Myrdal), 79–82
American Philosophical Association, 121
Amplifiers, 27–28, 95, 163, 164
"Amplifying feedback," 217–18
Annett, J., 243–44, 336
Anthropology, 3, 75, 85–86, 98
ARIMA time series analysis, 326, 348n.3
Aristotle, 8
Arithmetic, 6, 57
Arms and Insecurity (Richardson), 41
Arms race models, 38–41, 200, 217, 218, 338
Artificial intelligence, 148, 266, 329, 332, 338, 339
Ashby, W. Ross, 97, 99, 102–3, 162, 166–67n.2, 269n.23, 269–70n.28, 324, 330; *Design for a Brain*, 108–13; *Introduction to Cybernetics*, 113–18; Law of Requisite Variety, 113, 116, 173, 174, 322–23, 330, 331, 343; subsequent applications, 11, 41, 119, 123, 128, 188, 190, 233, 250, 251, 258, 278, 287, 290, 335; "ultrastability," 109–10, 171–72, 258, 306
Ataxia, 104
Athens, 64–65
Automation, 128, 264, 266
Autonomy, 101, 218–19, 220
Autopoesis, 265, 331
Autoregressive model, 348n.3
Aviation, 104

Balzac, Honoré de, 54–55
Bandwagon effect, 82–83
Bar-Hillel, Y., 166–67n.2
Bateson, Gregory, 106, 109, 166–67n.2, 200, 204, 217, 233; "schismogenesis," 3, 41, 85–90, 96
Bavelas, Alex, 166–67n.2
BCP psychology, 253, 255–58
Beer, Stafford, 97, 102–3, 198, 322, 327, 334–35, 340; "black box," 171, 173, 174, 182, 267n.2, 325–26; computer simulation, 194, 195; management cybernetics, 118, 170–85, 217, 220, 250, 263, 264–65, 267nn.1, 2, 3; Project Cybersyn, 185–89; and servomechanisms thread, 175, 183, 184, 185, 188, 189, 204, 266
Behavior: The Control of Perception (Powers), 241, 252
"Behavior, Purpose, and Teleology" (Rosenblueth, Wiener, and Bigelow), 94–95, 96, 97, 101, 104, 112, 120, 128
Behavioral psychology, 75, 190, 241–42, 244, 246, 259, 262
Behavior of Organisms (Skinner), 259
Behavior patterns, 336, 337, 341
Bernard, Claude, 49–50
Bhagavad-Gita, 271
Bibliographic methods, 10–12
Bigelow, Julian, 94, 104, 162
Biology, 17, 35, 46, 93, 94, 106
Birch, Herbert G., 166–67n.2
Birth control, 306
Black, Harold, 27–28
Black boxes, 116, 171, 173–74, 176, 179, 182, 184–85, 267n.2, 286, 325–26
Blood, 50–51
Bode, Heinrich W., 27–28, 31, 163, 164
Boolean algebra, 162, 175

Boulding, Kenneth, 118, 119, 123–27, 162, 167n.5
Bourgeois democracy, 73
Bowman, John R., 166–67n.2, 183
Brain, 47–48, 340; feedback models of, 109, 112, 171, 240, 241–43, 250, 253, 264–65, 324–25
Brain of the Firm (Beer), 176, 185–86
Brown, Gordon, 30–31, 149, 150, 161
Brussels school, 311, 312
Bureaucracies, 272–73
Burke, Edmund, 1
Business cycles, 2–3, 44–45, 70, 125, 129, 130, 131, 134, 135, 136, 141, 142, 159, 322
Business Cycles and National Income (Hansen), 130
Business Cycles in the United States (Tinbergen), 44

Calcium, 51
Cameron, S., 251
Campbell, Donald T., 237
Cannon, Walter, 48–53, 126
Cantor, Georg, 56–57, 91n.11, 103
Capitalism, 70, 84
Carnap, Rudolph, 162
Carroll, Lewis, 318
Causality, 2, 5, 106, 328, 337–38, 344–45; in complex systems, 174–75, 176, 300–301; cumulative, 79–80, 81; in human behavior, 241, 245–46; in social systems, 7–8, 212, 213
Causal loops, 3, 42, 58, 70, 298, 325, 328, 342, 343, 344; causal-loop diagrams, 7, 72, 110–12, 141, 203, 204, 231, 233, 234, 237, 265, 269n.22, 272–73, 302, 315, 316–17n.13; loop dominance, 33, 34, 35–36, 37–38, 46, 54–55, 90n.2, 213, 225, 235, 295, 300, 308, 310, 312, 314, 321, 339; loop polarity, 5–7, 46, 58, 90, 195, 199–200, 205, 211, 222, 229, 234, 247–48, 265, 276, 290, 293, 310, 323, 325, 330, 337, 345–46. *See also* Mutual causality
Cause maps, 235–38, 269n.24
Center for Advanced Study in the Behavioral Sciences, 126, 189
Centrifugal governor, 17–18, 19
Centrifugal pendulum, 18, 25
Chain reactions, 212–13
Chao, Yuen Ren, 166–67n.2

Chaos: Making a New Science (Gleick), 3
Checo, 185, 187
Chile, 176, 185–88
China, 21–22
Chomsky, A. Noam, 190, 198
Circular causality. *See* Causal loops; Mutual causality
Circular reasoning, 44–45, 54, 78–79, 131–32
Circular reflexes, 189
Cities, 201, 203, 204, 299–300
Clonus, 104
Closed systems, 122, 123, 133, 144, 148, 297–98, 316n.12, 329–30, 347
Club of Rome, 296
Cochrane, James L., 277, 278
Cognitive dissonance, 2–3, 227, 253–55
Cognitive functions, 198, 271–72, 339–40, 343
Cognitive maps, 269n.24, 316–17n.13
Collected Papers of Jay W. Forrester (Forrester), 149
Collins, John, 299
Combinatorics, 175
Communication, 92, 105, 222, 266, 284, 286, 324, 330–31, 337; engineering, 164–65, 168n.10, 227; networks, 101–2, 128, 264, 327, 335, 342, 343
Communications theory, 162, 163, 165, 228, 330–31
"Comparing stations," 253, 255, 257
Compensating feedback, 4, 15–16n.2, 306–8, 347
Complex systems, 112, 114, 116–17, 149, 166, 170–71, 186, 228, 299–306, 311, 325–26, 327, 334–35
Compound interest, 200
Comprehensive Dictionary of Psychological and Psycho-Analytical Terms, 48
Computers, 11, 53, 58, 98–99, 102, 162, 196, 266
Computer simulation, 4, 148, 315, 323–24, 329, 339–40; economic systems, 129–30, 136; human cognition, 241, 266, 271–72, 276; industrial dynamics, 149, 152, 153–55; management cybernetics, 186, 265; political systems, 211; TOTE unit, 190, 194–95
Conceptual distance, 342–43, 346
Conference on Teleological Mechanisms, 120
Consciousness, 101
Constructive criticism, 11, 265

Continuity and discreteness, 153, 213–14, 242, 323–24, 343–45, 346–47
Continuous point of view, 152–153, 157, 242–43, 247
Control, 3, 113; in cybernetics thread, 96, 128, 164–65, 188, 264, 340; economic, 136, 138, 277–78, 286; industrial dynamics, 170, 174; of positive loops, 85, 88–89; in servomechanisms thread, 160, 165, 315
Control theory, 30, 31–32, 46, 132, 136, 138, 152, 160, 163, 165, 315
Cooper, W. W., 145–46
Corporate systems, 145–46, 149–51, 171–75, 176–83, 185, 186, 188–89
Crane, Robert, 118
Crime, 307–8
Criticisms, 11, 235–36, 265
Culbertson, John, 278, 286–96, 315nn.3, 4, 315–16n.5, 323, 324
Culture, 215–16
Cumulative causality, 79–80, 81
Cybernetic factory, 171–72, 173
Cybernetics, 86, 116, 118, 242, 311, 324; feedback in, 41, 95, 96, 199, 264–65; Macy conferences, 95–96, 97–99, 102, 108–9, 207; management, 170–71, 174, 184, 188, 267nn.2, 3; social applications, 105–7, 189, 204, 217, 221, 227, 263, 266
Cybernetics: Control and Communication in the Animal and the Machine (Wiener), 3, 103–7, 112–13, 161–62, 207
Cybernetics and Management (Beer), 170–75, 184, 188
Cybernetics thread, 1–2, 10, 93–97, 128, 139–40, 160, 161–62, 169, 263–67, 333, 342–43; Ashby's brain model and ultrastability, 108–18; Beer's management science, 170–89; and complexity, 112, 170–72, 335; computers in, 339–40; Deutsch's political science, 214–27; deviation-amplifying feedback, 200–204; and dynamic behavior, 220, 221–22, 230, 233, 264, 276, 340–41; early social science writings, 99–103, 134, 159, 188; Easton's political science, 204–14; in economics, 277, 282, 284, 286; General Systems Theory, 118–28, 297; Macy conferences, 97–99; management and control, 340; negative feedback in, 94, 95, 96, 98, 99, 101, 128, 200, 264, 293; point of view, 164–65, 166, 263, 334, 338; positive feedback in, 95, 96, 114–16, 127–28, 200, 201–3; Powers's psychology, 240–63; self-reference in, 199, 219, 265, 311, 312; and structural change, 235, 266, 338; TOTE unit, 189–200; units of analysis, 337; units of description, 336, 341; weaknesses of, 321–28; Weick's organizational systems, 228–40; Wiener's *Cybernetics*, 103–8
Cybersyn project, 185–88

Darwin, Charles, 49, 70
Decision and Control (Beer), 175–76, 184
Decision-making theory, 149, 150, 155–59, 336
Democracy, 64–65, 73, 100
Depression, 234
Descartes, René, 47–48
Design for a Brain (Ashby), 108–12
Deterministic chaos, 309–10
Deterministic systems, 170–71, 329
Deutsch, Karl, 92, 96, 97, 99, 100–103, 108, 126, 133, 162, 196, 204–5, 209, 250, 264, 265, 322, 325, 336, 337; *Nationalism and Social Communication*, 214–27, 268–69n.18; *Nerves of Government*, 3, 205, 214–22, 268n.16
Deviation-amplifying processes, 200, 201–2, 203–4, 230, 231, 239, 278–79, 280, 283, 321, 324
Deviation-counteracting processes, 7, 203–4, 230, 231–32, 268n.9, 278–79, 283, 284, 321, 324
Dewey, John, 73–74, 75–77, 189–90, 192, 243, 343, 346
Dialectic, 57–58, 71–73, 134
Difference equations, 42–43, 44, 46, 286, 289–90, 293, 294–95, 299, 314, 323, 324, 337
Differential equations, 5, 46, 167–68n.6, 195, 337; cybernetics thread and, 96, 106; early feedback devices, 24–25, 26–27, 28–30; economic systems, 42, 109–10, 112, 113, 134, 135–36, 137, 293, 294; General Systems Theory, 119, 120, 121, 123; industrial dynamics, 152, 154, 157; political models, 222, 224–25; population dynamics, 32, 34, 37; servomechanisms thread and 165, 273–76, 286, 314, 323–24, 329
Discourse on Method (Descartes), 47

Discreteness, 195–96, 213–14, 322–24, 343–45, 346–47
Discrimination, 6–7, 79–81, 82, 85
Disraeli, Benjamin, 93
DO-loop, 53–54, 91n.10, 190
Drebbel, Cornelius, 22–23, 31
Drive-reduction theory, 73, 189
Drug addiction, 257
Dynamic behavior, 337, 340–41, 347; cybernetics thread and, 203–4, 211, 217, 220–22, 226, 266; economic models, 163, 287–88, 289; industrial dynamics, 150, 152; servomechanisms thread and, 92, 113, 126, 140, 159–60, 163, 166, 271, 284, 321–22, 335; system dynamics, 297, 306, 312–13
DYNAMO language, 154, 185, 187, 188, 329, 339

Easton, David, 118, 204, 205–14, 221, 266, 268nn.11, 13, 14, 336, 344
Ecology, 122–23
Econometrics, 5, 43–46, 82, 93, 134, 153, 313–15
Economic depressions, 284, 285
Economics, 215–16, 346; Chile, 185–88; complex systems, 112, 171, 265; Culbertson, 286–96; feedback mechanisms, 2–3, 17, 42–43, 46, 132–36; industrial dynamics, 149–59; Leijonhufvud, 278–86; market-growth model, 303–6; positive loops, 83, 201; production control, 143–48; in servomechanisms thread, 94, 129–32, 139–43, 159–61, 163, 310, 315; stabilization policy, 109, 136–39; Tustin and Phillips, 277–78; *Wealth of Nations*, 59–64, 65, 69–70
Ecosystems, 36–37
Eden, Colin, 316–17n.13
Electrical circuits, 26, 27–28, 95, 242
Elements of Physical Biology (Lotka), 36
Enactment, 237, 239
Endogenous point of view, 4, 16n.3, 38, 82, 333; economic, 130, 132, 150, 153, 278–79, 287, 289; in servomechanisms thread, 164, 166, 286, 287, 314, 329–30, 334; in system dynamics, 297, 298, 301, 303, 312–13
Engels, Friedrich, 134
Engineering, 12, 93, 162; behavioral applications, 82, 101–2, 104, 242–43; control and communication, 163–64, 165–66, 271; and cybernetics thread, 94–95, 128, 227; economic applications, 129–32, 136, 141–43; feedback devices, 17–32, 46, 53; and servomechanisms thread, 113, 161, 271, 329
ENIAC, 162
Entropy, 107–8; negative entropy, 99, 105, 201, 330
Environment, 205, 207, 237
Epidemic model, 35–36
Epimenides, 55–56, 57
Equifinality, 120, 121
"Equilibrium," 49, 80, 220–21
Error signal, 31, 76, 116–17, 253, 258
Eskimos, 108
Europe, 22, 23, 88
Events, 336, 337, 338, 341–42
Evolution, 36, 49, 53, 70, 149, 200, 201
Exceedingly complex systems, 97, 170–71, 265, 335
Exogenous point of view, 16n.3, 117–18, 153, 188, 288, 314, 333–34
Exponential smoothing, 29
Extrapolation, Interpolation, and Smoothing of Stationary Time Series (Wiener), 161–62, 326

Farm machinery, 228
Federalist papers, 64–66
Feedback devices, 17–30, 59, 63, 69, 149
Feedback loop concept: in circular processes, 77–90; complex systems, 149, 171, 173, 174–75, 228–33, 276, 299–306; conceptual threads of, 12–14, 15, 17, 92–94, 161, 163, 164–66, 187–88, 232, 233, 240–41, 332–47; cybernetics thread and, 94–95, 96, 101, 102–4, 106–7, 108–9, 112, 114–15, 117–18, 128, 170, 175, 200, 263–67, 321–28; decision making, 145, 155–59; defining, 3–4, 5–9, 123; Dewey's psychology, 75–77; dynamic behavior, 203–4, 220–21, 222, 226; economic systems, 41–43, 59–64, 129, 130, 132–43, 277–96; engineering servomechanisms, 17–32, 329; in General Systems Theory, 118–27; homeostasis and, 46, 48, 51–52, 53, 175, 321; industrial dynamics, 149–50, 152–53; logic loops, 53–58; mathematical models and, 23–26, 32–41, 46; nonlinearities and, 35–36,

212–13, 236–37, 308–12; political science, 64–74, 204–12, 218–19; Powers's four blunders, 240–41, 258–60; Powers's theory of behavior, 241–58, 268n.17; servomechanisms thread and, 129, 139–40, 148, 159–61, 169, 271, 313–15, 328–32; social science evolution, 1–2, 10–12, 14–15, 31, 38, 58, 59, 74–75, 90, 97, 99, 227, 233, 272, 296; social science applications, 2–3, 92–94, 318–20; system dynamics, 4, 296–99, 302–3, 306–8, 312–13; TOTE unit, 189, 190, 192–93, 195–96, 199–200
Festinger, Leon, 2–3, 227, 253–55
Finality, 119–20
Firm, theory of the, 145–46
Fitness of the Environment (Henderson), 49
Fluid control mechanisms, 18–21, 22
Foreign policy, 222
Forrester, Jay W., 3, 30–31, 82, 132, 160–61, 164–65, 185, 194, 195, 242, 248, 282, 336, 346, 348n.5; industrial dynamics, 3, 81, 149–59, 170, 171, 175, 183, 211; subsequent applications, 11, 187, 211, 213, 268nn.13, 14, 269–70n.28, 272, 345; system dynamics, 80, 247, 296–306, 315n.2, 316n.12, 331–32, 334–35, 342, 344; *Urban Dynamics*, 61, 299–300
Forrester, Judith, 348n.5
Forrester, Nathan B., 348n.2
Freedom, 101
Freud, Sigmund, 83
Frisch, Ragnar, 129, 133, 151
"Frontiers in Group Dynamics" (Lewin), 99–100
Functional causal imagery, 72

Gain, 221
Galanter, Eugene, 3, 189–200, 220, 263
Game theory, 162, 219–20, 268n.17
Gases, 311
General Equilibrium Theory, 277
General interchange model, 215–16
"Generalized Foreign Politics" (Richardson), 39, 41
General Systems Theory, 118–19, 120–21, 123–27, 182, 311
General Theory of Employment, Interest, and Money (Keynes), 133, 278, 283–86
Gerard, Ralph W., 118, 166–67n.2

Germany, 104
Gestalt psychology, 73, 75
Glasser, William, 3, 253, 255–58
Gleick, James, 3
Goal-seeking loops, 115–16, 147–48, 222
Gödel, Kurt, 57, 58, 91n.11, 178
Goodwin, Richard M., 129–32, 135, 139–40, 142, 143, 148, 149, 150, 151, 152, 153, 159, 160–61, 164, 278, 279, 314
Government, 52, 64–66, 73, 211, 212; Chile, 176; Deutsch's model, 214–22, 264
Graham, John A., 277, 278
Great Britain, 59, 64, 109
Greece, 18, 48
Grey-Walter, W., 166–67n.2

Hamilton, Alexander, 64
Harvard Business Review, 149
Harvard University, 49
Hebb, D. D., 243–44
Hegel, Georg Wilhelm Friedrich, 58, 71
Henderson, L. J., 49
Heroin, 257, 307–8
Heron, 21
Hicks model, 141, 142
Hippocrates, 48–49
Hofstadter, Douglas R., 56
Holt, E. B., 189
Homans, George, 148
Homeostasis: in cybernetics thread, 17, 128, 162, 166, 171, 184–85, 264, 321–22; in economic systems, 284; feedback influences, 3, 5, 12, 46, 48–53, 59, 93, 94, 104, 120, 122, 123; negative feedback in, 126, 175, 188
Homeostat, 258, 278, 281
Homeostatic equilibrium, 182
Homeostatic ultrastability, 173
Homomorphisms, 173, 267n.1
Hooke's law, 293–94
Households, 215–16
Housing, 300
Hull, C. S., 73
Human behavior, 3, 48; cybernetics and, 103–4, 106; in organizations, 276; Powers's feedback theory of, 240, 241, 243–58, 259–63, 269–70n.28; TOTE model, 194–95, 196–99
Human Use of Human Beings: Cybernetics and Society (Wiener), 107–8

Hume, David, 8, 59–60, 226
"Hunting," 23–24, 25, 62, 69, 131
Hurwitz, A., 113
Hutchinson, G. Evelyn, 166–67n.2

Iatmul tribe, 87, 89
Incompleteness theorem, 178
Industrial dynamics, 149–59, 171
Industrial Dynamics (Forrester), 3, 152–59, 160, 211, 296
"Industrial Dynamics: A Major Breakthrough for Decision Makers" (Forrester), 81
Industrial revolution, 23
Infant mortality, 63, 69, 306
Information: in cybernetics thread, 282, 284, 324–25; as feedback, 31–32, 45, 58, 122, 200, 204, 268n.17, 276, 335; as negative entropy, 99, 105, 201, 330; in servomechanisms thread, 330–31
Information theory, 98–99, 102, 106, 128, 171, 175, 212, 264
Input blunder, 258–59, 260
Input/output point of view, 98, 105, 127–28, 164–65, 207, 264, 328
Instincts, 197–98
Integral equations, 154, 157
Integrity, 101
Interdependence, 1, 3, 9, 230
International trade, 59
Interpersonal control, 252
Introduction to Cybernetics (Ashby), 109–10, 112, 113–18
Introduction to Servomechanisms (Porter), 170
Inventories, 144, 151, 154, 155, 157
Inverse feedback, 98
Investment, 44–45, 130, 131, 132, 134, 135, 141, 187, 200
Invisible hands, 5
Isomorphism, 267n.1

James, H. M., 130–31
James, William, 76
Jay, John, 64
"Jevon's homeostat," 281
Job-training programs, 300, 301

Kaldor, N., 129
Kalecki, M., 129, 141, 142
Kennedy, John F., 218

Keynes, John Maynard, 2, 131, 133, 183–84, 200, 278, 283–86, 287, 288, 289
Kinesthesia, 46–47, 48, 189
Kleene, Stephen, 183
Klein, Lawrence, 44, 45, 134, 135, 153, 277
Klir, Jiri, 97, 102–3
Knowledge of results, 3, 195, 200, 205, 209, 244, 265, 269–70n.28, 336
Koopmans, T. C., 43–44, 277
Kramer, Ernest, 74
Ktesibios, 18–21, 22, 23
Kubie, Lawrence S., 166–67n.2
Kuhn, Thomas, 327

Lags, 44–45, 212, 221
Laplace transforms, 26, 28, 30, 96, 106, 143
Law of Requisite Variety. *See* Ashby
Leads, 221
Learning, 101, 250–52, 253
Learning feedback, 244
Le Châtelier's principle, 119
Leibniz, Gottfried Wilhelm, 64
Leijonhufvud, Axel, 278–86, 287, 288
Lenz's rule of electricity, 119
Lewin, Kurt, 3, 92, 96, 99–100, 133, 162, 197, 227
Liar's paradox, 55–56
Licklider, J. C. R., 166–67n.2
Limits to Growth (Meadows), 296
Linearity, 211, 225
LISP language, 190, 266, 294, 329, 332, 339
Living Systems (Miller), 3, 127–28, 266
Load, 221
Logic, 12, 17, 53–58, 93, 94, 162, 190
"Logical Calculus of the Ideas Immanent in Nervous Activity" (McCulloch and Pitts), 96–97, 102, 173, 215, 240, 241, 322, 324
Logical types theory, 57, 58, 108, 178, 220
Logic Theorist, 190
Logistics equation, 314
Lorenz, Konrad, 198, 309–10
Lotka, Alfred, 36–38, 162, 183, 269n.23, 293, 314

McAnulty, J. C., 129–30
MacColl, L. A., 28

McCulloch, Warren S., 96, 97–99, 106, 122, 162, 173, 215, 240, 241, 322, 324
Machine analogy blunder, 258–59
MacIver, Robert M., 8
MacKay, Donald M., 166–67n.2
Macroeconomic Theory and Stabilization Policy (Culbertson), 287
Macy Foundation conferences, 95–96, 97–99, 102, 105, 108–9, 118, 128, 162, 166–67n.2, 204, 207, 328
Madison, James, 64
Malinowski, Bronislaw Kasper, 72
Malthus, Thomas R., 2, 63, 66–70
Management cybernetics, 170–83, 185–86
Management systems, 3, 118, 184–85, 264–65, 340
Man-machine blunder, 258–59, 260
Mannheim's paradox, 16n.5
March, J. G., 190, 272
"Market Growth as Influenced by Capital Investment" (Forrester), 303
Market growth model, 303–6
Markovian randomizers, 171
Marriage, 87–88
Maruyama, Magoroh, 200–204, 209, 229, 231–32, 265, 266, 268n.9, 283, 330
Marx, Karl, 2, 58, 71–73, 83, 84, 277
Massachusetts Institute of Technology: Servomechanisms Laboratory, 30–31, 149, 161; System Dynamics Group, 315n.2
Mathematical models, 17, 82, 165, 339; cybernetics thread and, 96, 103, 106, 112–13, 233; early feedback devices, 24–30; economic systems, 42–43, 46, 129, 134, 136–38, 140–43, 144, 145, 146–48, 277, 279–80, 289–92; industrial dynamics, 152, 154; management cybernetics, 175, 176; nonlinear dynamics, 35–36, 46, 308–12; organizations, 273–76; political systems, 39–41, 222–26; populations, 32–34, 37–38; psychology, 258, 260–63, 269–70n.28; servomechanisms thread and, 94, 140, 160, 231; TOTE unit, 195
Maxwell, James Clerk, 23–26, 30, 53, 62, 65, 261
Mayr, Otto, 18, 22, 23, 59, 60, 64, 70
Mead, George H., 73–74
Mead, Margaret, 3, 96, 98, 106, 166–67n.2

Mead, Thomas, 18
Meadows, Donella H., 302, 303, 305, 331–32
Means-end analysis, 148, 272, 345
Mechanism of Economic Systems (Tustin), 3, 132–33, 170, 183
Memory, 197, 198, 219
Merton, Robert King, 2–3, 72, 83–85, 209, 272–73, 301
Message loops, 58, 337, 342, 343, 344, 347
Meta-languages, 58, 178, 263
Metaplans, 198–99, 235
Mill, John Stuart, 2, 8, 78–79, 195, 204, 277
Miller, George A., 3, 189–200, 220, 263
Miller, James Grier, 3, 118, 127–28, 266
Modeling, 8, 102, 149, 153, 214
Models of Man (Simon), 183
"Monadology" (Leibniz), 64
Monetary policy, 138, 139
Morgenstern, Oscar, 162
"Morphogenesis," 201, 202, 266, 330
"Morphostasis," 200–201, 266
Motivation, 73–74
Multiplier/accelerator model, 2–3, 42–43, 44, 129–30, 136–37, 139, 152, 290, 291, 343
Multiplier effect, 140–41
Mutual (circular) causality, 1, 3, 5, 7, 8, 9, 38, 77–78, 123, 324, 326, 327, 330, 342; bandwagon effect, 82–83; cumulation, 79–82; in cybernetics thread, 114, 119, 122, 128, 169, 190, 204, 209, 227, 228, 230, 233, 237, 244, 264, 265, 268n.9; economic, 44, 45, 46, 277, 310; homeostatic, 46; in Macy conferences, 98, 106; positive loops, 59, 200; schismogenesis, 85–90; self-fulfilling prophecy, 83–85; in servomechanisms thread, 272, 276, 287, 312, 334, 335, 337, 340; social science applications, 2, 4, 12, 17, 319; vicious circle, 54, 78–79. *See also* Causal loops; Feedback loop concept
Myrdal, Gunnar, 2–3, 54, 79–82, 85, 86, 195, 200, 201, 204, 209, 265, 283, 313–14

Naines, J. B., 129–30
Nationalism, 222–27
Nationalism and Social Communication (Deutsch), 222–27, 268–69n.18

Naven (Bateson), 89
Needham, Joseph, 21–22
Negative entropy, 99, 105, 201, 330
Negative feedback, 3, 5–6, 7, 8–9, 15n.1, 58, 90, 203, 227, 308, 345; control of positive loops, 66, 88–89, 218; in cybernetics thread, 94, 95, 96, 98, 99, 101, 128, 200, 264; early feedback devices, 17, 22, 23, 27–28; in economic systems, 60, 62, 63, 70–71, 132, 140, 146–48, 279–80, 284, 290–94; in homeostasis, 50, 52, 53, 59, 126, 175, 188; in human behavior, 48, 73, 74, 76, 247–48, 251, 255, 262–63, 300; and organizational stability, 231, 232, 238–39; oscillations, 103, 104, 295; in population growth, 33–34, 37–38, 306–7; in servomechanisms thread, 160, 272; TOTE loop, 195, 199–200
Nerves of Government (Deutsch), 3, 103, 205, 214–22, 268n.16
Nervous system, 47, 50, 95, 102, 104, 241, 243, 248
Neural nets, 97, 102, 173, 215, 241, 264, 322, 324, 339–40
Neurophysiology, 75, 96–97, 162, 264
Newell, Allan, 148, 190, 194, 242, 272
New Guinea, 87
Newton, Sir Isaac, 12, 24, 293–94
Nichols, N. B., 130–31
Nonlinearity, 46, 82, 149; in economic systems, 129, 132, 134, 135–36; and loop dominance, 34, 35–36, 37, 235, 295; in political systems, 211, 212–13, 218, 225; in servomechanisms thread, 140, 160, 165, 235, 266, 308, 313, 339; and structural change, 233–36; system dynamics, 308–12, 314
Nyquist, H., 113

Objectification blunder, 258–60
"On the Application of Servomechanism Theory in the Study of Production Control" (Simon), 143
"On the Balance of Trade" (Hume), 59
"On Governors" (Maxwell), 23
On Keynesian Economics and the Economics of Keynes (Leijonhufvud), 283
Open systems, 120, 121–22, 123, 298, 329–30
Operations research, 170–71
Operator notation, 26, 28–30, 143

Order Out of Chaos (Prigogine and Stenger), 3
Organisms: computer simulations, 102, 194; drive-reduction, 73; feedback behavior, 95, 110, 241, 252; homeostasis, 48–52
Organizational studies, 228–33, 234, 272–76, 315
Organizations (March and Simon), 272–76
Oscillations: economic, 43, 62, 130, 131, 132, 134, 138, 151; feedback devices, 25, 53; negative feedback and, 103, 104, 293–94, 295, 321, 323; population growth, 37, 67–69

Paradoxes, 55–57
Parsons, Talcott, 215
Pask, Gordon, 183, 251
Pavlov, Ivan P., 74
Perception, 212, 250, 253
"Perceptual error," 253, 255–58
Phillips, A. W., 92, 129, 132, 149, 150, 161; economic stabilization, 136–40, 142, 143, 151, 159, 164, 168n.7, 277, 295, 343, 346; subsequent applications, 152, 153, 315
Phillips, R. S., 130–31
Philosophers, 8
Physiology, 94
PID. *See* Proportional, integral, and derivative control
Pitts, Walter H., 96, 162, 166–67n.2, 173, 215, 240, 242, 322, 324
Plans and the Structure of Behavior (Miller, Galanter, and Pribram), 3, 189–200, 240, 241, 247–48, 265, 267n.6, 336, 343
Platform for Change (Beer), 176, 185
Plato, 188–89, 267n.5
Platt, John, 64, 118
Pneumatics (Heron), 21
Point of view, 152, 333–34
Policy analysis, 4, 82, 140, 160, 288–89, 313
Policy resistance, 4, 302, 312, 326, 347
Policy structure, 158, 336
Political science, 3; Bateson's schismogenesis, 87, 88; Deutsch's *Nationalism*, 222–27; Deutsch's *Nerves of Government*, 214–22; Easton's political systems analysis, 204–14, 344
Polity, 215–16

Population dynamics: Deutsch's nationalism, 224, 226–27; Forrester's compensating feedback, 306–7; Lotka-Volterra predator-prey system, 36–38, 269n.23, 314; Malthusian, 66–70; Smith's labor supply, 63; Verhulst equation, 32–34, 36, 90n.2, 233
Porter, A., 170, 183
Positive feedback, 5–7, 8–9, 48, 53, 59, 90, 230, 234; bandwagon effect, 82–83; control of, 66, 88–89, 218; in cybernetics thread, 95, 96, 114–16, 127–28, 200, 201–3; in economic systems, 65, 132, 136, 140, 141, 188, 283, 290; in political systems, 213, 217–18, 224, 226; population growth, 32–34, 37–38, 66–67; schismogenesis, 86; self-fulfilling prophecy, 83–85; in servomechanisms thread, 160; vicious circle, 54, 78–79
Powers, William T., 11, 220; feedback behavior model, 240–58, 260–63, 264, 269nn.26, 28, 324, 343, 346; and feedback threads, 240–41, 263, 266–67; four blunders, 240, 258–60
Predator-prey ecosystem, 37–38, 269n.23, 314
Pribram, Karl, 3, 189–200, 220, 263
Prigogine, Ilya, 3, 311
"Priors," 10
Probabilistic systems, 170–71, 175
Problem solving, 145, 148, 198, 199, 241, 271–72, 345
Production control, 136, 138, 139, 144–45, 151, 155, 157, 176, 177, 183, 184, 284
Profit maximization, 44–45, 145–46, 147–48
Proportional, integral, and derivative control (PID), 136–39, 159, 168n.8
Proprioception, 46, 47, 48, 104, 123, 189
Proteins, 51
Protestantism, 84
Psychiatry, 88
Psychology: behaviorism, 75, 190, 240–42, 244, 246, 259, 262; Dewey and reflex arc, 75–77, 343; motivation, 73–74; Powers's control systems hierarchy, 240–58, 259–63; TOTE unit, 3, 189, 196–99
Psychotherapy, 3, 253–58
Purpose tremor, 104

"Quantitative Analysis of Purposive Systems" (Powers), 258–60
Quastler, Henry, 166–67n.2, 183

Race relations, 2–3, 79–80
Radcliffe-Brown, A. R., 72
Randomness, 308–10, 327–28, 340
Rapoport, Anatol, 118, 126
Rashevsky, Nicholas, 183
Rational soul, 47–48
Reality therapy, 3, 253, 255
Recursion, 3
Reflections on the Revolution of France (Burke), 1
Reflex arc, 75–76, 77, 189–90, 192, 244, 343
Regenerative loops, 97, 230
Regression equations, 334
Regulators, 23–25, 116–18, 163, 164, 330
Reinforcement, 74
Representation, 336–37
Retention, 237, 239
"Retroflex," 189
Revolution, 213, 344
Ricardo, David, 70
Richards, Ivor A., 166–67n.2
Richardson, Lewis F., 38–41, 200, 204, 217, 269n.23
Rich Lands and Poor (Myrdal), 81
Roosevelt, Theodore, 82–83
Rosenblueth, Arturo, 52, 94, 104, 162, 203
Russell, Bertrand, 8, 56–57, 58, 91n.11, 108, 162, 178, 220
Russell's paradox, 56–57

Samuelson, Paul A., 2–3, 42–43, 44, 45, 137, 200, 290, 291
"Schematic sowbug," 189
Schismogenesis, 3, 41, 85–90
Schizophrenia, 88, 200
Schön, Donald A., 337
"Second cybernetics," 200, 203, 204
Selection, 237, 239
Selective attention, 237
Self-fulfilling prophecies, 2–3, 5, 83–85
"Self-Fulfilling Prophecy" (Merton), 85
Self-organization, 219, 265, 311–12
Self-Organizing Systems: The Emergence of Order (Yates), 311–12

Self-reference, 3, 53–54, 55–58, 94, 199, 219, 220, 265, 279, 331–32
Self-regulation, 49, 59–60, 64, 65–66, 70, 171, 268n.13
Self-reinforcing loops, 7, 12, 41, 53, 90, 200, 338; bandwagon effect, 83; chain reaction, 212–13; discrimination, 6–7; population growth, 69; vicious circle, 54–55, 78–79. *See also* Positive feedback
Self-steering, 218–19, 268n.16
Sensory feedback, 189
Sensory receptors, 47
Servomechanisms engineering, 5, 129, 158, 162, 164, 168n.10, 259, 260
Servomechanisms thread, 1–2, 10, 13, 93–94, 118, 205, 222, 228, 271, 313–15, 333; Beer and, 175, 183, 184, 185, 188, 189, 204; and complexity, 112–13, 175, 299–306, 335; computers in, 339–40; continuity and discreteness, 195, 196, 213, 343; and dynamic behavior, 126, 159–61, 162–63, 166, 178, 220, 231, 232, 266, 284, 313, 340–41; economics in, 129–48, 162–63, 277–96; industrial dynamics, 149–59; management and control, 340; and nonlinearity, 140, 160, 165, 235, 266, 308, 313, 339; point of view, 164, 165, 204, 286, 297, 313, 314, 334, 338, 342; Powers and, 240, 248, 258; and self-organization, 311, 312; Simon and, 143, 271–76; and structural change, 339; system dynamics, 296–313; units of analysis, 337; units of description, 336, 341; weaknesses of, 328–32
Set theory, 56–57, 108
Shannon, Claude E., 102, 162, 166–67n.2
Shannon's Tenth Theorem, 331
Shaw, George Bernard, 169
Shaw, J. C., 190, 194, 242, 272
Simon, Herbert A., 8, 92, 149, 150, 152, 153, 160–61, 205, 282; human behavior, 3, 145, 147, 148, 241, 271–76, 315, 345; production control, 143–45, 147–48, 295; subsequent applications, 159, 183, 184, 190–91, 194
Skinner, B. F., 74, 189, 190, 259
Smith, Adam, 2, 59, 60–64, 65, 66, 67, 69–70, 226, 277, 284
"Social homeostasis," 52–53

Social planning, 99–100
Social Psychology of Organizing (Weick), 3, 228, 233
Social sciences: bandwagon effect, 82–83; causality in, 7–8, 212, 213; cybernetics thread, 96, 99–103, 105, 113, 169, 188, 263, 321–28; evolution of feedback concept, 1–3, 4, 9, 17, 46, 58, 59, 71, 74–75, 77–78, 90, 92–94, 95, 110, 117–18, 161–63, 164–66, 204, 227, 332–33; feedback applications, 5, 11, 12, 14, 15, 318–20; homeostasis and, 46–53, 123–26; mathematical models, 32–46, 148; Powers's four blunders, 258–60; schismogenesis, 85–90; self-fulfilling prophecy, 83–85; servomechanisms thread, 113, 129, 160, 271, 310–11, 328–32; vicious circle and, 54, 78–79, 82
Social Sciences Citation Index, 10, 11
Social systems, 268nn.13, 14, 336–37, 338, 339–40; causality in, 7–8, 212, 213; continuity in, 153, 344; cybernetics and, 106–7, 118, 219–20, 322, 324, 335; feedback view of, 2, 31, 38, 41, 82, 126, 152, 221, 272; nonlinearity in, 211, 310–11; policy resistance, 4, 302, 326; servomechanisms and, 160, 271, 312, 329–30, 335; system dynamics, 296–97, 299–300, 301–2
Sociology, 3, 75
Solow, Robert, 222, 224–26, 269n.19
Sommerhoff, Gert, 178
South-pointing chariot, 21–22
Soviet Union, 218
Speaking, 198
Speculation, 78
Stability: Ashby and, 109–10, 114–15; cybernetics thread and, 113, 118, 128, 230, 231, 233, 264, 266, 321–22; economic, 136–39, 143. *See also* Ultrastability
Stations of the Mind: New Directions for Reality Therapy (Glasser), 3, 253
Steady state dynamics, 25, 221
Steam engine governor, 17–18, 19, 24, 25, 31, 67, 69, 111–12, 123
STELLA, 339
Sterman, John D., 310
Stimulus-response-reinforcement theory, 74, 189, 190, 245–46, 252, 269–70n.28

Stinchcombe, Arthur L., 71–73, 75
Strange loops, 56
Strotz, R. J., 129–30
Stroud, John, 166–67n.2
Structural change, 82, 218–20, 233, 235, 265–66, 338–39
Suicidal prophecy, 84
Supply and demand, 61–63, 69, 279–80
Su Sung, 22
Symbolic logic, 195
System dynamics, 4, 80–81, 248, 296–99, 312–13, 314, 315, 323, 328, 329, 336, 339, 348n.5; compensating feedback, 306–8; complex systems, 265, 299–306; nonlinearity, 308–12
System of Logic (Mill), 8
Systems Analysis of Political Life (Easton), 213
Systems hierarchy, 126

Tâtonnement, 279, 284
Teleology, 94–95, 119, 120, 121
Tension-reduction theory, 189
"Tension system," 197
Thermodynamics, 201, 311
Thermostats, 19, 22–23, 123, 127, 149, 163
Thomas, W. I., 83
Through the Looking Glass (Carroll), 318
Time series, 106, 153, 161–62, 326
Tinbergen, Jan, 2–3, 43–45, 91n.8, 200, 277
Tinbergen, Nicholas, 197–98
Tolman, Edward C., 73, 189
Tolstoy, Leo, 344–45, 346, 348n.5
Totalitarianism, 100
TOTE unit, 3, 75, 189, 191–200, 222, 242, 247–48, 265, 322, 330, 343
Toynbee, Arnold J., 268n.16
Tracking experiment, 244–47, 262
Traité de l'Homme (Descartes), 47
Tranquilizers, 257
Transfer functions, 26, 158
Transformational grammar, 198–99
Troland, L. T., 189
Turing, Alan, 183
Tustin, Arnold, 92, 149, 150, 151, 152, 153, 159, 161, 164; economic servomechanisms, 3, 132–36, 139–40, 143, 314, 322, 343, 346; subsequent applications, 170, 183–84, 277, 295, 310, 315

Ultrastability, 109–10, 171–72, 173, 175, 184–85, 251, 258, 306
"Unanticipated Consequences of Purposive Social Action" (Merton), 84, 301
Unemployment, 283, 285
UNESCO, 185
United States, 6, 79, 218
United States Constitution, 64–65
Units of analysis, 337
Units of description, 336, 341–42
University of Chicago, 118
University of Michigan, 118
"Uptight variables," 233–34
Urban Dynamics (Forrester), 61, 299–300

Valach, Miroslav, 97
Values, 101
Van der Pol equation, 167–68n.6
Variables, reversal of, 233, 234–35
Variety, 258, 279, 286; Ashby's law, 113, 116, 173, 174, 322–23, 330, 343; in Beer's management cybernetics, 171–72, 173, 174, 175, 176, 182, 184–85, 189, 267n.1
Venn, John, 84
Verbal Behavior (Skinner), 190
Verhulst, P. F., 32–34, 36, 90n.2, 233, 294
Viable systems, 186–87
Vicious circles, 2–3, 5, 53, 54–55, 56, 78–80, 82, 85, 89, 200, 228, 230, 234
Vickers, Geoffrey, 128
Virtuous circles, 234
Vivian Grey (Disraeli), 92
Volterra, Vito, 36–38, 122–23, 269n.23, 293, 314
Von Bertalanffy, Ludwig, 99, 118, 119–23, 126, 127, 162, 167n.4, 252, 267n.4, 297, 298, 316n.12
Von Foerster, Heinz, 98, 162, 166–67n.2
Von Neumann, John, 162

Wallace, Alfred Russel, 70, 200, 201
Wang P'u, 22
War and Peace (Tolstoy), 344–45
Water clock, 20–21, 22
Watt, James, 17–18, 24, 25, 31, 67, 111–12, 123
Wealth of Nations (Smith), 59, 60–64, 70
Weick, Karl, 3, 228–40, 265, 266, 269nn.21, 23, 24, 335, 338, 344
Weizenbaum, Joseph, 272, 345

Wender, P. H., 234
Werner, Heinze, 166–67n.2
White boxes, 176, 177, 178, 325–26
Whitehead, Alfred North, 108
Wiener, Norbert, 52, 58, 92, 94, 98, 99, 133, 161, 166–67n.2; *Cybernetics*, 3, 103–7, 112–13, 207; *Extrapolation*, 161–62, 326; *Human Use of Human Beings*, 107–8; negative entropy concept, 105, 201, 330; subsequent applications, 11, 120, 126, 159, 190, 203, 204, 205, 207, 240–41, 244, 258, 260, 263, 277

Will, 101
Windmills, 18
Wisdom of the Body (Cannon), 50–51, 52
Wisdom of the Fathers (*Pirke About*), 17
World Dynamics (Forrester), 149
World War I, 39, 41, 161
World War II, 104, 109, 161, 164

Yates, F. Eugene, 311
Young, J. Z., 166–67n.2
Yovitz, M., 251

Zero population growth, 306–7

This book was set in Baskerville and Eras typefaces. Baskerville was designed by John Baskerville at his private press in Birmingham, England, in the eighteenth century. The first typeface to depart from oldstyle typeface design, Baskerville has more variation between thick and thin strokes. In an effort to insure that the thick and thin strokes of his typeface reproduced well on paper, John Baskerville developed the first wove paper, the surface of which was much smoother than the laid paper of the time. The development of wove paper was partly responsible for the introduction of typefaces classified as modern, which have even more contrast between thick and thin strokes.

Eras was designed in 1969 by Studio Hollenstein in Paris for the Wagner Typefoundry. A contemporary script-like version of a sans-serif typeface, the letters of Eras have a monotone stroke and are slightly inclined.

Printed on acid-free paper.

FEEDBACK THOUGHT
in Social Science and Systems Theory
George P. Richardson

Feedback Thought in Social Science and Systems Theory is an original investigation in the history of an idea and a way of thinking in the social sciences—the loop concept underlying the notions of feedback and circular causality. After tracing its historical roots, George P. Richardson argues that modern usage of feedback thinking in the social sciences divides rather dramatically into two main lines of development, which proceeded from 1945 through at least the 1970s in considerable ignorance of each other. Richardson makes extensive use of analysis of citations and texts from many branches of the social sciences to document this split and to trace its development and implications.

The presumption underlying the work is that feedback thinking is one of the most penetrating patterns of thought in all social science. The work itself tends to demonstrate that by the length and diversity of the examples it provides; it is also a testament to the depth and richness of circular thinking in the social sciences. Usually implicit, sometimes explicit, feedback thought is embedded in the very foundations of social science and systems theory. It is a building block. Great social scientists are feedback thinkers; great social theories are feedback thoughts. Part of the purpose of this intellectual history is to illuminate the significance of feedback thinking in social science and social policy—current as well as classical.

Feedback Thought in Social Science and Systems Theory will be of interest to intellectual historians, students of the sociology of knowledge, social science scholars in a very wide range of disciplines, and students and scholars in various systems sciences.

GEORGE P. RICHARDSON is Associate Professor of Public Administration and Public Policy at the Nelson A. Rockefeller College of Public Affairs and Policy at the State University of New York at Albany. He is the author of *Introduction to System Dynamics Modeling with Dynamo*.

UNIVERSITY OF PENNSYLVANIA PRESS
418 Service Drive
Philadelphia, PA 19104-6097

Cover design: Adrianne Onderdonk Dudden   ISBN 0-8122-1332-7